Lecture Notes in Physics

Volume 882

Founding Editors

W. Beiglböck
J. Ehlers
K. Hepp
H. Weidenmüller

Editorial Board

B.-G. Englert, Singapore, Singapore
P. Hänggi, Augsburg, Germany
W. Hillebrandt, Garching, Germany
M. Hjorth-Jensen, Oslo, Norway
R.A.L. Jones, Sheffield, UK
M. Lewenstein, Barcelona, Spain
H. von Löhneysen, Karlsruhe, Germany
M.S. Longair, Cambridge, UK
J.-F. Pinton, Lyon, France
J.-M. Raimond, Paris, France
A. Rubio, Donostia, San Sebastian, Spain
M. Salmhofer, Heidelberg, Germany
S. Theisen, Potsdam, Germany
D. Vollhardt, Augsburg, Germany
J.D. Wells, Geneva, Switzerland

For further volumes:
www.springer.com/series/5304

The Lecture Notes in Physics

The series Lecture Notes in Physics (LNP), founded in 1969, reports new developments in physics research and teaching—quickly and informally, but with a high quality and the explicit aim to summarize and communicate current knowledge in an accessible way. Books published in this series are conceived as bridging material between advanced graduate textbooks and the forefront of research and to serve three purposes:

- to be a compact and modern up-to-date source of reference on a well-defined topic
- to serve as an accessible introduction to the field to postgraduate students and nonspecialist researchers from related areas
- to be a source of advanced teaching material for specialized seminars, courses and schools

Both monographs and multi-author volumes will be considered for publication. Edited volumes should, however, consist of a very limited number of contributions only. Proceedings will not be considered for LNP.

Volumes published in LNP are disseminated both in print and in electronic formats, the electronic archive being available at springerlink.com. The series content is indexed, abstracted and referenced by many abstracting and information services, bibliographic networks, subscription agencies, library networks, and consortia.

Proposals should be sent to a member of the Editorial Board, or directly to the managing editor at Springer:

Christian Caron
Springer Heidelberg
Physics Editorial Department I
Tiergartenstrasse 17
69121 Heidelberg/Germany
christian.caron@springer.com

Hans Paetz gen. Schieck

Nuclear Reactions

An Introduction

 Springer

Hans Paetz gen. Schieck
Institut für Kernphysik
University of Cologne
Köln, Germany

ISSN 0075-8450   ISSN 1616-6361 (electronic)
Lecture Notes in Physics
ISBN 978-3-642-53985-5   ISBN 978-3-642-53986-2 (eBook)
DOI 10.1007/978-3-642-53986-2
Springer Heidelberg New York Dordrecht London

Library of Congress Control Number: 2014931767

© Springer-Verlag Berlin Heidelberg 2014
This work is subject to copyright. All rights are reserved by the Publisher, whether the whole or part of the material is concerned, specifically the rights of translation, reprinting, reuse of illustrations, recitation, broadcasting, reproduction on microfilms or in any other physical way, and transmission or information storage and retrieval, electronic adaptation, computer software, or by similar or dissimilar methodology now known or hereafter developed. Exempted from this legal reservation are brief excerpts in connection with reviews or scholarly analysis or material supplied specifically for the purpose of being entered and executed on a computer system, for exclusive use by the purchaser of the work. Duplication of this publication or parts thereof is permitted only under the provisions of the Copyright Law of the Publisher's location, in its current version, and permission for use must always be obtained from Springer. Permissions for use may be obtained through RightsLink at the Copyright Clearance Center. Violations are liable to prosecution under the respective Copyright Law.
The use of general descriptive names, registered names, trademarks, service marks, etc. in this publication does not imply, even in the absence of a specific statement, that such names are exempt from the relevant protective laws and regulations and therefore free for general use.
While the advice and information in this book are believed to be true and accurate at the date of publication, neither the authors nor the editors nor the publisher can accept any legal responsibility for any errors or omissions that may be made. The publisher makes no warranty, express or implied, with respect to the material contained herein.

Printed on acid-free paper

Springer is part of Springer Science+Business Media (www.springer.com)

To Sybille, Annette, Birgit

Preface

Nuclei and nuclear reactions are a playground (or a laboratory) of three of the four (in special cases four) fundamental interactions in nature (the reactions have predominantly to do with the hadronic strong and the electromagnetic interactions). Figure 1 shows the characteristic features of the relevant forces. Nuclei are complex many-particle systems of nucleons (hadrons). These proved to be—mainly by performing scattering experiments with leptons (electrons, muons, and neutrinos)—extended objects with complex internal structure: Constituent quarks, gluons, whose exchange binds the quarks together, sea-quarks (quark-antiquark pairs, into which the gluons transform and vice versa) and—to the outside—virtual mesons, surrounding an inner nuclear region (the *bag*). Figure 2 depicts schematically this internal structure of the nucleons and their interactions via exchange of virtual bosons (especially mesons). The virtual mesons clouds surrounding the nucleons are—by their mutual exchange of mesons— the cause of the nucleon-nucleon interaction. This exchange is therefore responsible for the existence of nuclei, their rich structure, and the variety of their interactions. Especially the spin structure of the nucleons, i.e. the way the spins and orbital angular momenta of the constituents act together to form the nucleon spin has been of great concern ("Spin Crisis") and prompted many experimental investigations, often using spin-polarized particles. It turned out, however, that the internal structure of the nucleons has comparatively little influence on the behavior of the nucleons in nuclei, i.e. nuclear structure and nuclear reactions. Thus, nuclear physics—and especially nuclear reactions—is a field of science in its own right, even without much recourse to subnuclear degrees of freedom.

Historical Remark

Elastic scattering of particles off each other is a special class of nuclear reactions, i.e. one without change of the identity of the particles involved. In this sense the famous Rutherford/Geiger/Marsden scattering experiment started the field of nuclear reactions around 1911 in Manchester. Energetic particles from radioactive sources were

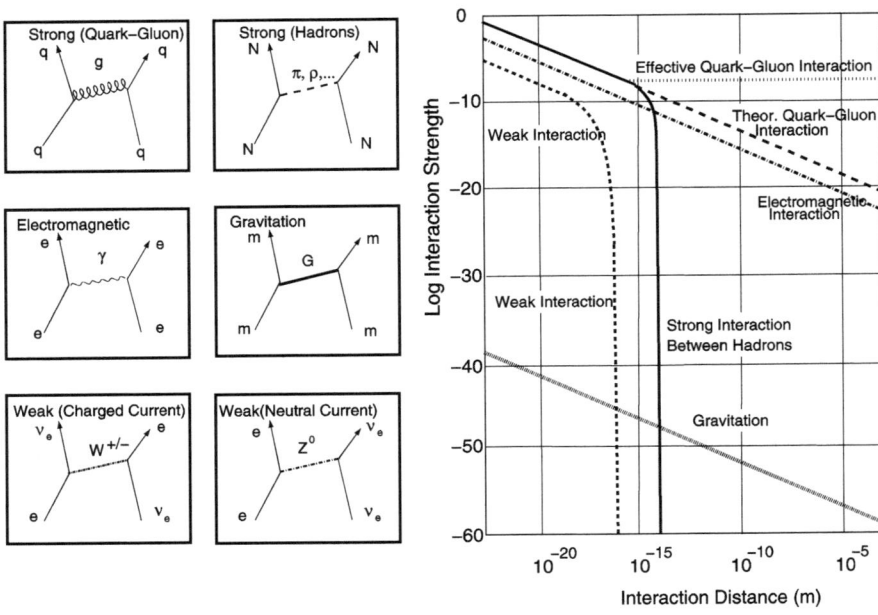

Fig. 1 Unified view of the fundamental forces of nature. All can be described as caused by the exchange of bosons between the constituent particles where massless exchange bosons lead to forces of infinite range (the photon, the graviton, and—in principle—the gluon). However, the strong interaction is special in two ways: the confinement of QCD causes the finite range of the "effective" quark-gluon interaction, and the interaction takes place also between the exchange bosons, the gluons. The masses of the other exchange bosons determine the range of the interactions via Heisenberg's uncertainty relation: ≈ 1.4 fm for the nuclear force due to the meson masses of a few hundred MeV, $\approx 10^{-18}$ m for the weak interaction due to the high masses of the W or Z bosons. Although the cornerstone of the standard model and responsible for the particle masses, the *Higgs boson*, has recently been found in agreement with the standard model the values of the masses of all particles as well as the very different strengths of the forces are still unexplained leaving the standard model an incomplete theory

used as projectiles (α's from heavy elements such as "radium emanation" ($^{222}_{86}$Rn) with sufficiently high energies and intensities). Figure 3 shows the original setup used by Geiger and Marsden displaying already all features of a modern scattering chamber, e.g. a thin-foil target, a detector registering single scattering events, the rotation of the detector around the target for angular distributions, the collimation of the projectiles for a good definition of the z axis etc. The scattering experiments were tedious: A MBq (in 4π solid angle) source corresponds to an incident "beam" current into a solid angle, small enough to define a reasonable scattering geometry, of only $\approx 10^{-6}$ nA. Single scintillation events had to be counted by observing them on a ZnS screen in the dark. The first "true" nuclear reaction (i.e. one with transmutation into different particles) was discovered by Rutherford in 1919 (after

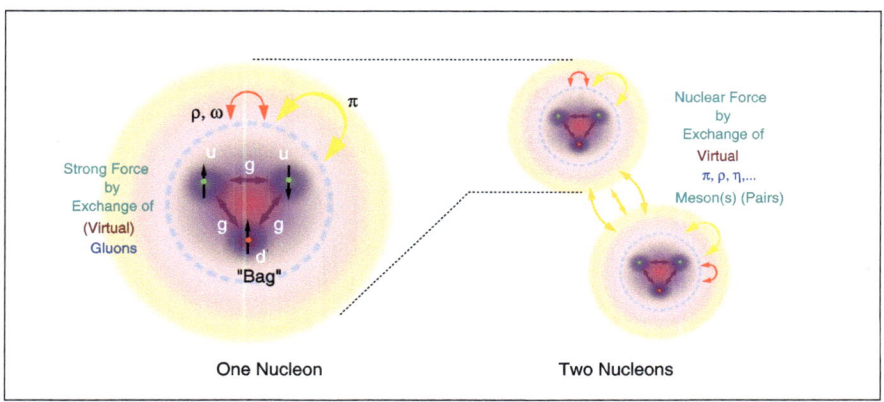

Fig. 2 Structure of the nucleons (*left*, here for a proton with u, u, and d quarks—for a neutron it would be d, d, and u). The core ("bag") is surrounded by virtual meson clouds of different ranges, depending on their masses. The picture of the quark-gluon structure inside a nucleon depends on the resolution with which we observe it; here we have rather a low-energy picture with three *constituent* quarks interacting via gluons, no sea-quarks (quark-antiquark pairs), no heavier quarks, and no gluon-gluon interactions. At much better resolution their importance grows, see Sect. 2.3. The coupling of he quark spins to form the nucleon spin as shown is a crude simplification because the role of the spin $J^P = 1^-$ of the gluons as well as of the orbital angular momenta is still somewhat unclear and under investigation using spin-polarized probes. In any case, the quark-spin contribution is much smaller than suggested by the simple picture, and the gluon-spin contribution also appears too small to overcome what has been called "*Nucleon-Spin Crisis*". On the *right* the *nuclear force* is depicted as resulting from the exchange of virtual mesons or meson pairs. For two protons, in addition, the electromagnetic force is mediated by the exchange of virtual photons. Although the nuclear force is predominantly a two-body interaction (a *saturation* property), multinucleon forces cannot be excluded that are defined as "simultaneous" interactions between more than two nucleons not accounted for by a sum of two-nucleon forces (see Chap. 9). In principle, also the exchange of Z^0 and $W^\pm$ bosons mediating the weak interaction must also be considered, e.g. in the parity-violating part of the interaction, which is, however, weaker by a factor of $\approx 10^{-7}$

earlier work together with Ernest Marsden), in which particles with larger range, ^{1}H nuclei, than that of the α's in scattering from different targets were observed:

$$\alpha + {}^{14}\text{N} \rightarrow {}^{17}\text{O} + p \tag{1}$$

using 6 MeV α's from a radioactive source, observed with an apparatus quite similar to the above mentioned setup. Figure 4 shows such an event in a cloud chamber, including also an event of "Rutherford" scattering of the recoil ^{17}O nucleus on an ^{14}N nucleus [BLA25,GEN40]. The cloud chamber, which is still unsurpassed as an instrument for visualizing such events but also cosmic rays etc. was invented by Charles Wilson following 1911, but developed for practical use by Blackett only since 1921, was not yet used by Rutherford. Consequently the ^{1}H nuclei were identified as part of all nuclei and Rutherford coined the term "proton". However, the still remaining basic puzzles about the true structure of nuclei were only resolved after the neutron was discovered by James Chadwick in 1932 (after Rutherford had already speculated about neutrons in nuclei and others had mistakenly interpreted

Fig. 3 The original setup of E. Rutherford, H.W. Geiger, and E. Marsden for α scattering from a gold foil [RUT11,GEI13]

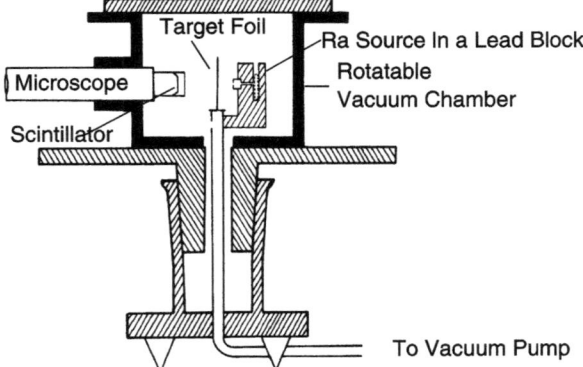

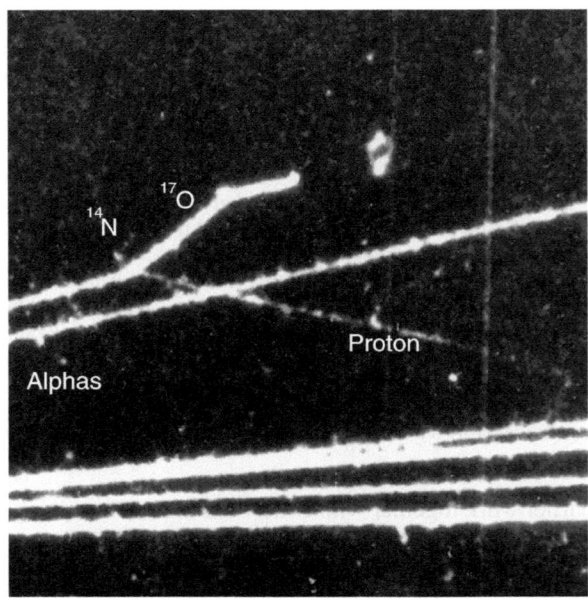

Fig. 4 Cloud chamber photograph by Blackett of the first nuclear reaction $\alpha + {}^{14}\text{N} \rightarrow {}^{17}\text{O} + p$ observed by Rutherford in 1919 [RUT19,BLA25]

the neutron radiation from the reaction $\alpha + {}^{9}\text{Be} \rightarrow {}^{12}\text{C} + n$ (with α's from a polonium source) to be an energetic γ radiation). It is evident that the use of radioactive sources imposed severe restrictions: fixed or very limited energy range and extremely low intensities. It is clear that the field of nuclear reactions could only progress with the invention of particle accelerators. The first accelerator prototype important for nuclear physics was the linear accelerator ("LINAC") developed and published in 1929 by Ralf Wideröe at the Aachen Institute of Technology, also laying the ground for the betatron, which was realized by Kerst and Serber in 1940, and the cyclotron by Lawrence in 1931. Wideröe's ideas also included the synchrotron

Preface   xi

and storage-ring schemes. The first nuclear reaction initiated with accelerated beams was the reaction

$$p + {}^7\text{Li} \rightarrow 2\alpha \qquad (2)$$

by Cockroft and Walton in 1932 at the Cavendish Laboratory at Cambridge using a DC high voltage across several accelerating gaps and produced by the Delon/Greinacher voltage multiplication scheme. This and the ensuing developments in nuclear and particle physics up to the present energies of up to 14 TeV (at the Large Hadron collider LHC at CERN/Geneva) are intimately connected with the achievements in accelerator physics and technology. Likewise the development of detector technologies—from the first scintillators, later equipped with photomultipliers, to the cloud and the bubble chambers, the ionization chamber, Geiger-Müller counter, multiwire ionization chambers, and the large field of solid-state detectors—was essential. Not unjustifiably accelerators have been called "tools of our culture" or "Engines of Discovery" (so the title of an opulent book by Sessler and Wilson [SES07]). Their impact reaches now into social applications such as tumor diagnosis and therapy, materials identification and modification, age and provenience analyses in archaeology, geology, arts, environmental science etc., see Part III. Many nuclear-physics textbooks deal also with applications, a comprehensive text is e.g. Ref. [HER99].

The—possibly somewhat underestimated—importance of accelerators and their developments for nuclear and particle physics will be stressed in this text by chapters on their principles, about ion sources, ion optics and other important features, see Chap. 16.

With the discovery of the *neutron* by Chadwick (1932) another branch of nuclear physics and especially nuclear reactions opened up that only partly depends on accelerators. Not only was the discovery of the neutron the keystone to the fundamental structure of nuclei removing all kinds of inconsistencies about e.g. about nuclear isotopes, but immediately it incited Heisenberg to formulate the idea of charge independence of the nuclear interaction and the fundamental symmetry of *isospin*. The neutrality of the neutron facilitates the description of nuclear reactions. On the other hand, production of neutrons for nuclear reactions as well as the detection methods are more complicated (see Sect. 17.4). Normally, except when neutrons from nuclear reactions are used, the choice or selection of specific neutron energies requires additional methods such as moderation by elastic collisions with light nuclei and/or chopper and time-of-flight facilities. Much of neutron work relies on neutrons from fission in reactors (an example is the high-flux 660 MW research reactor with a thermal flux of $>1 \cdot 15 \text{ s}^{-1} \text{cm}^{-2}$, at the Institute Laue-Langevin (ILL Grenoble)) or on spallation neutron sources where intense proton beams in the GeV and mA range incident on (liquid) metal targets release many (up to 30) neutrons per proton with high energies (a typical research center is the LANSCE facility with a proton LINAC, originally designed as *meson factory* at Los Alamos, New Mexico, another the spallation neutron source (SNS) at Oak Ridge, Tennessee, with 1.4 MW beam power and $4.8 \cdot 16$ neutrons/s.) The neutron has also fundamental properties in its own right that have been studied:

- β decay.
- The internal (quark + gluon) structure and charge and magnetic-moment distributions. They have been studied e.g. by elastic and inelastic electron scattering where deuterons and especially ^{3}He served as neutron targets. Polarized ^{3}He is an almost pure polarized neutron target. The charge and magnetic-moment distributions inside the neutron are proof of its inner structure.
- The possible electric dipole moment and thus time-reversal and parity violations were studied where the absence of the Coulomb force is experimentally advantageous.
- The wave nature of neutrons of low energies was studied in reflection, diffraction, and interference experiments.
- Especially ultracold neutrons offer many interesting properties and applications, e.g. its interaction with the gravitational field or that of its magnetic moment with magnetic fields.

Observables

In this text the term *observable* is used repeatedly. In quantum mechanics the most widely used definition is

Definition of "Observable" An observable is an operator in Hilbert space corresponding to a physically measurable quantity that must be hermitean in order to have real eigenvalues (the measurable quantities).

This definition lacks uniqueness in the sense that only a small fraction of all hermitean operators corresponds to measurable quantities. Without going into the details of the still ongoing discussion of different interpretations of quantum mechanics (e.g. the "Copenhagen" interpretation, which includes the interaction between a macroscopic measuring apparatus with the microscopic quantum world, or the question of *decoherence* etc.) the term *observable* will be used more loosely in the sense of *measurable quantity*. Thus, anything that can be measured is an observable in this sense, and a state vector Ψ is not an observable. A special role is played by the measurable quantity *time* to which no operator can be assigned. On the other hand energy (cf. Hamilton operator), coordinates of space and momentum, components of angular momentum, also spin, and spin operators are observables, also in the strict sense of the above definition.

About This Book

This book contains the essential material that was presented in nuclear-physics courses for graduate students at the University of Cologne. Therefore, the references in this book, but especially the list of general references in the front part of the book contain also selected textbooks in German that may not be accessible easily to an international readership.

The author, an experimental nuclear physicist, considers the intertwining of experimental facts, experimental methods, and tools, with basic theoretical knowledge a good method to teach the subject.

The problems attached to each chapter serve rather to elucidate and detail physical ideas that could not be presented in full detail in this text, than to give an ample collection for classroom use which would exceed the space available. Many good (older) books contain such collections.

Subjects such as (relativistic or non-relativistic) kinematics will therefore not be treated extensively. However, a basic knowledge of both is a prerequisite for understanding nuclear reactions and the connections to particle and high-energy physics.

Acknowledgments

The author acknowledges many fruitful discussions, not only on few-body problems, with his friend Walter Glöckle, who passed away, much too early, in 2012. He gave few-body theory an entirely new impact and scope and, thus, incited also many important experiments.

Thanks go to Claus Kiefer, Cologne, for an enlightening discussion on different aspects of the term "Observable".

The author is very grateful to Morton Hjorth-Jensen, Oslo, for reading the manuscript very carefully and suggesting many improvements.

References

[BLA25] P.M.S. Blackett, Proc. R. Soc. A **107**, 349 (1925)
[GEI13] H. Geiger, E. Marsden, Philos. Mag. **25**, 604 (1913)
[GEN40] W. Gentner, H. Maier-Leibnitz, W. Bothe, *Atlas Typischer Nebelkammerbilder (Atlas of Typical Expansion Chamber Photographs)* (Springer, Heidelberg, 1940). Reprint (Springer, Berlin, 2013)
[HER99] W.T. Hering, *Angewandte Kernphysik* (Teubner, Leipzig, 1999)
[RUT11] E. Rutherford, Philos. Mag. **21**, 669 (1911)
[RUT19] E. Rutherford, Philos. Mag. **37**, 537, 581 (1919)
[SES07] A. Sessler, E. Wilson, *Engines of Discovery—A Century of Particle Accelerators* (World Scientific, Singapore, 2007)

About the Author

Born 1938 in Coburg, Germany. Studies of Physics, Mathematics, and Philosophy at Universities at Stuttgart, Hamburg, and Basel. Physics Diploma (1964) and Dr. phil. (1966) from the University of Basel. 1967–1970 Postdoc at Basel and Ohio State University, Columbus, OH. 1970 Visiting Assistant Professor at OSU. Since 1971 at the University of Cologne. Habilitation 1973. Apl. Professor 1978, University Professor since 1983. Retired since 2004. Member of the German Physical Society DPG, member and fellow of the American Physical Society APS. Main fields of interest: Low-energy nuclear reactions, especially few-body and fusion reactions, spin physics, and polarization methods.

Contents

Part I Nuclear Reactions

1 Introduction: Role of Nuclear Reactions in Nuclear and Particle Physics 3
  1.1 Nuclear Reactions in Nuclear Spectroscopy 3
  1.2 Extension of Nuclei 3
  1.3 Typical Energies 5
    1.3.1 Binding Energies of Nuclei 5
    1.3.2 Coulomb Barrier 6
    1.3.3 "Optical" Argument 7
  1.4 Nuclear Reaction Models 8
    1.4.1 Hierarchy of Excitations 8
  1.5 Exercises 10
  References 11

2 Classical Cross Section 13
  2.1 Deflection Function 13
  2.2 Rutherford Scattering 15
    2.2.1 Rutherford Scattering Cross Section 15
    2.2.2 Minimal Scattering Distance d 16
    2.2.3 Trajectories in the Point-Charge Coulomb Field 17
    2.2.4 Consequences 18
    2.2.5 Consequences of the Rutherford Experiments and Their Historic Significance 19
    2.2.6 Quantum-Mechanical Derivation of Rutherford's Formula 19
    2.2.7 Deviations from the Rutherford Formula 20
  2.3 Scattering, Density Distributions, and Nuclear Radii 21
    2.3.1 Nuclear Radii from Deviations from Rutherford Scattering 21
    2.3.2 Coulomb Scattering from an Extended Charge Distribution 21

		2.3.3	Ansatz for Models	26
		2.3.4	Expansion into Moments	26
		2.3.5	Results of Hadron Scattering	28
	2.4	Electron Scattering		29
		2.4.1	Matter-Density Distributions and Radii	33
		2.4.2	Special Cases—Neutron Skins and Halo Nuclei	33
	2.5	Exercises		36
	References			37
3	**Role of Conservation Laws and Symmetries in Nuclear Reactions**			39
	3.1	Generalities		39
		3.1.1	Discrete Transformations	41
		3.1.2	Continuous Transformations	41
	3.2	Conserved Quantities in Nuclear Reactions		42
		3.2.1	Energy Conservation	42
		3.2.2	Momentum Conservation	42
		3.2.3	Reaction Kinematics	43
		3.2.4	Conservation of Angular Momentum	43
		3.2.5	Conservation of Parity	44
		3.2.6	Nuclear Reactions Under Parity Conservation	45
		3.2.7	Nuclear Reactions Under Parity Violation	46
	3.3	Isospin in Nuclear Reactions		48
		3.3.1	Formalism	48
		3.3.2	Isospin as Conserved Quantity	50
		3.3.3	Isospin Breaking	51
	3.4	Exchange Symmetry in Nuclear Reactions of Identical Particles		52
		3.4.1	Identical Bosons with Spin $I = 0$	54
		3.4.2	Identical Fermions with Spin $I = 1/2$	54
	3.5	Time-Reversal Invariance		54
		3.5.1	Time Reversal, Reciprocity, and Detailed Balance	56
		3.5.2	Other Nuclear Observables of Possible Time Reversal Violation TRV	58
	3.6	Exercises		59
	References			59
4	**Cross Sections**			61
	4.1	General Appearance of Cross Sections		61
		4.1.1	Neutral Particles—Elastic Scattering (n, n)	61
		4.1.2	Neutral Particles—Inelastic Neutron Scattering	62
		4.1.3	Neutral Particles—Exothermic Reactions with Thermal Neutrons $(n, \gamma), (n, p), (n, f)$ etc.	63
		4.1.4	Neutral Incident Particles—Endothermic Reactions $(Q < 0)$ with Charged Exit Channel	63
		4.1.5	Charged Particles in the Entrance and Exit Channels	64
		4.1.6	Threshold Effects	64
		4.1.7	Other Phenomena	65

	4.2	Formal Description of Nuclear Reactions	67
		4.2.1 Wave Function and Scattering Amplitude	67
		4.2.2 Scattering Amplitude and Cross Section	68
		4.2.3 Schrödinger Equation	69
		4.2.4 The Optical Theorem	73
	4.3	Remark and Exercise	74
	References		75
5	**Polarization in Nuclear Reactions—Formalism**		**77**
	5.1	Polarization Formalism	77
	5.2	Expectation Value and Average of Observables in Measurements	78
	5.3	Density Operator, Density Matrix	78
		5.3.1 General Properties of ρ	79
		5.3.2 Density Matrix of the General Mixed State	79
		5.3.3 Examples for Density Matrices	80
		5.3.4 Complete Description of Spin Systems	84
		5.3.5 Expansions of the Density Matrix, Spin Tensor Moments	85
	5.4	Rotations, Angular Dependence of the Tensor Moments	97
		5.4.1 Generalities	97
		5.4.2 The Description of Rotations by Rotation Operators	97
		5.4.3 Rotation of the Density Matrix and of the Tensor Moments	99
		5.4.4 Practical Realization of Rotations	101
		5.4.5 Coordinate Systems	101
	5.5	Exercises	102
	References		103
6	**Nuclear Reactions of Particles with Spin**		**105**
	6.1	General	105
	6.2	The M Matrix	106
	6.3	Types of Polarization Observables	108
	6.4	Coordinate Systems	110
		6.4.1 Coordinate Systems for Analyzing Powers	110
		6.4.2 Coordinate Systems for Polarization Transfer	111
		6.4.3 Coordinate Systems for Spin Correlations	112
	6.5	Structure of the M Matrix and Number of "Necessary" Experiments	115
	6.6	Examples	118
		6.6.1 Systems with Spin Structure $1/2 + 0 \longrightarrow 1/2 + 0$	118
		6.6.2 Systems with Spin Structure $1/2 + 1/2 \longrightarrow 1/2 + 1/2$	119
		6.6.3 Systems with Spin Structure $\vec{\tfrac{1}{2}} + \vec{1}$ and Three-Nucleon Studies	120
		6.6.4 Systems with Spin Structure $\vec{1} + \vec{1}$ and $\vec{\tfrac{1}{2}} + \vec{\tfrac{1}{2}}$ and the Four-Nucleon Systems	121

		6.6.5 Practical Criteria for the Choice of Observables	121

- 6.7 Exercises 122
- References 122

7 Partial Wave Expansion 125
- 7.1 Neutral Particles 125
- 7.2 Charged Particles 128
- 7.3 Exercises 129
- References 130

8 Unpolarized Cross Sections 131
- 8.1 General Features 131
- 8.2 Inelasticity and Absorption 133
- 8.3 Low-Energy Behavior of the Scattering 136
  - 8.3.1 Scattering Length a 137
  - 8.3.2 Analytically Solvable Models for the Low-Energy Behavior 139
- 8.4 Exercises 141
- References 143

9 The Nucleon-Nucleon Interaction 145
- 9.1 The Observables of the NN Systems 145
  - 9.1.1 NN Observables 146
  - 9.1.2 NN Scattering Phases 148
  - 9.1.3 NN Interaction as Exchange Force 149
- 9.2 Few-Nucleon Systems 152
  - 9.2.1 The Two-Nucleon System 153
  - 9.2.2 The Three-Nucleon System 154
  - 9.2.3 Elastic Scattering in the Three-Nucleon System ... 154
  - 9.2.4 Kinematics of Three-Nucleon Breakup Reactions ... 154
  - 9.2.5 Results for the Three-Nucleon Breakup Reaction ... 159
  - 9.2.6 Recent Progress in Few-Nucleon Reactions ... 159
  - 9.2.7 Other Few-Nucleon Systems 162
- 9.3 Exercises 162
- References 165

10 Models of Reactions—Direct Reactions 167
- 10.1 Generalities 167
- 10.2 Elastic Scattering 167
- 10.3 Optical Model 168
- 10.4 Direct (Rearrangement) Reactions 169
- 10.5 Stripping Reactions 171
- 10.6 T Matrix and Born Series 174
  - 10.6.1 Integral Equations 174
- 10.7 Born Approximation 175
  - 10.7.1 First Born Approximation = PWBA = Plane Wave Born Approximation 175

	10.7.2 Distorted Wave Born Approximation = DWBA	176
10.8	Details of the Born Approximations	176
10.9	Exercises	179
	References	180

11 Models of Reactions—Compound-Nucleus (CN) Reactions 181
- 11.1 Generalities 181
- 11.2 Theoretical Shape of the Cross Sections 182
- 11.3 Derivation of the Partial-Width Amplitude for Nuclei (s Waves Only) 184
- 11.4 Role of Level Densities 185
  - 11.4.1 Induced Nuclear Fission 187
- 11.5 Single Resonances 188
  - 11.5.1 General Features 188
  - 11.5.2 Alternative Description of Resonances 189
  - 11.5.3 Overlapping Resonances 192
  - 11.5.4 Ericson Fluctuations 192
  - 11.5.5 Analysis of Ericson Fluctuations 196
  - 11.5.6 Results for Level Widths 197
- 11.6 Complete Averaging over the CN States 197
  - 11.6.1 Generalities 197
  - 11.6.2 Hauser-Feshbach Formalism 199
- 11.7 Exercises 202
- References 204

12 Intermediate Structures 207
- 12.1 Heavy-Ion Scattering and Molecular Resonances 207
- 12.2 Structures in Neutron Reactions 209
  - 12.2.1 Neutron-Nucleus IS 209
  - 12.2.2 Single-Particle Neutron Resonances 209
- 12.3 Giant Resonances 210
- 12.4 Fission Doorways 213
- 12.5 Isobaric Analog Resonances (IAR) 213
- 12.6 Exercises 217
- References 218

13 Heavy-Ion (HI) Reactions 221
- 13.1 General Characteristics of HI Interactions 221
- 13.2 Semi-Classical Phenomena and Description 223
  - 13.2.1 Elastic Scattering 224
  - 13.2.2 Other HI Reaction Models 225
  - 13.2.3 Molecular Resonances 225
  - 13.2.4 Heavy-Ion Reactions and Superheavies SH 225
- 13.3 Nuclear Spectroscopy and Nuclear Reactions 226
- 13.4 Heavy-Ion Reactions as Special Tools for Nuclear Spectroscopy . 226
  - 13.4.1 Coulomb Excitation 226

		13.4.2 Fusion-Evaporation Reactions	227
	13.5	Exercises	228
	References		229

14 Nuclear Astrophysics 231
    14.1 Reaction Rates 231
    14.2 Typical S-Factor Behavior 232
        14.2.1 Calculation of Reaction Rates in Plasmas 234
    14.3 Exercises 239
    References 240

15 Spectroscopy at the Driplines, Exotic Nuclei, and Radioactive Ion Beams (RIB) 241
    15.1 Use of RIB 241
        15.1.1 Nuclear Radii and Neutron vs. Proton Distributions . . . 241
        15.1.2 Nuclear Models for Exotic Nuclei 242
        15.1.3 Giant Resonances of Exotic Nuclei 242
    15.2 Production of Radioactive-Ion Beams 242
        15.2.1 The ISOL Principle 242
        15.2.2 IFF and Post-Accelerating Schemes 243
    15.3 Nuclear Reactions and the Way to Superheavies 244
    References 246

Part II Tools of Nuclear Reactions

16 Accelerators 251
    16.1 Electrostatic Accelerators 251
        16.1.1 The Cockroft-Walton Accelerator 251
        16.1.2 The Van-de-Graaff (VdG) Accelerator 252
    16.2 RF Accelerators 255
        16.2.1 The Linear Accelerator (LINAC) 255
        16.2.2 The Cyclotron 255
        16.2.3 Betatron and Synchrotron 258
    16.3 Beam Forming and Guiding Elements 260
        16.3.1 Electrostatic Lenses 262
        16.3.2 Magnetic Lenses 263
        16.3.3 Strong Focusing 265
    16.4 Ion Sources 266
        16.4.1 Unpolarized Beams 266
        16.4.2 Polarized Beams and Targets 269
    16.5 Physics and Techniques of the Ground-State Atomic Beam Sources ABS 271
        16.5.1 Production of H and D Ground-State Atomic Beams . . . 271
        16.5.2 Dissociators, Beam Formation and Accommodation . . . 272
        16.5.3 State-Separation Magnets—Classical and Modern Designs 273
        16.5.4 RF Transitions 275

Contents

	16.6	Ionizers	279
		16.6.1 Ionizers—Electron-Bombardment and Colliding-Beams Designs	279
		16.6.2 Sources for Polarized ^{6,7}Li and ^{23}Na Beams	283
		16.6.3 Optically Pumped Polarized Ion Sources (OPPIS)	283
	16.7	Physics of the Lambshift Source LSS	284
		16.7.1 The Lamb Shift	284
		16.7.2 Level Crossings and Quench Effect	285
		16.7.3 Enhancement of Polarization	286
		16.7.4 Production and Maximization of the Beam Polarization	288
	16.8	Spin Rotation in Beamlines and Precession in a Wien Filter	293
		16.8.1 Spin Rotation in Beamlines	294
		16.8.2 Spin Rotation in a Wien Filter	295
	16.9	Exercises	297
	References		299
17	**Detectors, Spectrometers, and Electronics**		301
	17.1	Ionization Chambers	301
	17.2	Scintillation Detectors	302
	17.3	Solid-State Detectors	303
		17.3.1 Si Detectors	304
		17.3.2 Particle Identification	306
		17.3.3 Magnetic Spectrographs	306
		17.3.4 Ge Detectors	307
	17.4	Neutrons	309
		17.4.1 Production of Neutrons	309
	17.5	Neutron Detectors	310
		17.5.1 Neutron-Induced Reactions	311
		17.5.2 Neutrons as Reaction Products	311
		17.5.3 Different Neutron Detection Methods Depending on Neutron Energies	311
	17.6	Polarized Neutrons	313
		17.6.1 Magnetized Materials	313
		17.6.2 Polarized ^{3}He as Spin Filter	314
	17.7	γ Spectroscopy	315
	References		318

Part III   Applications of Nuclear Reactions and Special Accelerators

18	**Medical Applications**		321
	18.1	Particle (Hadron) Tumor Therapy	321
	18.2	Isotope Production	323
	18.3	Exercises	324
	References		325

19	**Nuclear-Energy Applications**	327
	19.1 Fusion-Energy Research	327
	19.1.1 Fusion Basics	327
	19.1.2 Nuclear Cross Sections	327
	19.2 Five-Nucleon Fusion Reactions	328
	19.2.1 "Polarized" Fusion	329
	19.3 Four-Nucleon Reactions	331
	19.3.1 Suppression of Unwanted DD Neutrons	332
	19.3.2 Possible Reaction-Rate Enhancement for the DD Reactions by Polarization?	334
	19.4 Other Fusion Reactions	334
	19.5 Present Status of "Polarized" Fusion	334
	19.6 New Calculations for Few-Body Systems	335
	19.6.1 Theoretical Approaches	336
	19.7 New Aspects of Polarized Fusion	337
	19.7.1 Effect of Electron Screening	337
	19.7.2 Rate Enhancement and Electron Screening	337
	19.7.3 Pellet Implosion Dynamics	338
	19.7.4 Technical Questions	338
	19.7.5 Preservation of Polarization on Injection	339
	19.8 Future of Polarized Fusion	339
	19.9 Transmutation of Nuclear Waste	339
	19.10 Exercises	340
	References	341
20	**Other Important Applications of Nuclear-Reaction Techniques**	345
	20.1 Archaeology, Geology, and Art	345
	20.1.1 Archaeology and Age Determination	345
	20.2 Materials Analysis and Modification	346
	20.2.1 RBS	347
	20.2.2 PIXE	347
	20.2.3 Neutron Radiography	348
	20.2.4 Materials Modifications	349
	20.3 Exercises	349
	References	350
21	**Trends and Future Developments of Nuclear Reactions**	351
	21.1 Exotic Nuclei	351
	21.2 Low-Cross Section Reactions	352
	21.3 Accelerator Developments	352
	21.4 Theoretical Progress	353
	References	353
22	**Appendices**	355
	22.1 Appendix A—Tables of Useful Numbers and Relations	355
	22.2 Appendix B—Practical Units	356

    22.3 Appendix C—Angular-Momentum Recoupling Coefficients . . . 357
          22.3.1 Coupling of Two Angular Momenta 357
          22.3.2 Coupling of Three Angular Momenta 357
          22.3.3 Coupling of Four Angular Momenta 358
          22.3.4 Spherical Harmonics 358
    22.4 Appendix D—General Resources 359
    References 359

Index 361

Part I
Nuclear Reactions

Chapter 1
Introduction: Role of Nuclear Reactions in Nuclear and Particle Physics

1.1 Nuclear Reactions in Nuclear Spectroscopy

We define nuclear spectroscopy as the science of learning all about the properties of the thousands of nuclides, each with individual and also collective properties. Aside from early studies of radioactive decays, nuclear reactions have been the tool to investigate the action of nuclear forces (in the sense of an interplay of the strong interaction proper, the electromagnetic, and the weak force). In high-density situations, e.g. in neutron stars, even the gravitational force enters the stage via the density dependence of the nuclear interactions. The aim of modern nuclear spectroscopy is now moving away from stable nuclei, from deformed highly excited nuclei with high angular momenta on to the investigation of nuclei in the regions near the limits of existing nuclei with either high neutron excess, high neutron deficiency, or the region of new elements, the superheavy nuclei. They can be characterized by their isospin $T = (N - Z)/A$ (for details see Chap. 12. Z is the charge number of the nucleus considered, determining its position in the periodic table and thus its chemical properties, N the neutron number of the specific isotope, and A its mass number). Thus a three-dimensional plot with the coordinates excitation energy, angular momentum, and isospin charts the nuclear landscape to be explored. The latest developments encompass therefore nuclear reactions with *exotic* beams with radioactive (or: rare) ion-beam facilities (RIB) exploring the *drip-line limits* of the nuclear chart, the astrophysical chains of reactions such as the r-process, and the changes of nuclear models such as the shell model with varying isospin. Figure 1.1 illustrates this nuclear parameter space. In the following chapters details of the different nuclear reaction models and tools will be discussed.

1.2 Extension of Nuclei

Since the epoch-making scattering experiments of Geiger, Marsden, and Rutherford around 1912 and their interpretation the basic structure of the atom is known. It

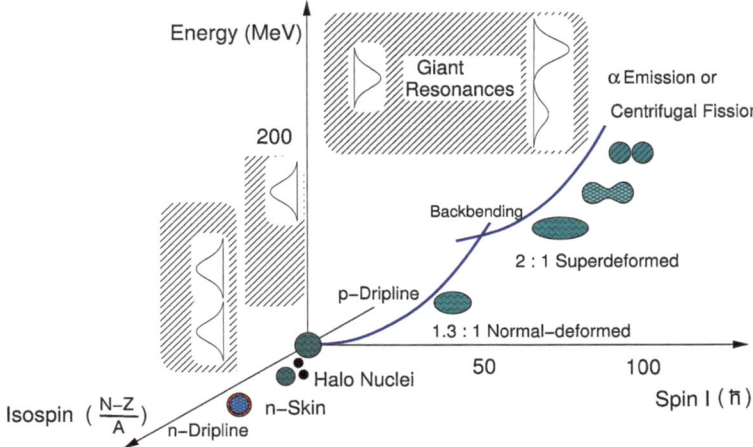

Fig. 1.1 Nuclear parameter space with phenomena classified according to excitation energy, angular momentum, and isospin

consists of a compact (i.e. small and massive) nucleus, which carries the charge Ze (Z = nuclear charge number = element number of the periodic table) and nearly the entire mass of the atom. This evidence formed the basis of Bohr's (and later Sommerfeld's) atomic-structure models of electrons circulating around the nucleus like planets around the sun (with some unsatisfactory ad-hoc assumptions that became obsolete only around 1924 by the new *quantum mechanics*). The explanation of the existence of isotopes and the correct placement of them in the *chart of nuclides* (Z vs. N) required the discovery of the neutron in 1932.

Already Rutherford could—by comparing the measured scattering angular distribution of α particles on gold with his Ansatz of a point-Coulomb interaction—conclude that the nucleus is an object smaller than the scattering distances (order of magnitude: 1 fm = $1 \cdot 10^{-15}$ m). The very fact that scattering at backward angles occured, showed that the scattering center had to be heavier than the α (this is pure kinematics). The electron cloud relative to this is very large (order-of-magnitude radius: 1 Å = $1 \cdot 10^{-10}$ m) and carries the charge $-Ze$ such that the atom is exactly neutral. After the invention of accelerators the use of α particles of much higher energies with penetration into the target nucleus was possible and the extension (the radius) of nuclei could be obtained by the onset of deviations from the point-Coulomb scattering. A key role is played here by the *charge form factor* and its Fourier transform, the *charge-density distribution*. It expresses how strongly the Coulomb potential of an extended (often simply assumed to be homogeneous) charge distribution in the nuclear interior deviates from that of a point charge or what the influence of the (hadronic) nuclear interaction on the observables is, see Fig. 2.6.

Using charged leptons as probes, which have no measurable extension and do not feel the strong interaction, charge (and current) distributions in nuclei and nucleons have been determined. At higher momentum transfer (i.e. at high energies and large scattering angles via *inelastic or quasi-elastic scattering*) excited states of the nucleon and later-on, (via *deep-inelastic scattering*), substructures of the nucleons (*partons*) were discovered that had all the properties of quarks: 1/3 charges, spin $1/2\hbar$, color charge and confinement, characteristics of truly elementary particles (point shape, no internal structure), and they proved to be sources of the strong, electromagnetic, and weak interactions, also by probing them with neutrinos.

1.3 Typical Energies

The order-of-magnitude of the energies relevant in nuclear reactions will be shown in three examples:

1.3.1 Binding Energies of Nuclei

To construct a spherical nucleus with an assumed homogeneous charge distribution and a charge radius R from all its nucleons (protons) all protons (total charge $Q_{tot} = Ze$ with Z the charge number) must be brought together from infinity against their mutual rejection and, therefore, at the expense of the electrostatic binding energy. The increase of Coulomb energy between a charge element dQ and a partially filled sphere with radius $r < R$ and charge $Q_r = Q_{tot}(r^3/R^3)$, and $dQ_r = Q_{tot}(3r^2/R^3)dr$ is

$$dE_{\text{Coul}} = \int_\infty^r \frac{Q_r dQ_r}{x^2} dx = 3 \cdot \frac{Q_{tot}^2}{R^6} r^4 dr. \tag{1.1}$$

From this by integration the required total energy is obtained:

$$E_{\text{Coul}} = \int_\infty^R dE_{\text{Coul}} = \frac{3}{5} \frac{(Ze)^2}{R}, \tag{1.2}$$

which is numerically

$$E_{\text{Coul}} = 0.72 \frac{Z^2}{A^{1/3}} \tag{1.3}$$

(with $R = R_0 A^{1/3}$, $R_0 \approx 1.2$ fm the radius constant, see Chap. 2, A the mass number of the nucleus). For a heavy nucleus such as Pb this amounts to $\approx$819 MeV (about 4 MeV/nucleon). The binding energies of the nuclei must therefore be larger than these values. A nuclear reaction in which the projectile should probe the interior of the target nuclei must therefore proceed at appropriately high energies. This explains the necessity of developing and using accelerators because the energies of the "beams" from radioactive sources are very limited to a few MeV. Figure 1.2 shows the development of the accelerator energies since 1930, and it is interesting

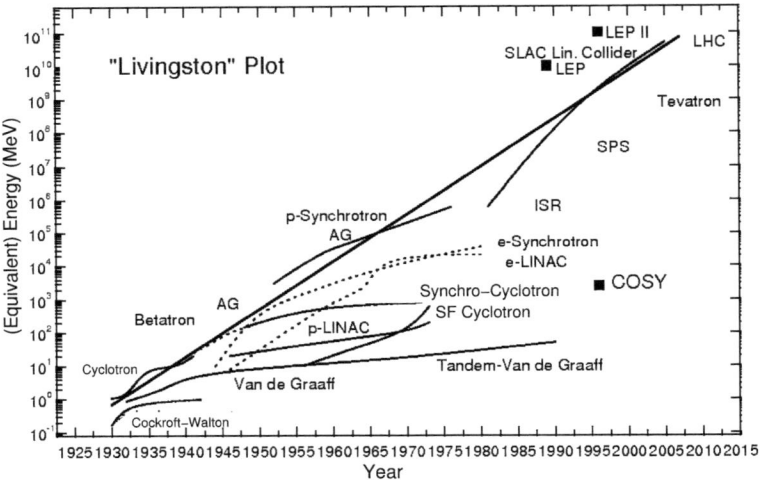

Fig. 1.2 Time development of the accelerator energies. The *straight line* corresponds to an energy doubling time of about seven years. This type of plot was first created by M.S. Livingston [LIV54] who co-invented the first cyclotron with E.O. Lawrence and also was instrumental in the invention of "strong focusing", see Sects. 16.2.2 and 16.3.3

to see the onset of new technologies every time when an older line of development became "saturated" (Livingston plot). Parallel to the progress in accelerators were the developments in detection methods, detector technologies, electronics and computing. The principles of the main tools of nuclear reactions are discussed in Part II. Applications in many other fields are continuing to gain importance, especially in medicine, art, archaeology, materials analysis and modification, technology etc., see Part III.

1.3.2 Coulomb Barrier

The height of the Coulomb barrier

$$V_C = \frac{Z_1 Z_2 e^2}{R} \approx 1.44 \frac{Z_1 Z_2}{R(\text{fm})} (\text{MeV}) = 1.44 \frac{Z_1 Z_2}{R_0(A_1^{1/3} + A_2^{1/3})(\text{fm})} (\text{MeV}) \quad (1.4)$$

is decisive for a charged particle with charge number Z_1 to be able to penetrate into a nucleus with charge number Z_2 and to initiate a reaction or not (this energy is the one calculated in the "c.m. system", i.e. that of the relative motion). For illustration see Figs. 2.4 and 13.1.

1.3 Typical Energies

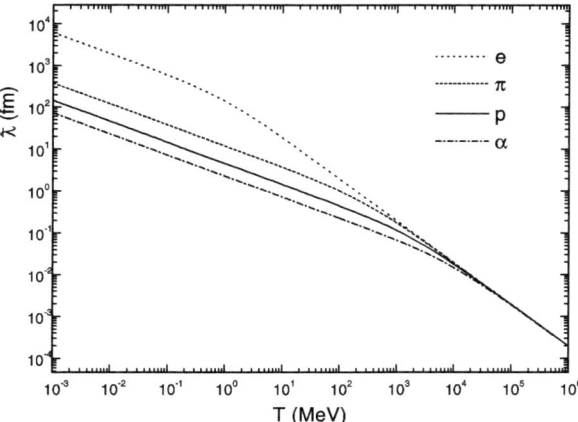

Fig. 1.3 The reduced de Broglie wavelength ƛ of different particles as functions of their kinetic energy T

1.3.3 "Optical" Argument

The de-Broglie wavelength of a particle beam is determined by its momentum:

$$\lambdabar_{\text{de Broglie}} = \frac{\hbar}{p} = \frac{\hbar c}{\sqrt{T^2 + 2(m_0 c^2) T}} = \frac{197 (\text{MeV})}{\sqrt{T^2 + 2(m_0 c^2) T}} (\text{fm}). \qquad (1.5)$$

T is the kinetic energy of the particles and is often used in a relativistic context when kinetic and total energy, i.e. including the rest energy have to be distinguished. At low energies, i.e. in low-energy nuclear physics and in this book E is used for the kinetic energy.

Analogous to light optics the resolution of an device or experiment is limited by diffraction such that only objects can be observed separately, which have at least a distance from each other of about one wavelength. Details of the spatial nuclear structure can only be resolved with electrons with energies $\gg 197$ MeV (there $\lambdabar_{\text{de Broglie}} = 1$ fm) or with protons with energies $\gg 1$ MeV. The argument is somewhat imprecise: more precisely the wavelength of the radiation exchanged in the interaction should be taken as measure, i.e. of the virtual photons in the electron-nucleus scattering, or that of the exchanged mesons for the nuclear interaction, and that of the gluons in the strong interaction. The transferred momentum $\hbar \vec{K}$, which has to be used here depends on the projectile momentum and also on the scattering angle. For elastic scattering ($|\vec{k}_{\text{in}}| = \hbar |\vec{k}_{\text{out}}|$) with the law of cosines one obtains

$$\hbar |\vec{K}| = \hbar |\vec{k}_{\text{in}} - \vec{k}_{\text{out}}| = 2\hbar |\vec{k}_{\text{in}}| \sin(\theta/2). \qquad (1.6)$$

Figure 1.3 shows the reduced de Broglie wavelength of electrons, pions, protons, and α particles as functions of the kinetic energy T. In the range of highly relativistic energies ƛ does not depend on the particle mass and decreases with T^{-1} (instead of non-relativistically with $T^{-1/2}$).

1.4 Nuclear Reaction Models

Lacking a complete and fundamental theory of the nuclear interaction (the application of the quantum chromodynamics (QCD) as a fundamental theory of the strong interaction meets unsurmountable difficulties in the non-perturbative regime) in nuclear physics, i.e. also for the description of nuclear reactions, there is no complete Hamiltonian. Very promising steps have recently been successful in describing nuclear structures of light nuclei and reactions between them in the framework of EFT ("effective-field" theories, i.e. low-energy approximations of the QCD, also χPT, "chiral-perturbation" theory) and even lattice-gauge calculations. But, in general, one has to rely on models of nuclei and the nuclear interactions with their naturally limited range of validity. The different nuclear reaction models can be classified in different ways, but a classification according to times scales of interactions is especially intuitive with the limiting cases of direct (fast) reactions on the one hand, (slow) compound nuclear reactions on the other. Typical models for fast reactions are e.g.:

- The optical model of elastic scattering (OM).
- DWBA ("Distorted Waves Born Approximation"), CCBA ("Coupled Channels Born Approximation") for direct inelastic scattering or rearrangement reactions such as stripping or pickup reactions.

Slow processes via intermediate states e.g. isolated resonances are described with the

- R-matrix and Breit-Wigner theories for compound-nucleus (CN) resonances

Between both limits there are e.g.

- Hauser-Feshbach and statistical theories for reactions via strongly overlapping compound-nucleus (CN) states.
- Chaos theories describing distribution functions of resonances and Ericson fluctuations.
- Descriptions of "intermediate" structures by "doorway" states (examples are: isobaric-analog resonances (IAR), neutron single-particle resonances, giant resonances, and fission doorways).
- Incomplete fusion and pre-compound nuclear reactions.

In following chapters of this book the most important of these models are discussed and references for further and in-depth reading are given there.

1.4.1 Hierarchy of Excitations

The classification of structures in excitation functions reveals that there seems to exist a certain hierarchy, i.e. the widths of the observed structures (equivalent to the lifetimes of the corresponding states via Heisenberg's uncertainty relation $\Gamma = \hbar/\tau$)

1.4 Nuclear Reaction Models

Fig. 1.4 Hierarchy of excitations. The figure illustrates the spreading of simple structures into more complicated ones

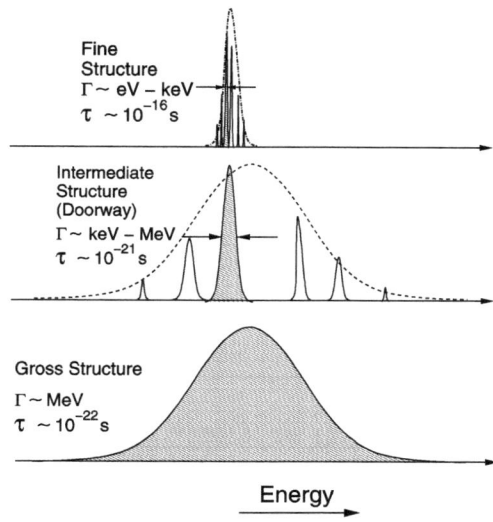

are sufficiently different to assign them to at least three groups: *Fine structure, intermediate, and gross structure*. Very schematically the fine-structures can be associated with compound-nucleus states, which are complicated n-particle-n-hole configurations (in the language of the shell model). The gross structures are simple excitations, e.g. 1-particle-1 hole states, and intermediate structures can be associated with few-particle-few-hole states, e.g. 2-particle-2-hole configurations, often called *doorways*. It is characteristic that the gross and intermediate structures may decay directly ("escape") as well as via narrower (longer-living) structures ("spreading"). Thus, their widths are the sum of the two (sometimes three) widths.

$$\Gamma_{\text{tot}} = \Gamma^\uparrow(\text{escape}) + \Gamma^\downarrow(\text{spreading}). \tag{1.7}$$

The shape of the fine structure state distribution, i.e. its strength and energy distribution normally in first order is Lorentzian. Figure 1.4 illustrates this schematically. Several examples of at least three distinct levels of structures have been identified in nuclear physics. The finest structures are associated with true compound states and, in the limit, participation of many or all nucleons. The extreme gross structures are connected to single (quasi)-particle excitations, whereas the intermediate structures stem from few particle-few-hole excitations such as th giant resonances.

Experimentally, with sufficient energy resolution, the fine structures have been resolved (examples are the isobaric-analog resonances or the giant resonances). The location of the intermediate or gross structure is then given by the centroid of the fine-structure and its width from that of a Gaussian-shaped envelope. In the case of insufficient experimental resolution or, after deliberate averaging over fine structures, intermediate or gross structures are the only remaining visible structures.

In the reference list of this chapter—besides a few specific references—a number of general textbooks of general interest, covering the subject of nuclear reactions, is given without citations in the text (clearly the selection is a matter of taste of the author and not meant to be an evaluation). In the later chapters the references are more specific for the content of the respective chapter.

1.5 Exercises

1.1. Make yourself acquainted with the relativistic relations between total energy E, kinetic energy T, velocity v or $\beta = v/c$, 3-momentum $\vec{p}$, and rest mass of particles. Show the conditions for approximate non-relativistic relations as used in low-energy nuclear physics. What does the value of

$$\gamma = \frac{1}{\sqrt{1-\beta^2}} = 1 + \frac{T}{m_0 c^2} \tag{1.8}$$

tell us?

1.2. Using non-relativistic kinematics, show how in elastic scattering the mass ratio m_1/m_2 (with m_1 the mass of a projectile, m_2 that of the target nuclei at rest) enters the range of allowed scattering angles in the lab. and c.m. systems. (For the case of equal masses think of the billiard balls.) Discuss the case of a heavier nucleus scattered elastically from a lighter one, apply this to deuterium-hydrogen scattering ^{1}H(d, p)^{2}H and give the relations of lab. vs. c.m. scattering angles, as well as energies of the outgoing deuterons and protons vs. lab. angles.

1.3. The semi-classical theory of energy loss of heavy charged particles (mass M) in matter assumes that the main effect of energy degradation is caused by many collisions with momentum transfer to atomic electrons (mass m) at statistically varying impact parameters. This is the basis of the Bethe-Bloch theory of stopping power, see Sect. 17.3.1. How does the extreme mass ratio M/m determine the appearance of a trace in a Wilson cloud chamber, cf. Fig. 4?

1.4. Verify the main part of the Bethe-Boch formula semi-classically by calculating the energy transferred on an atomic electron for a given impact parameter b, see Eq. (17.1) and Sect. 2.2.2.

   Hints: Calculate first the momentum $p_\perp(b)$ transferred to one electron at collision parameter b, assume a statistical (uniform) distribution of b for many electrons of charge density NZe, then integrate over b from a minimum $b_{\min}$ determined from the ionization energy I to a maximum $b_{\max}$ (minimal momentum transfer) from the classical kinematical limit. Employ the *Sudden Approximation* in which only the momenta perpendicular to the particle's path contribute, the longitudinal components cancel in the integral.

1.5. Summarize, along which lines of thought Rutherford's conclusions about the nuclear atom, with a compact and heavy nucleus as source of a point Coulomb field, can be understood? Is there a "proof" (in the mathematical sense) of Rutherford's model (think of Popper's "falsification only!" [POP34])—or why do we think this model is right? Apply this reasoning to the scientific method in general: For how long is a model considered "good". (Example: N. Bohr's model of the atom; why have his "ad hoc" assumptions not been satisfactory?)

1.6. For the dynamics of nuclear reactions the "energy, available in the c.m. system ($\tilde{E}$)" is one decisive parameter (the other is the scattering angle or, equivalently, the momentum transfer, see also the section on direct reactions (10.5)). Compare this quantity (simplification: $m_1 = m_2 = m$) as function of the lab. energy (energies) for the two cases: (a) Fixed target, (b) Colliding beams. Which lab. energy of a proton beam on a fixed target would be necessary to obtain the same "available" energy of 14 TeV as in the LHC?

1.7. The W and Z bosons are responsible for transmitting the weak interaction. What is its approximate range? Which features of the neutron decay $n \to p + e^- + \bar{\nu}$ follow from this range?

1.8. Until 1932, when the neutron was discovered, some nuclear-structure models assumed electrons in the nuclei to compensate for the positive proton charge. Why is this conjecture wrong?

References

[BER04] C.A. Bertulani, P. Danielewicz, *Introd. to Nucl. Reactions*. IOP Graduate Student Series in Physics (IOP, Bristol, 2004)
[BET00] K. Bethge, G. Walter, B. Wiedemann, *Kernphysik (Eine Einführung)*, 2nd edn. (Springer, Berlin, 2000)
[BLA25] P.M.S. Blackett, Proc. R. Soc. A **107**, 349 (1925)
[BLA52] J.M. Blatt, V.F. Weisskopf, *Theoretical Nuclear Physics* (Wiley, New York, 1952)
[BOD79] E. Bodenstedt, *Experimente der Kernphysik und ihre Deutung I–III* (Bibliographisches Institut, Mannheim, 1978)
[BOH69] N. Bohr, B. Mottelson, *Nuclear Structure* (Benjamin, New York, 1969)
[deS92] A. de Shalit, H. Feshbach, *Theor. Nucl. Phys. II, Nuclear Reactions* (Wiley, New York, 1992)
[FRA86] H. Frauenfelder, E.M. Henley, *Nuclear and Particle Physics, A: Background and Symmetries*. Lect. Notes and Suppl. in Phys., 2nd printing (Benjamin/Cummings, Reading, 1986)
[GEI13] H. Geiger, E. Marsden, Philos. Mag. **25**, 604 (1913)
[GEN40] W. Gentner, H. Maier-Leibnitz, W. Bothe, *Atlas Typischer Nebelkammerbilder (Atlas of Typical Expansion Chamber Photographs)* (Springer, Heidelberg, 1940). Reprint (Springer, Berlin, 2013)
[GOT86] K. Gottfried, V.F. Weisskopf, *Concepts of Particle Physics* I + II (Oxford University Press, New York, 1986)
[HEN07] E.M. Henley, A. Garcia, *Subatomic Physics*, 3rd edn. (World Scientific, Singapore, 2007)
[HOD71] P.E. Hodgson, *Nucl. Reactions and Nucl. Structure* (Clarendon Press, Oxford, 1971)
[JAC70] D.F. Jackson, in *Nuclear Reactions* (Methuen, London, 1970)
[KRA88] K.S. Krane, *Introductory Nuclear Physics* (Wiley, New York, 1988)
[LIV54] M.S. Livingston, *High-Energy Accelerators* (Interscience, New York, 1954)
[MAR70] P. Marmier, E. Sheldon, *Physics of Nuclei and Particles*, vol. II (Academic Press, New York, 1970), 811 ff.
[MAY02] T. Mayer-Kuckuk, *Kernphysik*, 7th edn. (Teubner, Stuttgart, 2002)
[PER87] D.H. Perkins, *Introd. to High Energy Physics* (Addison-Wesley, Reading, 1987)

[POP34] K. Popper, *Logik der Forschung. Zur Erkenntnistheorie der modernen Naturwissenschaft* (Springer, Wien, 1934). English: The Logic of Scientific Discovery (1995)
[RUT11] E. Rutherford, Philos. Mag. **21**, 669 (1911)
[RUT19] E. Rutherford, Philos. Mag. **37**, 537, 581 (1919)
[SAT90] G.R. Satchler, *Introd. Nucl. Reactions*, 2nd edn. (McMillan, London, 1990)
[SEG83] E. Segrè, *Nuclei and Particles* (Benjamin, Reading, 1983)
[SES07] A. Sessler, E. Wilson, *Engines of Discovery—A Century of Particle Accelerators* (World Scientific, Singapore, 2007)

Chapter 2
Classical Cross Section

2.1 Deflection Function

The question whether physical phenomena can be described "classically" or need a microscopic treatment, i.e. need quantum mechanics for their description is fundamental and still under intensive discussion. Ever larger objects (e.g. heavy ions, see below in Sect. 3.4, large molecules like the fullerenes) have been shown to have quantum properties. They exhibit e.g. interference. This touches upon basic concepts such as *Schrödinger's Cat* or *decoherence* of quantum states by the environment. For our purpose it may suffice to assume a fundamentally quantum-mechanical description that in special cases, i.e. when the relevant de Broglie wavelengths are small, may be approximated by classical methods.

Cross sections are the central observable of nuclear and particle physics. In some areas of nuclear physics (e.g. heavy-ion physics) nuclear reactions (scattering) are often treated semi-classically. Likewise, the historically important Rutherford scattering can be treated classically. The classical description implies that particles and their trajectories are localized. However, in each case it must be checked whether a classical description is valid. A criterion for classicity is (like in geometrical light optics) the wavelength of the radiation used is small as compared with some characteristic object dimension d. In agreement with Heisenberg's uncertainty relation this means

$$\lambda_{\text{deBroglie}} = \hbar/p \ll d. \tag{2.1}$$

When choosing for a typical object dimension half the distance of the trajectory turning point d_0 for a central collision the Sommerfeld criterion for classical scattering is obtained (with $\alpha = e^2/\hbar c$ Sommerfeld's fine-structure constant and β a short notation for v/c)

$$\eta_S = \frac{Z_1 Z_2 e^2}{\hbar v} = Z_1 Z_2 \frac{e^2}{\hbar c} \cdot \frac{c}{v} = Z_1 Z_2 \cdot \frac{\alpha}{\beta} \gg 1 \tag{2.2}$$

or numerically (for a very heavy target, thus at rest in the c.m. system)

$$\eta_S \approx 0.16 \cdot Z_1 Z_2 \sqrt{\frac{A_{\text{proj}}}{E_{\text{lab}}(\text{MeV})}} \gg 1 \qquad (2.3)$$

with A_{proj} and E_{lab} the projectile mass number and its kinetic energy in the lab. system, respectively.

There exist more refined criteria, which take into account that the wave nature of the radiation leads to diffraction phenomena, especially where the scattering potential changes strongly, e.g. at the nuclear surface. Therefore, a requirement is postulated that the de Broglie wavelength not change substantially by the potential gradient. For Coulomb scattering this provides a scattering-angle dependent criterion [NOE76]

$$\eta_S^2 \gg \eta_{\text{crit}}^2 = \left[\frac{\sin^2 \frac{\theta}{2}}{\cos \frac{\theta}{2}(1 - \sin \frac{\theta}{2})} \right]^2. \qquad (2.4)$$

Thus, a classical description is always possible at $\theta = 0°$ but never at $\theta = 180°$.

In the case of scattering of identical particles exchange symmetry and its ensuing interference effects entirely forbid any classical description (see below).

Here we present a complete definition of the (classical) cross section, which can be easily translated into a quantum-mechanical definition.

Definition of "Cross Section" The (differential) cross section is the number of particles of a given type from a reaction, which, per target atom and unit time, are scattered into the solid-angle element $d\Omega$ (formed by the angular interval $\theta \ldots \theta + d\theta$ and $\phi \ldots \phi + d\phi$), divided by the incident particle flux j (a current density = number of particles passing a unit area per unit time).

In the following we assume azimuthal (i.e. ϕ) independence of the scattering (e.g. valid for particles without spins or particles with spins, but with spin-independent interactions). The classical scattering situation is characterized by a definite trajectory and a unique relation between each particle incident from $r \to -\infty$ at a definite perpendicular distance b (the *impact parameter*) from the z axis and its (asymptotic) scattering angle at $r \to +\infty$ after the interaction. This definition yields the classical formula for the cross section. With the number of particles per unit time $j d\sigma = j \cdot 2\pi b db$ one obtains

$$\left(\frac{d\sigma}{d\Omega} \right)_{\text{class}} = \frac{2\pi b db}{2\pi \sin\theta d\theta} = \frac{b}{\sin\theta} \cdot \left| \frac{db}{d\theta} \right|. \qquad (2.5)$$

$b = b(\theta, E)$ contains the influence of the interaction (the dynamics). $\theta(b)$ for obvious reasons is called *deflection function*. Its knowledge determines the scattering completely.

Fig. 2.1 Classical Rutherford scattering. b is the *impact parameter*, (r, ϕ) are the polar coordinates of the projectile, θ the polar scattering angle, d is the distance of closest approach, and d_0 its minimum for a central collision

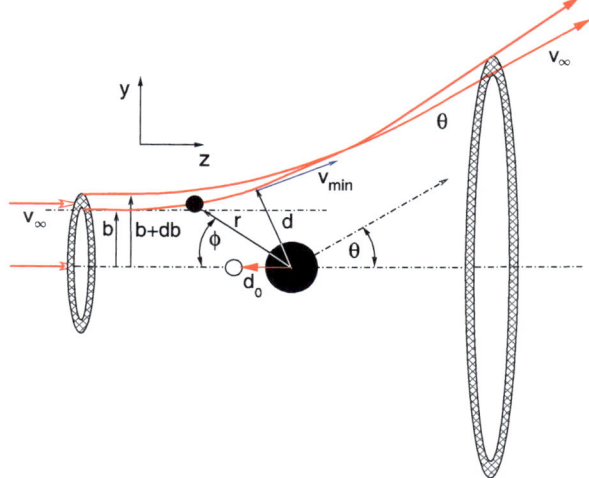

2.2 Rutherford Scattering

2.2.1 Rutherford Scattering Cross Section

The situation for the case of Rutherford scattering (for a repulsive Coulomb force between equal charges) is shown in Fig. 2.1. For the derivation of the Rutherford scattering cross section we assume:

- The projectile and the scattering center (target) are point particles (with Gauss's law it can be proved that this is also fulfilled for extended particles as long as the charge distribution is not touched upon).
- The target nucleus is infinitely heavy (i.e. the laboratory system coincides with the c.m. system).[1]
- The interaction is the purely electrostatic Coulomb force (more precisely: the monopole term of this force)[2]

$$F_C = \pm \frac{1}{4\pi\epsilon_0} \cdot \frac{Z_1 Z_2 e^2}{r^2} = \frac{C}{r^2} \tag{2.6}$$

with the Coulomb potential $V_C = \pm C/r$.

The deflection function is most simply determined by applying angular-momentum conservation and the equation of motion in one coordinate (y) (v_∞, E_∞, and

---

[1] In all scattering/reaction problems the projectile mass $m = m_a$ is correct for an infinitely heavy target or in the lab. system of coordinates with the target (mass m_A) at rest. In the c.m. system used for theoretical considerations m has to be understood as the *reduced mass* $\mu = m_a m_A/(m_a + m_A)$.

[2] The term $1/4\pi\epsilon_0$ is correct in SI units. Throughout the remainder of this book—as usual in the nuclear physics literature—it is set equal to 1 (Gaussian system of units).

p_∞ are the asymptotic (i.e. valid or prepared at $r \to \pm\infty$) quantities: projectile velocity, kinetic energy, and momentum).

$$L = mv_\infty b = mr^2\dot\phi = mv_{\min}d \tag{2.7}$$

and from this

$$dt = r^2 d\phi/v_\infty b \tag{2.8}$$

$$m\Delta v_y = \int F_y dt$$

$$v_\infty \sin\theta = \frac{C}{mv_\infty b}\int_{-\infty}^{\infty} \dot\phi \sin\phi \, dt$$

$$= \frac{C}{mv_\infty b}\int_0^{\pi-\theta} \sin\phi \, d\phi = \frac{C}{mv_\infty b}(1+\cos\theta). \tag{2.9}$$

After transformation to half the scattering angle the deflection function is

$$\cot(\theta/2) = mv_\infty^2 b/C = v_\infty L/C \tag{2.10}$$

and

$$b = \frac{C}{2E_\infty}\cdot\cot\left(\frac{\theta}{2}\right) \tag{2.11}$$

and

$$\frac{db}{d\theta} = \frac{C}{2mv_\infty^2}\cdot\frac{1}{\sin^2(\theta/2)} = \frac{C}{4E_\infty}\cdot\frac{1}{\sin^2(\theta/2)} \tag{2.12}$$

and thus for the Rutherford cross section

$$\frac{d\sigma}{d\Omega} = \left(\frac{Z_1 Z_2 e^2}{4E_\infty}\right)^2 \cdot \frac{1}{\sin^4(\theta/2)}. \tag{2.13}$$

Numerically:

$$\frac{d\sigma}{d\Omega} = 1.296\left(\frac{Z_1 Z_2}{E_\infty(\text{MeV})}\right)^2 \cdot \frac{1}{\sin^4(\theta/2)}\left[\frac{\text{mb}}{\text{sr}}\right]. \tag{2.14}$$

Figure 2.2 shows the strong angle dependence of this cross section together with the original data of Ref. [GEI13], adjusted to the theoretical curve shown.

2.2.2 Minimal Scattering Distance d

For this quantity one needs additionally the energy-conservation law:

$$\frac{mv_\infty^2}{2} = \frac{mv_{\min}^2}{2} + \frac{C}{d}. \tag{2.15}$$

2.2 Rutherford Scattering

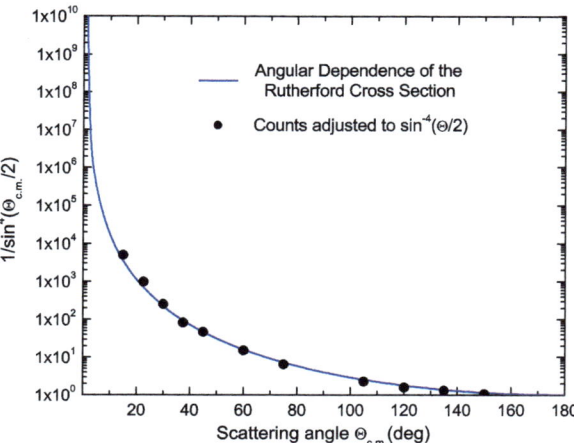

Fig. 2.2 The *curve* shows the angular dependence of the theoretical Rutherford cross section $\propto \sin^{-4}(\theta/2)$. The *points* are the original data (that consisted of tabulated numbers of counts with no error bars, and not transformed into cross-section values) of Ref. [GEI13], adjusted to the theoretical curve, giving a nearly perfect fit (Nowadays data with at least an error estimate or, better, error bars are mandatory)

The absolutely smallest distance d_0 is obtained in central collisions with:

$$E_\infty = \frac{mv_\infty^2}{2} = \frac{C}{d_0}. \tag{2.16}$$

From this and the angular-momentum conservation Eq. (2.7) the relation

$$b^2 = d(d - d_0) \tag{2.17}$$

is obtained with the solution:

$$\begin{aligned} d &= \frac{C}{2E_\infty}\left(1 + \sqrt{1 + b^2 \frac{4E_\infty^2}{C^2}}\right) \\ &= \frac{d_0}{2}\left(1 + \frac{1}{\sin\theta/2}\right). \end{aligned} \tag{2.18}$$

The classical scattering distance in relation to the minimum distance d_0 as function of the scattering angle is shown in Fig. 2.3.

2.2.3 Trajectories in the Point-Charge Coulomb Field

For the motion in a central-force field with a force $\propto r^{-2}$ classical mechanics shows that the trajectories are conic sections (for scattering, i.e. positive total energy, these are hyperbolae). To derive this one needs again the conservation laws of angular momentum and energy (with the Coulomb potential):

$$L = mr^2\dot{\phi} = \text{const} \tag{2.19}$$

Fig. 2.3 Minimal scattering distance d (in units of d_0) as function of the c.m. scattering angle

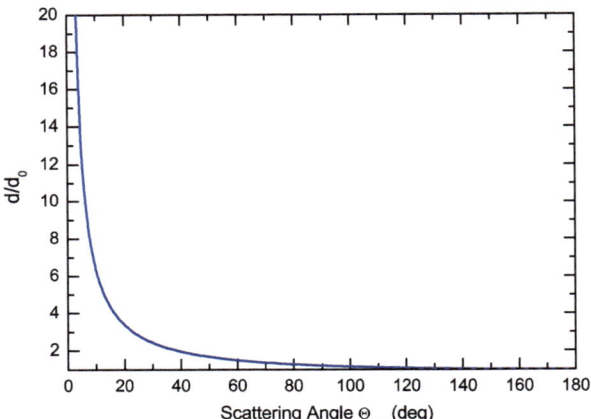

$$E = \frac{m\dot{r}^2}{2} + \frac{L^2}{2mr^2} + \frac{C}{r}. \tag{2.20}$$

In these equations dt can be eliminated. The integration of

$$d\phi = -\frac{L}{mr^2}\left[\frac{2}{m}\left(E - \frac{C}{r} - \frac{L^2}{2mr^2}\right)\right]^{-1/2} dr \tag{2.21}$$

results in

$$r = \frac{L^2}{mC} \cdot \frac{1}{1 - \epsilon\cos\phi} \tag{2.22}$$

with $b = L/\sqrt{2mE}$. With $k = L^2/mC$ and $\epsilon = \sqrt{1 + \frac{4E^2 b^2}{C^2}}$ (the eccentricity) the standard form of conic sections is obtained

$$\frac{1}{r} = \frac{1}{k}(1 - \epsilon\cos\phi). \tag{2.23}$$

2.2.4 Consequences

There is now a connection between impact parameter b, scattering angle θ, and (quantized) orbital angular momentum $L = \ell\hbar$

$$b = \frac{1}{2}d_0 \cot\frac{\theta}{2} = \frac{\ell\hbar}{p_\infty}. \tag{2.24}$$

Because of the quantization of L the orbital angular-momentum quantum numbers $l = 0, 1, 2, \ldots$ correspond to annular zones around the z direction. Larger scattering angles belong to smaller impact parameters and smaller orbital angular momenta.

2.2.5 Consequences of the Rutherford Experiments and Their Historic Significance

Rutherford and his collaborators Geiger and Marsden (later also Chadwick) used α particles from radiactive sources as projectiles. Their energies were so small that for all scattering angles the minimum scattering distances d were large compared with the sum of the two nuclear radii of projectiles and targets. The complete agreement between the results of the measurements and the (point-)Rutherford scattering cross section formula shows this in accordance with Gauss's law of electrostatics, which can be used to prove that a finite charge distribution in the external space beyond the charges cannot be distinguished from a point charge with an r^{-1} potential. In addition, the mere occurrence of backward-angle scattering events proves uniquely by simple kinematics that the target nuclei were heavier than the projectiles. Thus the existence of the atomic nucleus as a compact (i.e. very small and heavy object) was established (and Thomson's idea of a "plum-pudding" of negative charges from distributed electrons, in which the positive charges of ions were suspended, was refuted).

The dependence of the Rutherford cross section on the atomic charge number Z has been used to check on the correct assignment of Z to chemical elements and their positioning in the periodic table—complementing the use of the characteristic X-ray spectra together with Moseley's law.

2.2.6 Quantum-Mechanical Derivation of Rutherford's Formula

The Rutherford cross section may be derived quantum-mechanically by solving the Schrödinger equation with the point (or extended) Coulomb potential as input and with suitable boundary conditions. This equation has the form of a hypergeometric differential equation.

$$-\frac{\hbar^2}{2\mu}u'' + \left(\frac{C}{r} + \frac{\hbar^2}{2\mu}\frac{\ell(\ell+1)}{r^2} - \frac{\hbar^2 k^2}{2\mu}\right)u = 0. \quad (2.25)$$

After decomposition into partial waves (with ℓ designating the angular momentum of each), this equation may be written in its "normal" form with the Sommerfeld parameter η_S, as defined in Eq. (2.2), and $\rho = kr$ with k the c.m. wave number of the projectile:

$$\frac{d^2 u_\ell(\rho)}{d\rho^2} + \left(1 - \frac{\ell(\ell+1)}{\rho^2} - 2\frac{\eta_S}{\rho}\right)u_\ell(\rho) = 0, \quad (2.26)$$

where $u_\ell(\rho)$ is the radial wave function solving the equation. *Asymptotically* the solutions are the regular and irregular Coulomb Functions with the Coulomb phases $\sigma_\ell = \arg \Gamma(\ell + 1 + i\eta_S)$:

$$F_\ell \longrightarrow \sin(kr - \ell\pi/2 - \eta_S \ln 2kr + \sigma_\ell), \quad (2.27)$$

$$G_\ell \longrightarrow \cos(kr - \ell\pi/2 - \eta_S \ln 2kr + \sigma_\ell). \tag{2.28}$$

With the usual partial-wave expansion with incident plane waves the Coulomb scattering amplitude of the outgoing wave results:

$$\Psi_S \longrightarrow \frac{1}{r} e^{i(kr - \eta_S \ln 2kr)} f_c(\theta), \tag{2.29}$$

$$f_c(\theta) = -\eta_S \frac{e^{2i\sigma_0} \cdot e^{i\eta_S \ln \sin^2 \theta/2}}{2k^2 \sin^2 \theta/2}. \tag{2.30}$$

The amplitude squared $f_C \cdot f_C^*$ provides the Rutherford cross section, which is identical to the classically derived equation. However, for all applications where there is interference the Rutherford amplitude has to be used including its (logarithmic) phase, responsible for the long range, and the s-wave Coulomb phase σ_0. Typical cases are that of identical particles (see Sect. 3.4) or of interference with nuclear (hadronic) amplitudes (see e.g. Sect. 7.2). In these cases the partial-wave expansion cannot be truncated at low partial waves.

2.2.7 Deviations from the Rutherford Formula

According to the previous discussion deviations from the point Rutherford cross section are expected in the following cases:

- Modifications of the point Coulomb potential by the screening effects of the atomic electrons, which must be described by a screened Coulomb potential. These effects should show up especially at forward angles. Details will be discussed below when the Rutherford cross section is derived using the Born approximation (Sect. 10.7).
- Extended charge distribution and sufficiently high incident energy such that the projectile "dives" into the nuclear volume. Leptonic projectiles can probe the charge distribution without interfering strong-interaction effects and with no volume of their own. For hadronic charged projectiles one expects strong effects from the nuclear interaction whereas neutrons see the matter-density distribution only. With many such scattering experiments, besides the different density distributions, charge and matter radii of the nuclei and nucleons and their systematics with the nuclear mass number A were determined (see below). This principle was also applied in the deep-inelastic scattering of very high-energy leptons (electrons, muons, and neutrinos) from nucleons, which led to the evidence of substructures (partons, which finally turned out to be the quarks) inside the nucleons and to the determination of all their properties such as spin, masses, charges etc.

2.3 Scattering, Density Distributions, and Nuclear Radii

2.3.1 Nuclear Radii from Deviations from Rutherford Scattering

Already without detailed knowledge of the density distribution and of the potential some quite precise statements about nuclear radii by scattering of charged projectiles from nuclei were possible. One condition for this is, however, that the potential, which is responsible for the deviations from the point cross section is of short range, i.e. the charge distribution has a relatively sharp edge.

Most impressively these deviations from the point cross section appear with diminishing distances between projectile and target in a suitable plot. Because the Rutherford cross section itself is strongly energy and angle dependent one may choose to plot the ratio

$$\left(\frac{d\sigma}{d\Omega}\right)_{\text{exp}} \bigg/ \left(\frac{d\sigma}{d\Omega}\right)_{\text{point, theor.}} \tag{2.31}$$

as function of the minimum scattering distance d. Thus data at very different energies and angles can be directly compared (see Fig. 2.7 in Sect. 2.3.5). If, in addition, one wants to check on the assumption of the systematics of nuclear radii to follow $r = r_0 A^{1/3}$ a universal plot for all possible scattering partners by plotting the above ratio against $d/(A_1^{1/3} + A_2^{1/3})$ is useful. The experimental results show the extension of the charge distribution and the rather sudden onset of (hadronic) absorption (provided the interaction has a strong absorption term, which is typical for $A \geq 4$).

2.3.2 Coulomb Scattering from an Extended Charge Distribution

Here the quantum-mechanical derivation of the Rutherford-scattering cross section for a homogeneous charge distribution is useful. Starting points are

- Fermi's *Golden Rule* of perturbation theory.
- The first Born approximation.

For a "sufficiently weak" perturbation Fermi's *Golden Rule* gives the transition probability per unit time W:

$$W = \frac{2\pi}{\hbar} |\langle \Psi_{\text{out}} | H_{\text{int}} | \Psi_{\text{in}} \rangle|^2 \rho(E)$$

$$= \frac{Vmpd\Omega}{4\pi^2 \hbar^4} \cdot |H_{if}|^2. \tag{2.32}$$

The density of final states $\rho(E) = dn/dE$, which enters the calculation can be obtained from the ratio of the actual to the minimally allowed phase-space volumes:

$$\frac{dn}{dE} = \frac{V 4\pi p^2 dp \frac{d\Omega}{4\pi}}{(2\pi\hbar)^3 dE}, \tag{2.33}$$

$E = p^2/2m$ and $dp/dE = m/p = E/c^2 p$. Thus

$$\rho(E) = \frac{dn}{dE} = V\frac{pmd\Omega}{(2\pi\hbar)^3}$$
$$= V\frac{pEd\Omega}{(2\pi\hbar)^3 c^2}. \tag{2.34}$$

W becomes the cross section according to the definition on p. 14 with the incident particle-current density $j = v/V = p/mV$:

$$d\sigma = \frac{W}{j} = \frac{W}{(\frac{p}{mV})} = \frac{V^2 m^2 d\Omega}{4\pi^2 \hbar^4} \cdot |H_{if}|^2. \tag{2.35}$$

The 1st Born approximation consists in using only the first term of the Born series (see Sect. 10.7) with plane waves in the entrance and exit channels:

$$\Phi_{\text{in}} = \frac{1}{\sqrt{V}} e^{i\vec{k}_i \vec{r}} \quad \text{and} \quad \Phi_{\text{out}} = \frac{1}{\sqrt{V}} e^{i\vec{k}_f \vec{r}}. \tag{2.36}$$

If $H_{\text{int}} = U(r)$ signifies a small time-independent perturbation then, with $\vec{K} = \vec{k}_f - \vec{k}_i$

$$|H_{if}| = \left| \frac{1}{V} \int e^{i\vec{K}\vec{r}} U(r) d\tau \right| \tag{2.37}$$

and

$$\frac{d\sigma}{d\Omega} = \left(\frac{m}{2\pi\hbar^2}\right)^2 \left| \int e^{i\vec{K}\vec{r}} U(r) d\tau \right|^2 = |f(\theta)|^2. \tag{2.38}$$

Inserting the Coulomb potential $U(r) = C/r$ the classically calculated formula for the Rutherford scattering cross section is obtained. The cross section is (with the constant $Z_1 Z_2 e^2/16$ and the substitution $u = iKr\cos\theta$ and $du = -\sin\theta\, d\theta(iKr)$)

$$\frac{d\sigma}{d\Omega} = \text{const} \cdot \left| \int e^{i\vec{K}\vec{r}} \cdot \frac{1}{r} d\tau \right|^2$$

$$= \text{const} \cdot \left| \int \int \frac{1}{r} e^{iKr\cos\theta} 2\pi \sin\theta\, d\theta r^2 dr \right|^2$$

$$= \text{const} \cdot 2\pi \left| \int \int \frac{r}{iKr} e^u\, du\, dr \right|^2$$

$$= \text{const} \cdot \left(\frac{2\pi}{iK}\right)^2 \left| \int_r \left(e^{iKr\cos\pi} - e^{iKr\cos 0} \right) dr \right|^2$$

$$= \text{const} \cdot \left(\frac{2\pi}{iK}\right)^2 \left| \int_r \left(e^{-iKr} - e^{iKr} \right) dr \right|^2$$

2.3 Scattering, Density Distributions, and Nuclear Radii

$$= \text{const} \cdot \left(\frac{2\pi \cdot 2i}{iK}\right)^2 \left|\int_0^\infty \sin Kr \, dr\right|^2. \tag{2.39}$$

The integral is undefined. This is circumvented by a *screening Ansatz* after Bohr, which corresponds to the real situation of the screening of the point Coulomb potential by the electrons of the atomic shell, with the screening constant α. With

$$\int_0^\infty e^{-\alpha r} \sin Kr \, dr = \frac{K}{K^2 + \alpha^2} \tag{2.40}$$

one obtains

$$\left(\frac{d\sigma}{d\Omega}\right)_{R,s} = \left[\frac{2\mu Z_1 Z_2 e^2}{\hbar^2[\alpha^2 + 4k^2 \sin^2(\theta/2)]}\right]^2 \tag{2.41}$$

with the momentum transfer $K = 2k \sin(\theta/2)$ for elastic scattering. This cross section is finite for $\theta \to 0°$. By letting the screening constant go to zero a cross section results, which is identical with that from the classical derivation:

$$\left(\frac{d\sigma}{d\Omega}\right)_R = \lim_{\alpha \to 0}\left(\frac{d\sigma}{d\Omega}\right)_{R,s}$$

$$= \left(\frac{Z_1 Z_2 e^2}{4E_{\text{kin}}}\right)^2 \cdot \frac{1}{\sin^4(\theta/2)}. \tag{2.42}$$

The extension of the derivation of the Rutherford cross section to an extended (especially a homogeneous and spherically-symmetric) charge distribution is simple and leads to the fundamental concept of the *form factor*. Such a distribution is suggested by the approximately constant nucleon density with r in most nuclei (see Fig. 2.8), but cannot represent the density behavior around R very well.

We start with the Coulomb potential of such an extended homogeneous spherical charge distribution with radius R (see Fig. 2.4). It is calculated with Gauss's theorem of electrostatics (for units see the footnote 2 in Sect. 2.2.) At $r = R$ the interior and exterior potential must be suitably matched.

$$V(r) = \begin{cases} ze^2 \frac{1}{r} & \text{for } r > R \\ ze^2 \frac{1}{2R}(3 - \frac{r^2}{R^2}) & \text{for } r \le R. \end{cases} \tag{2.43}$$

In the exterior space the potential is identical with that of a point charge, continues at $r = R$ to a parabolic shape in the interior of the distribution. It is therefore to be expected that in the scattering with sufficiently high energy the scattering cross section would strongly deviate from the Rutherford cross section as soon as the nuclear surface is touched. In addition, the onset of the short-range strong interaction will influence the scattering, especially by absorption. For the calculation of the cross section an integral over the contributions from all charge elements $dq = Ze\rho(\vec{r})d\tau$

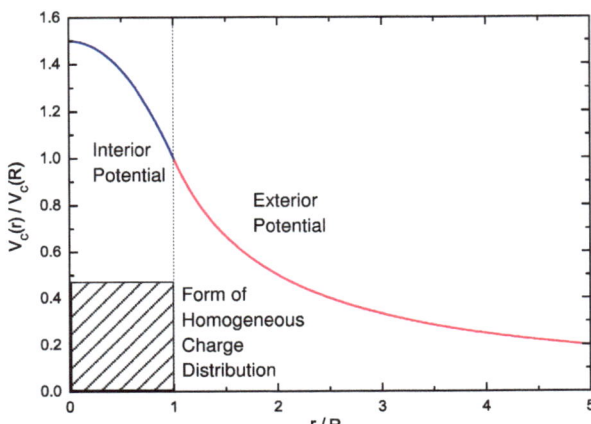

Fig. 2.4 Coulomb potential of a spherical homogeneous charge distribution in relation to the point-Coulomb potential that is valid outside the charge distribution, normalized to $V(r = R) = 1$

to the potential $U(\vec{r}) = -\frac{Z_1 Z_2 e^2}{R} \cdot e^{-\alpha R} \rho(\vec{r}) d\tau$ has to be performed.

$$U(\vec{r}') = -Z_1 Z_2 e^2 \int \rho(\vec{r}') \frac{e^{-\alpha R}}{R} d\tau. \tag{2.44}$$

By inserting this into the Born approximation Eq. (10.25) (with $d\vec{R} = d\vec{r}'$ and $\vec{R} = \vec{r}' - \vec{r}$) one obtains:

$$\frac{d\sigma}{d\Omega} = \left(\frac{Z_1 Z_2 e^2 m}{2\pi \hbar^2}\right)^2 \cdot \left[\int \rho(\vec{r}) e^{i\vec{K}\vec{r}} d\tau \cdot \int \frac{e^{-\alpha R}}{R} e^{i\vec{K}\vec{R}} d\vec{R}\right]^2$$

$$= \left[F(\vec{K}^2) \frac{K}{K^2 + \alpha^2}\right]^2. \tag{2.45}$$

The cross section factorizes into two parts, one of which (after a transition to the limit $\alpha \to 0$) results again in the point cross section, the other in the *form factor*:

$$\frac{d\sigma}{d\Omega} = \left(\frac{d\sigma}{d\Omega}\right)_{\text{point nucleus}} \cdot |F(\vec{K}^2)|^2. \tag{2.46}$$

This separation is characteristic for the interaction between extended objects and signifies a separation between the interaction (e.g. the Coulomb interaction) and the structure of the interacting particles.

For rotationally-symmetric problems the form factor has a simplified interpretation:

$$F(K) = \int \rho(r) \exp(i \vec{K} \vec{r}) 2\pi r^2 dr \sin\theta d\theta. \tag{2.47}$$

On substitution $u = iKr\cos\theta$ and $du = -iKr\sin\theta d\theta$ this becomes

2.3 Scattering, Density Distributions, and Nuclear Radii

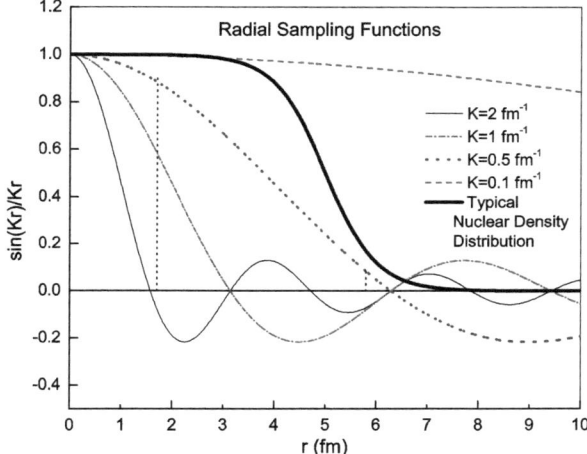

Fig. 2.5 Sampling functions for different momentum transfers show that in order to sample details of a given structure (e.g. the shape around the radius of a nuclear density (charge or mass) distribution) the momentum transfer (given by the incident energy and the scattering angle) has to have an appropriate intermediate value. In the example shown the value of $K = 0.5$ fm^{-1} is suitable for sampling the region around the nuclear radius of 5.0 fm. The *vertical dotted lines* indicate a 10 to 90 % sampling region

$$F(K) = 2\pi \int \rho(r) e^u r^2 dr \frac{du}{-iKr}$$
$$= \int \rho(r) 4\pi r^2 dr \cdot \underbrace{\left(\frac{\sin(Kr)}{Kr}\right)}_{\text{purely real}}. \qquad (2.48)$$

Thus the form factor is a folding integral of the density with the sampling function (in parentheses). This function is oscillatory and its oscillation "wavelength" $1/K$ (which depends on the energy of the transferred radiation) has to be adjusted to the rate of change of the density. If the oscillation is too frequent the integral results in ≈ 0 revealing no information on ρ. If it is too slow the sampling function is $\approx$constant, and the integral results in just the total charge Ze. Figure 2.5 illustrates this for different momentum transfers on a given nuclear density distribution. Experimentally the form factor is obtained as the ratio

$$\left(\frac{d\sigma}{d\Omega}\right)_{\text{experimental}} \bigg/ \left(\frac{d\sigma}{d\Omega}\right)_{\text{point, theor.}}. \qquad (2.49)$$

The charge distribution (or more generally: the density distribution e.g. of the hadronic matter) is obtained by *Fourier inversion* of the form factor F:

$$\rho_c(\vec{r}) = \frac{1}{(2\pi)^3} \int_{0 \to \infty} F_c(\vec{K}^2) e^{(-i\vec{K}\vec{r})} d\vec{K}. \qquad (2.50)$$

This means that (in principle) for a complete knowledge of $\rho(\vec{r})$ F must be known for all values of the momentum transfer. Since $\rho(\vec{r})$ for small $\vec{r}$ is governed by the high-momentum transfer components of $\vec{K}$ this cannot be achieved in practice. For this reason the following approximations may be used:

- Model assumptions are made for the form of the distribution: e.g. homogeneously charged sphere, exponential, Yukawa, or Woods-Saxon behavior.
- The model-independent method of the expansion of $e^{i\vec{K}\vec{r}}$ into moments.

2.3.3 Ansatz for Models

It is useful to get an impression of the Fourier transformation of different model density-distributions as shown in Fig. 2.6: It is a general observation that "sharp-edged" distributions lead to oscillating form factors (and therefore cross sections), and smooth distributions to smooth form factors. In agreement with our Ansatz a δ distribution (characteristic for a point charge or mass) corresponds to a constant form factor (this is called "scale invariance").

2.3.4 Expansion into Moments

With the power-series expansion of $e^{i\vec{K}\vec{r}}$ the form factor becomes

$$F(\vec{K}^2) \propto \int \rho(\vec{r}) \left[1 + i\vec{K}\vec{r} - \frac{(\vec{K}\vec{r})^2}{2!} \pm \cdots \right] d\tau. \qquad (2.51)$$

By assuming a spherically symmetric distribution (with pure r dependence only) and with a normalization such that for a point object the constant form factor is 1, we have:

$$F(\vec{K}^2) = 1 - \text{const} \cdot K^2 \int_{0 \to \infty} r^2 \rho(r) d\tau \pm \cdots . \qquad (2.52)$$

The second term contains the average squared radius $\langle r^2 \rangle = r_{rms}^2$. For small values of $K^2 \langle r^2 \rangle$ one gets in a model-independent way (i.e. for arbitrary form factors):

$$F(\vec{K}^2) \approx 1 - \frac{1}{6} K^2 \langle r^2 \rangle. \qquad (2.53)$$

Of course this approximation is becoming worse with smaller r (because one needs higher moments), i.e. if one wants to resolve finer structures.

2.3 Scattering, Density Distributions, and Nuclear Radii

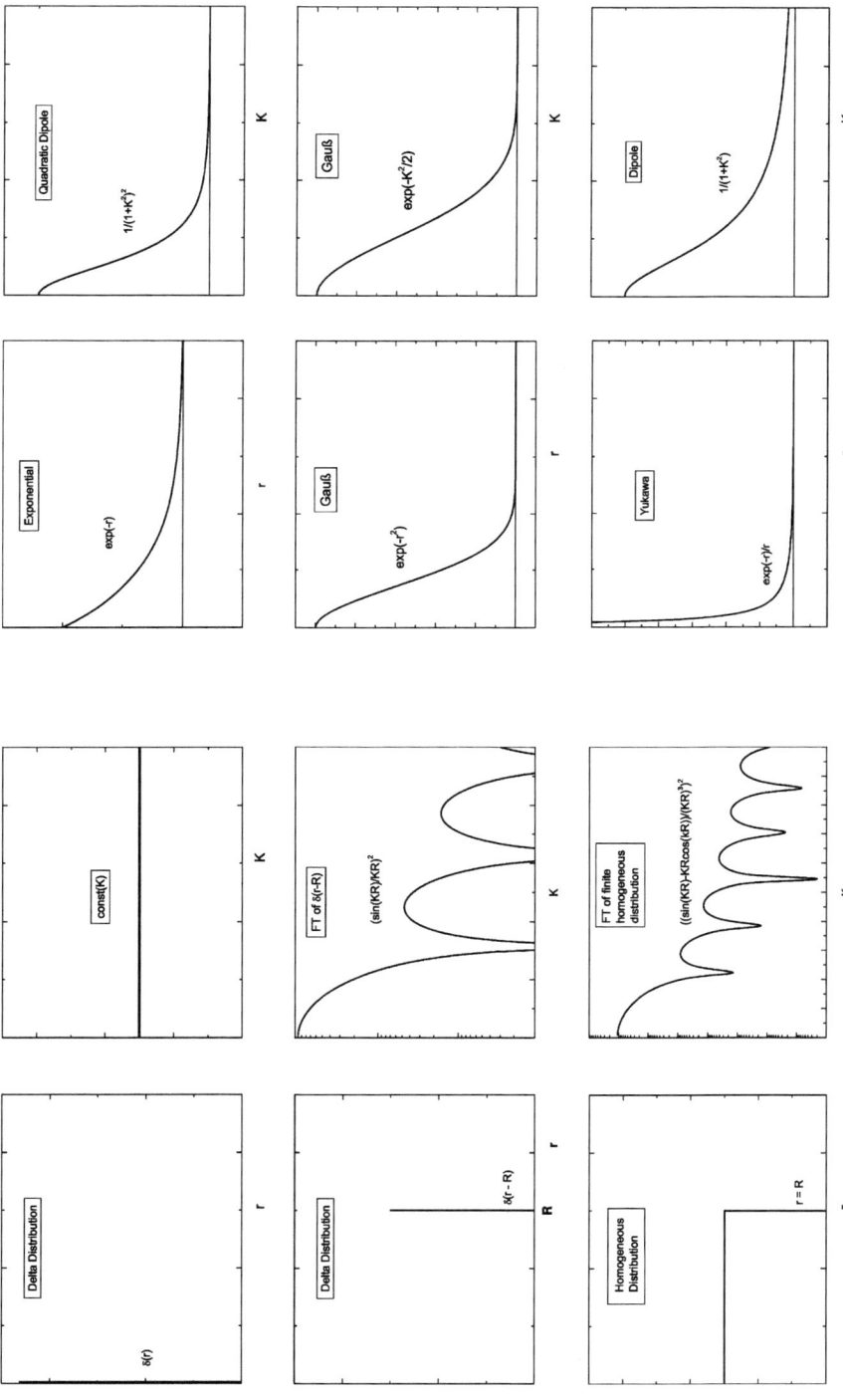

Fig. 2.6 *Squares* of the Fourier transforms—basically the form factors determining the shapes of the cross sections—of different charge-density distributions

2.3.5 Results of Hadron Scattering

After accelerators were available charge and matter density distributions of nuclei and their radii could be investigated by probing the distributions with hadronic projectiles. With light as well as with heavy ions, but also with neutrons as projectiles it is evident that they are extended and possess structure. The consequence is that detailed statements about the density distributions are difficult to make and may need the deconvolution of the contributions from projectile and target nuclei. However, statements about nuclear radii are possible, even with quite simple semi-classical assumptions such as absorption between nuclei setting in sharply at a well-defined distance and pue Coulomb scattering beyond that distance. Systematic α scattering studies on many nuclei (where we already have strong absorption at the nuclear surfaces) revealed good $A^{1/3}$ systematics for the nuclear radii. A dependence of

$$\sigma_{\alpha,\alpha} = R_0 \left(A^{1/3} + 4^{1/3} \right) \tag{2.54}$$

was fitted to the data, assuming a sharp-cutoff model for the cross sections and taking into account the finite radii of both nuclei. It yielded a radius constant of

$$R_0 = 1.414 \text{ fm}. \tag{2.55}$$

However, when considering the range of the nuclear force for both nuclei of about 1.4 fm a radius constant of ≈ 1.2 fm resulted.

Heavy-ion scattering experiments with a great number of different pairs of collision partners yielded a very good systematics shown in Fig. 2.7 that becomes evident when the relative cross sections were plotted against the distance parameter d, for which an assumed $A^{1/3}$ dependence of the radii of both collision partners was applied

$$d = D_0 \left(A_1^{1/3} + A_2^{1/3} \right)^{-1} \tag{2.56}$$

with D_0 the distance of closest approach, as calculated from energies and scattering angles. A well-defined sharp distance parameter of $d_0 = 1.49$ fm for the onset of absorption resulted. This corresponds to a universal radius parameter of $r_0 = 1.1$ fm if the range of the nuclear force is set to 1.5 fm. The simple model applied was to assume

- Pure point-Rutherford scattering outside the range of nuclear forces,
- Ratio of elastic to Rutherford cross section

$$\frac{d\sigma}{d\sigma_R} = 1 + P_{\text{abs}}(D) \tag{2.57}$$

and

$$P_{\text{abs}}(D) = \begin{cases} 0 & \text{for } D \geq D_0, \\ 1 - \exp(\frac{D-D_0}{\Delta}) & \text{for } D < D_0, \end{cases} \tag{2.58}$$

with $P_{\text{abs}}(D)$ the probability of absorption out of the elastic channel, D_0 the interaction distance, and Δ the "thickness" of the transition region.

2.4 Electron Scattering

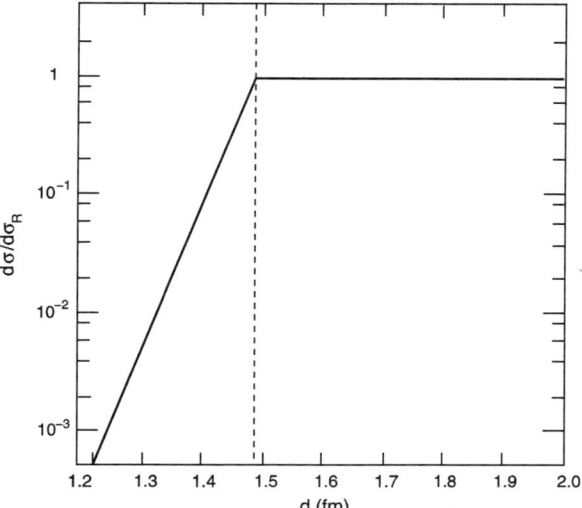

Fig. 2.7 Plot of the scattering cross sections (relative to the Rutherford cross section) of many different HI pairings vs. the distance parameter d in fm. The data used for the fit are in Ref. [OGA78], see also [CHR73]

The latter depends on the A of the nuclei involved and could be determined with good accuracy to be e.g. $\Delta \approx 0.33$ fm for scattering of nuclei near ^{40}Ca from ^{208}Pb.

Here we will discuss the results obtained with charged projectiles, for which the interaction is exactly known (e.g. electrons, which do not interact via the strong force) whereas e.g. for neutrons one needs nuclear scattering models (e.g. the optical model, see Sect. 10.3). The assumption that neutron and proton radii of nuclei are about equal has proved too simple with the evidence of neutron-halo and neutron-skin nuclei, see Sect. 2.4.2.

2.4 Electron Scattering

Since all electrons (and all leptons) are considered to be point-particles they are—as long as not the hadronic interaction region proper shall be probed—the ideal projectiles. They "see" the electromagnetic (and weak) structure of the nuclei. Of course, the treatment must be relativistic. Instead of the Rutherford- (point-Coulomb) approach one has to use the proper theory.

Besides the relativistic treatment differences to the (classical) Rutherford cross section come about by the lepton spin. The derivation of the correct scattering cross section relies on the methods of Quantum Electrodynamics (QED) and techniques such as the Feynman diagrams. Here only the results will be presented. The electromagnetic interaction between the electron and a hadron is mediated by the exchange of virtual photons, which is accompanied by a transfer of energy and momentum. The wavelength of these photons derives directly from the momentum transfer $\hbar K = 2(h\nu/c)\sin(\theta/2)$ to be

$$\lambda_{\text{de Broglie}} = \frac{\hbar}{\hbar K} = 1/K. \tag{2.59}$$

The argument of diffraction limitation may also be formulated in the complementary time picture; it may be said that at long wavelengths, due to the uncertainty relation, one needs long measurement times, in which the projectile sees only a time-averaged picture of the object considered while small wavelengths allow measurement times equivalent to snapshots of the object or its substructures (partons).

In contrast to low-energy Rutherford scattering, in which only the electric charges interact, in charged lepton scattering at higher (relativistic) energies there is also a magnetic interaction, and in neutrino scattering only the weak interaction is acting. Principally in lepton scattering at higher energies three distinct regions of momentum transfer can be distinguished:

- Elastic scattering at small momentum transfer is suitable to probe the shape of the hadrons. The resulting two form factors F_E (electric) and F_M (magnetic) produce again the charges and current (magnetic moment) distributions and the radii of the hadrons by Fourier inversion.
- Weakly inelastic scattering at higher momentum transfer leads to excitations of the hadrons (e.g. Delta- or Roper excitations (resonances) of the nucleons). The form factors are quite similar to those from the elastic scattering, which means that we have some excited state of the same nucleons.
- Deep-inelastic scattering is the suitable method to see partons inside the hadrons. In this way in electron and muon scattering the quarks bound in nucleons and their properties (spin, momentum fraction) and also the existence of sea quarks (*s* quark/anti-quark pairs) were identified. Especially the pointlike character of these constituents was shown by the near constancy of the form factors (here called: *structure functions*) with the momentum transfer (*Bjorken scaling*).

Here only elastic scattering will be discussed in detail. In QED theory for the differential cross section the Rosenbluth formula was deduced:

$$\frac{d\sigma}{d\Omega} = \left(\frac{d\sigma}{d\Omega}\right)_{\text{point}} \cdot \left(\frac{F_E^2 + bF_M^2}{1+b} + 2bF_M^2 \tan^2\frac{\theta}{2}\right). \tag{2.60}$$

The point cross section $(d\sigma/d\Omega)_{\text{point}}$ is a generalized Rutherford cross section and is calculable with the methods of QED (e.g. using Feynman diagrams). The most general form of this cross section (the Dirac scattering cross section) contains as main part the electrostatic scattering, a contribution from the magnetic (spin-dependent) interaction, which depends on the momentum transfer, and a correction for the nuclear recoil:

$$\left(\frac{d\sigma}{d\Omega}\right)_{\text{Dirac}} = \frac{\alpha^2}{4p_0^2\sin^4(\theta/2)}\left[1 + \frac{2p_0}{M}\sin^2\frac{\theta}{2}\right]\left(\cos^2\frac{\theta}{2} + \frac{q^2}{2M^2}\sin^2\frac{\theta}{2}\right). \tag{2.61}$$

2.4 Electron Scattering

For small energies or momentum transfers the cross section simplifies to:

$$\left(\frac{d\sigma}{d\Omega}\right)_{\text{Mott}} = \frac{[2e^2(E'c^2)]^2}{q^4} \cdot \frac{E'}{E} \cos^2\frac{\theta}{2}. \tag{2.62}$$

The symbols used here mean: q = four-momentum transfer, $b = -q^2/(4m^2c^2)$, E', and E the energies of the outgoing and incoming electrons. F_E and F_M are the electric and magnetic form factors of the nucleons. Experimentally they are obtained from the measured data by least-squares fitting of the parameters of the theory, graphically through the *Rosenbluth plot*, i.e. by plotting $(d\sigma/d\Omega)_{\text{exp}}/(d\sigma/d\Omega)_{\text{point}}$ against $\tan^2(\theta/2)$.

In analogy to the Rutherford cross section here the form factors (or structure functions) are Fourier transforms of the charge and current-density distributions (or: distributions of the (anomalous) magnetic moments). Like there, these distributions result from Fourier inversion of the form factors, and at the same time quantitative values of the shape and size of the nucleons are obtained.

The measured form factors as functions of q^2 are normalized such that for $q \to 0$ they become the static values of the electric charge and magnetic moments. Except for the electric form factor of the neutron all others are well described by the dipole Ansatz corresponding to a density distribution of an exponential function.

An early model for the charge-density distribution was—besides the homogeneously charged sphere with only one parameter, its radius—a modified Woods-Saxon distribution with three parameters, because, besides the radius parameter r_0 and the surface thickness a, also the central density ρ_0 must be adjustable because it varies especially in light nuclei:

$$\rho_c(r) = \frac{\rho_0}{1+e^{\frac{r-r_{1/2}}{a}}}. \tag{2.63}$$

The surface thickness $t = 4\ln 3 \cdot a$ signifies the 10 to 90 % thickness range centered around $r_{1/2}$. From this parametrization an electromagnetic radius constant of $r_{1/2} = 1.07$ fm, a surface-thickness parameter of $a = 0.545$ fm, and a central density of $\rho_N = 0.17$ nucleons/fm^3 or $1.4 \cdot 10^{14}$ g/cm^3 for nuclei with $A > 30$ have been derived. The description of "modern" density distributions is not so simple because the nuclei have individual and more complex structure even if the essential features such as the three parameters do not vary too much. The detailed structure information is obtained from model-independent approaches such as Fourier-Bessel expansions. Radii are given as rms radii or converted into the equivalent radii R_0. R_0 is the radius of a homogeneously charged sphere of equal charge using the relation

$$r_{\text{rms}} = \sqrt{3/5}\, R_0. \tag{2.64}$$

The definition of the (model-independent) Coulomb rms radius is

$$r_{\text{rms}} = \langle r^2 \rangle^{1/2} = \left[\frac{1}{Ze}\int_0^\infty r^2 \rho_C(r) 4\pi r^2 dr\right]^{1/2}. \tag{2.65}$$

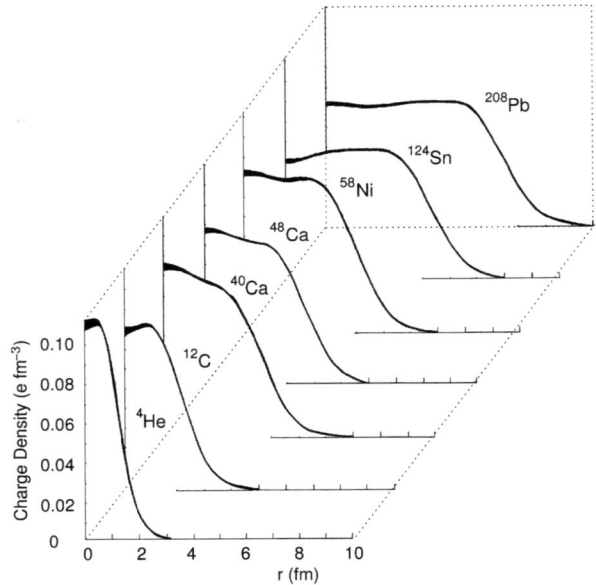

Fig. 2.8 Charge density distributions of different doubly closed-shell nuclei. The central density is only weakly changing over the nuclear chart whereas the radii increase with $A^{1/3}$. After [FRO87]

Nuclear radii from muonic atoms are often more precise than those from lepton scattering but they are in a way complementary in relation to the radius region probed they measure different moments). Thus, the results of both methods can be combined (Fig. 2.8). The distributions are quite well reproduced by "mean-field" calculations, see e.g. [FRO87, DEC68]. The salient results of these investigations are:

- From the distributions a central density is derived, which for heavier nuclei is constant in first approximation. This and the systematics of radii are characteristic for nuclear forces; their properties are: short range, saturation and incompressibility of nuclear matter, and suggest the analogy to the behavior of liquids, which led to the development of collective nuclear models (liquid-drop models, models of nuclear rotation and vibration).
- The radii follow more or less a simple law $r = R_0 A^{\frac{1}{3}}$. For the radius parameter $R_0 = 1.24$ fm is a good value. From Coulomb-energy differences of mirror nuclei a value of $R_0 = (1.22 \pm 0.05)$ fm has been derived.
- The surface thickness of all nuclei is nearly constant with a 10–90 % value of $t = 2.31$ fm corresponding to $a = t/4\ln 3 = 0.53$ fm. This is explained by the range of the nuclear forces independent of the nuclear mass number A.
- The nucleons have no nuclear surface. The charge and current as well as the matter densities of the proton follow essentially an exponential distribution. For the neutron the charge distribution is more complicated because volumes of negative and positive charges must compensate each other to zero notwithstanding some complicated internal charge distribution that originates from its internal quark-gluon structure.

2.4 Electron Scattering

- The rms radii for the current distributions of protons and neutrons and the charge distribution of the protons are 0.88 fm. Recently, with increased experimental precision an unresolved discrepancy between values from lepton scattering and muonic-atom work has been published. The rms charge radius of the neutron it is 0.12 fm, which means that there must be positive and negative charges distributed differently over the nuclear volume.
- Thus, nucleons are not "elementary", but have complicated internal structures.

2.4.1 Matter-Density Distributions and Radii

The matter density—apart from and independent of the charge or current distributions—can be investigated only by additional hadronic scattering experiments because neutrons and protons in principle need not have the same distributions in nuclei.

Hadronic Radii from Neutron Scattering The total cross sections of 14 MeV neutron scattering under simple assumptions have been shown to also follow a $A^{1/3}$ law, see e.g. Ref. [SAT90], p. 32, cited from Ref. [ENG74]. The assumptions were that the sharp-edged range of the nuclear force was 1.2 fm and the total cross section σ_{tot} follows $2\pi(R + \lambda)^2$ with R the nuclear (hadronic) radius, i.e. the nuclei are considered to be black (totally absorbent) to these neutrons, which is not exactly fulfilled, as the structures in this dependence show. These can be explained with the optical model, see below. The radius constant extracted from this systematics is

$$R_{\text{hadr}} = 1.4 \text{ fm}. \tag{2.66}$$

In addition, there have been attempts to extract the neutron radius of ^{208}Pb from parity-violating electron scattering [ABR12].

2.4.2 Special Cases—Neutron Skins and Halo Nuclei

Neutron Skins Different from halos skins are a volume effect expected and appearing in heavier nuclei with increasing neutron excess (and densities over proton densities). The question of a neutron skin in these nuclei is interesting, and only recently such a thin skin was consistently shown to exist, see e.g. [TSA12] and references therein. Among the hadronic probes used have been protons, α's, heavy ions, antiprotons, and, recently, also pions e.g. on ^{208}Pb, ^{48}Ca and others. The extraction of rms radii requires some model assumptions concerning the reaction mechanism and the interplay of hadronic and Coulomb interactions. The pion results are derived from two sources: pionic atoms (in analogy to the derivation of the electromagnetic radii from muonic atoms) and total reaction cross sections of π^+ [FRI12].

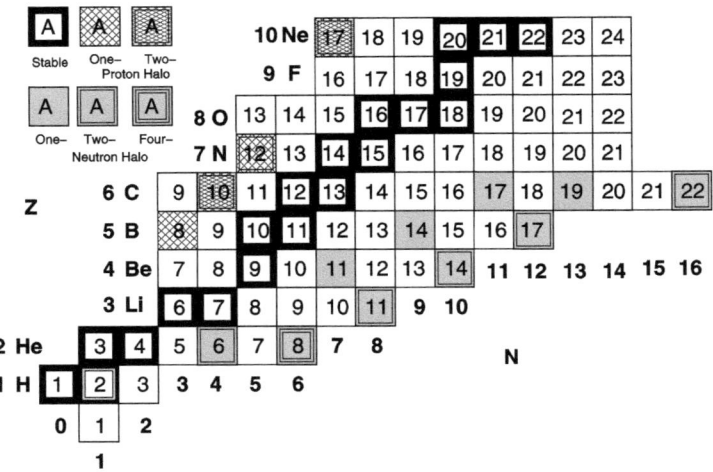

Fig. 2.9 Halo nuclei at the driplines of the chart of nuclides

The neutron skin is related to the symmetry energy, which plays a role in the mass formula of Bethe and Weizsäcker for the binding energies of nuclei, especially for "asymmetric" nuclei with strong neutron excess, but also for astrophysics and nuclear-matter calculations. The radius of neutron stars is closely related to the symmetry energy value in high-density nuclear matter, see e.g. Ref. [TSA12].

Usually the quantity

$$\delta R_{np} = \langle r^2 \rangle_n^{1/2} - \langle r^2 \rangle_p^{1/2} \tag{2.67}$$

is taken as a measure of skin thickness. The experimental values deduced from different experiments are on the order of $\delta R_{np} \approx 0.2$ fm.

Halo Nuclei At the "rims" of the valley of stability (the neutron or proton *driplines*) there are a number of nuclei that have much larger radii than expected from the systematics. ^{11}B has about the same radius as ^{208}Pb. Also the deuteron has an rms radius of about 3.4 fm. In all cases the nuclei seem to have a halo of weakly bound neutrons (or protons), which surrounds a more strongly bound core. Different cores are possible, i.e. besides the strongly bound α making ^{6}He and ^{8}He one- and two-neutron halo nuclei and ^{8}B and ^{10}C seem to form one- and two-proton halos, also ^{9}Be forms some type of core. Indications of halo structures were—among others—the exceptionally large cross sections in heavy-ion reactions, a narrower momentum distribution of the nucleons in the nuclei, and larger radii, as compared to the $A^{1/3}$ law. Figure 2.9 shows the low-mass portion of the chart of nuclides where halo nuclei have been found. The scientific interest in halo nuclei is manifold. They were among the first where the driplines have been reached. The results show that the shell structures established for the valley of stability can be extended to "exotic" nuclei, but with modifications of the closed shells, i.e. with new magic

Fig. 2.10 Coat of arms and symbol of the Renaissance Borromean family (and other north Italian families like the Sforzas) at their castle on the Borromean island Isola Bella in the Lago Maggiore, Italy

numbers emerging. The low mass numbers invite application of microscopic theories such as Faddeev-(Yakubowsky), no-core shell models, Green's function Monte Carlo (GFMC), and other approaches to test nuclear forces, e.g. three-body forces, or effective-field (EFT) approaches. Impressive results have been obtained by such "ab initio" calculations, see e.g. Ref. [PIE01, DEA07]. A special role is played by the so-called *Borromean* nuclei, i.e. those that consist of a core plus two weakly (un)bound neutrons at large radii, and for which any of the two-particle subsystems are unbound (Example: ^{4}He $+ n + n$). They can be treated by well-established three-body methods; see also Chap. 9.2. Their name is derived from the three intertwined Borromean rings that fall apart when one ring is removed and hold together only when united, see Fig. 2.10. Many nucleosynthesis processes pass through nuclei that are neutron rich or neutron poor and are not well known. Thus, for astrophysics, a better understanding of all these reactions and their reaction rates is essential.

Since we deal with unstable (radioactive) nuclei the "radioactive-ion beams (RIB)" facilities, many of which are being developed, are especially suited for their investigation (for details see Chap. 15). These facilities collect, focus and accelerate nuclear reaction products in order to use them as projectiles in reactions. Figures 2.11 and 2.12 show the properties typical for halo nuclei:

- They have radii, which are larger than predicted from the usual $A^{1/3}$ systematics.
- Their density distributions reach further out than usual.
- In agreement with this they show narrower momentum distributions of the breakup fragments of the halo nuclei (one example: ^{19}C $\rightarrow$ ^{18}C $+ n$, compared to ^{17}C $\rightarrow$ ^{16}C $+ n$).
- The halo structure has to be distinguished from a neutron skin structure (see the preceding subsection). The latter is a volume effect caused by the increasing numbers of neutrons over protons with increasing A, whereas in light nuclei,

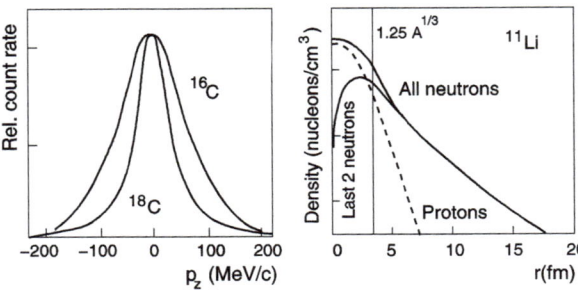

Fig. 2.11 Schematic fragment-momentum distributions from breakup reactions (*left*) and density distributions in halo nuclei, see e.g. Ref. [DOB06]

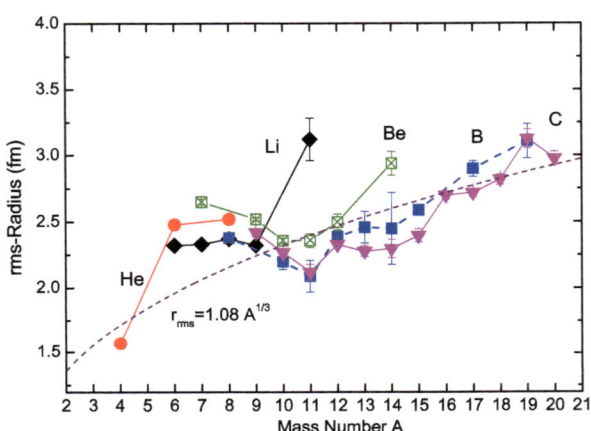

Fig. 2.12 Radii of halo nuclei. There exist (slightly varying, depending on extraction methods from experiments) numbers from different sources, especially from the earliest systematics from Ref. [TAN85]. Here they were taken from Refs. [OZA01] and [KRI12]. For comparison a plausible rms radius function for "normal" nuclei is shown

in approaching the n (or p) driplines, the binding energy of additional neutrons (or protons) approaches zero, making the nuclei nearly unbound which results in large radii.

The latest discovery at present of a halo nucleus is that of ^{22}C [TAN10] which showed an increased reaction cross section and an rms radius of $r_{\mathrm{rms}} = 5.4 \pm 0.9$ fm, both larger than expected from the usual systematics.

2.5 Exercises

2.1. (a) Show how kinematical arguments lead to Rutherford's conclusions on a (compact and) heavy target nucleus (gold).
(b) In the semi-classical theory of Bethe and Bloch (see Eq. (17.1)) the slowing-down of charged particles is explained with their being scattered by atomic electrons. What is different and what are the consequences from kinematics in this case: effects on the α projectiles and on the recoils electrons?

2.2. The "first" artificial nuclear reaction induced by an accelerated beam (Cockroft/Walton, 1932) was $^{7}_{3}\mathrm{Li}(p,\alpha)^{4}_{2}\mathrm{He}$ at an incident energy of 150 keV.

(a) Calculate the Q value of the reaction from the known masses of the particles (p: 1.007289 u, α: 4.002425 u, and $^{7}_{3}$Li: 7.014907 u).
(b) What are the energies of the two α's under the lab. angles of 0° and 90°?
(c) Does the *inverse*, i.e. time-reversed reaction have an energy threshold, and at which α lab. energy?

2.3. Chadwick discovered the neutron in 1932 by correctly identifying the energetic radiation emitted from the reaction $\alpha + {}^{9}_{4}\text{Be} \rightarrow {}^{12}_{6}\text{C} + {}^{1}_{0}n$ (induced by α's from a Po source). The recoil energies transferred to the protons and ^{14}N nuclei of the filling gas of the ionization chamber were measured to be 5.7 MeV and 1.6 MeV, respectively (masses: ${}^{9}_{4}$Be: 9.011348 MeV; ^{14}N: 14.002863 u).

(a) Which value of the neutron mass (in u) was obtained by Chadwick?
(b) Other nuclear physicists (among them the Curies) had erroneously interpreted the energetic radiation as γ radiation. How high would their energy have to be if they had transferred their energy by elastic Compton scattering (see Eq. (17.7)) on the protons or ^{14}N nuclei? Could such energies of γ transitions occur in nuclei?
(c) What exactly is their error of reasoning?

2.4. An ${}^{16}_{8}$O nucleus is scattered elastically at 80 MeV and 150° from a ${}^{179}_{79}$Au nucleus at rest in the lab. system.

(a) How close to each other (in a semi-classical picture) do the centers of the two nuclei get?
(b) Do the two nuclei "feel" the (hadronic) nuclear force, if a radius constant $R_0 = 1.25$ fm and as range of the force the value resulting from pion exchange + Heisenberg's uncertainty relation are assumed?

2.5. To resolve substructures ("partons") in nuclei (or nucleons), e.g. by elastic electron scattering the wavelength of the probing radiation must be chosen sufficiently small, more precisely: the wavelength of the exchanged (virtual) radiation.

(a) What electron energy is required to probe the surface of a ${}^{208}_{82}$Pb nucleus, of a ^{4}He nucleus, the rms-radius range of a proton, or quark structures inside nucleons in the $\leq 1 \cdot 10^{18}$ m range?
(b) What about using muons?

References

[ABR12] S. Abrahamyian et al. (PREX Collaboration), Phys. Rev. Lett. **108**, 112502 (2012)
[CHR73] P.R. Christensen, V.I. Manko, F.D. Becchetti, R.J. Nickles, Nucl. Phys. A **207**, 33 (1973)
[DEA07] D.J. Dean, Phys. Today. **November**, 48 (2007)
[DEC68] J. Dechargé, D. Cogny, Phys. Rev. C **21**, 1568 (1968)
[DOB06] A.V. Dobrovolsky et al., Nucl. Phys. A **766**, 1 (2006)

[ENG74] J.B.A. England, *Techniques in Nuclear Structure Physics* (Halstead, New York, 1974)
[FRI12] E. Friedman, Nucl. Phys. A **896**, 46 (2012)
[FRO87] B. Frois, C.N. Papanicolas, Annu. Rev. Nucl. Part. Sci. **37**, 133 (1987)
[GEI13] H. Geiger, E. Marsden, Philos. Mag. **25**, 604 (1913)
[KRI12] A. Krieger et al., Phys. Rev. Lett. **108**, 142501 (2012)
[NOE76] W. Nörenberg, H.A. Weidenmüller, *Introd. Theory of Heavy Ion Collisions*. Lecture Notes in Physics, vol. 51 (Springer, Heidelberg, 1976)
[OGA78] Y.T. Oganessian, Y.E. Penionzhkevich, V.I. Man'ko, V.N. Polyansky, Nucl. Phys. A **303**, 259 (1973)
[OZA01] A. Ozawa, T. Suzuki, I. Tanihata, Nucl. Phys. A **693**, 32 (2001)
[PIE01] S.C. Pieper, R.B. Wiringa, Annu. Rev. Nucl. Part. Sci. **51**, 53 (2001)
[SAT90] G.R. Satchler, *Introd. Nucl. Reactions*, 2nd edn. (McMillan, London, 1990)
[TAN85] I. Tanihata, H. Hamagaki, O. Hashimoto, Y. Shida, N. Yoshikawa, K. Sugimoto, O. Yamakawa, T. Kobayashi, N. Takahashi, Phys. Rev. Lett. **55**, 2676 (1985)
[TAN10] K. Tanaka et al., Phys. Rev. Lett. **104**, 062701 (2010)
[TSA12] M.B. Tsang, J.R. Stone, F. Camera, P. Danielewicz, S. Gandolfi, K. Hebeler, C.J. Horowitz, J. Lee, W.G. Lynch, Z. Kohley, R. Lemmon, P. Möller, T. Murakami, S. Riordan, X. Roca-Maza, F. Sammarruca, A.W. Steiner, I. Vidaña, S.J. Yennello, Phys. Rev. C **86**, 015803 (2012)

Chapter 3
Role of Conservation Laws and Symmetries in Nuclear Reactions

3.1 Generalities

Besides the classical conservation laws in nuclear and particle physics a number of non-classical conservations laws (e.g. that of parity conservation) or symmetries, respectively (e.g. that of time-reversal invariance or the exchange symmetry of identical particles) are important. Historically, the discoveries of violations of laws, which so far were considered valid without any doubt were extremely spectacular and equaled small revolutions of our view of the world. It was shown that the validity of conservation laws or symmetries depends on the type (or strength) of the fundamental interaction, which is considered and that their number increases with increasing strength of the interaction. Table 3.1 shows this for the most prominent symmetries.

If **F** is the operator of a conservation quantity and is not explicitly dependent on the time t, then this is equivalent to the commutativity of **F** with the Hamilton operator **H** and there exist eigenfunctions of **H**, that are simultaneous eigenfunctions of **F**. (If **F** depends explicitly on time it is additionally required that also $\partial \mathbf{F}/\partial t = 0$.)

In simple cases the commutativity may be shown directly. Examples are: the Hamilton operator and the conservation of the total energy E lead to the statement that the physical process considered must not be time-dependent; analogously for momentum conservation and space independence, and angular momentum conservation and angle independence in rotations.

From this it is immediately clear that conservation laws are connected to symmetries, which is expressed [NOE18] by the famous.

Noether Theorem (1918) *If a physical law is invariant under a symmetry transformation then there exists a corresponding conservation law.*

The operator **U** is a symmetry operator if $\mathbf{U}|\Psi\rangle$ fulfills the same Schrödinger equation as $|\Psi\rangle$:

$$i\hbar \frac{d}{dt}(\mathbf{U}|\Psi\rangle) = \mathbf{H}\mathbf{U}|\Psi\rangle \qquad (3.1)$$

Table 3.1 Conservation quantities and their violation, and fundamental interactions: conservation: +, violation: −

Conservation quantity or symmetry	Strong Interaction	EL-Mag	Weak
Mass m/Energy E	+	+	+
Momentum p			
Angular Momentum L, S			
Charge Q	+	+	+
Isospin T	+	−	−
Strangeness S	+	+	−
Charm C			
Beauty B, Topness T			
Parity P	+	+	−
Charge Conjugation C	+	+	−
Baryon Number B	+	+	+
Lepton Number (s)		+	+
Hypercharge Y	+	+	−
Time Reversal T	+	+	−
Charge Parity CP	+	+	−
CPT	+	+	+

Then also, together with $\mathbf{UU}^{-1} = \mathbf{1}$

$$i\hbar \frac{d}{dt}|\Psi\rangle = \mathbf{U}^{-1}\mathbf{HU}|\Psi\rangle, \qquad (3.2)$$

whence, together with $i\hbar d/dt|\Psi\rangle$

$$[\mathbf{H}, \mathbf{U}] = 0. \qquad (3.3)$$

If the explicit commutativity cannot be shown directly because e.g. the potential in the Hamilton operator is not or insufficiently known in analytic form (which is a case typical for nuclear physics) the behavior under the corresponding symmetry operation must be investigated. In the case of time-reversal invariance there is, because of the properties of the time-reversal operator, no conservation quantity, but only a symmetry.

Another important distinction is that between *discrete* and *continuous* transformations. The discrete transformations have *multiplicative*, the others *additive* quantum numbers. Typical discrete transformations are the parity operation or the exchange-symmetry operation. Continuous operations are rotations in space (the conservation quantity is the angular momentum) or isospin space (the conservation quantity is the isospin). Translations in space and time have the conservation quantities of momentum and energy. A detailed treatment of symmetries, invariances, conservation laws, etc. can be found in Ref. [FRA86].

3.1 Generalities

3.1.1 Discrete Transformations

The operator of a symmetry transformation must be unitary: $\mathbf{U}^\dagger = \mathbf{U}^{-1}$. The twofold application of the transformation leads back to the original state. The eigenvalues of the transformation operator are therefore ± 1. For the parity operation (see below) this means

$$P\Psi = \pi\Psi \quad \text{and} \quad P^2\Psi = \Psi. \tag{3.4}$$

In order for a symmetry operator to represent an *observable*, i.e. to have real eigenvalues it must be, in addition, hermitean: $\mathbf{U} = \mathbf{U}^\dagger$. For discrete transformations this is fulfilled and the symmetry operator belongs to a conserved quantity (example: parity operator, parity conservation).

3.1.2 Continuous Transformations

The symmetry operator connects continuously to (i.e. it is only infinitesimally different from) the unit (identity) operation and is generated by a *generator* $\mathbf{F}$:

$$\mathbf{U} = e^{i\epsilon\mathbf{F}} = 1 + i\epsilon\mathbf{F} + \frac{(i\epsilon\mathbf{F})^2}{2!} + \cdots \tag{3.5}$$

with ϵ real and <1. The condition for unitarity of $\mathbf{U}$ is (for infinitesimal transformations)

$$\mathbf{U}^\dagger\mathbf{U} \approx \left(1 - i\epsilon\mathbf{F}^\dagger\right)(1 + i\epsilon\mathbf{F}) \tag{3.6}$$

$$= 1 - i\epsilon\left(\mathbf{F}^\dagger - \mathbf{F}\right) + O\left[(i\epsilon\mathbf{F})^2\right] = 1 \tag{3.7}$$

i.e.

$$\mathbf{F}^\dagger = \mathbf{F}, \tag{3.8}$$

which shows the hermiticity of $\mathbf{F}$. Thus, $\mathbf{F}$ is the observable belonging to $\mathbf{U}$. If $\mathbf{U}$ is a symmetry operator, which commutes with $\mathbf{H}$, thus also with $\mathbf{F}$:

$$\mathbf{H}(1 + i\epsilon\mathbf{F}) - (1 + i\epsilon\mathbf{F})\mathbf{H} = 0, \tag{3.9}$$

i.e.

$$[\mathbf{H}, \mathbf{F}] = 0. \tag{3.10}$$

The quantity $\mathbf{F}$ is the conservation quantity belonging to the symmetry transformation $\mathbf{U}$.

Conservation laws lead to quantum numbers, selection rules (forbiddenness), and branching ratios, which may determine the behavior of nuclear reactions.

3.2 Conserved Quantities in Nuclear Reactions

In nuclear reactions, especially those mediated by the strong and electromagnetic interactions, a number of conserved quantities are taken for granted, e.g. the existence and the conservation of the electric charge or the baryon number. It can be shown (but will not be discussed here in detail) that these belong to a class of quantities with additive quantum numbers that are conserved because of *gauge-transformation* properties of wave functions or field operators. Charge conservation is thus a consequence of the invariance of the S-matrix against gauge transformations of the electromagnetic field. This gauge invariance is a basic property of all field theories such as QCD.

3.2.1 Energy Conservation

The conservation of the total energy (rest + kinetic energy) plays an important role in the dynamics and in the kinematics of nuclear reactions. Here only the kinematics will be mentioned briefly. Basically it is the application of the laws of conservation of energy and momentum to nuclear reactions. For the planning and interpretation of experiments kinematics is immensely important and deserves more space than can be provided here. It should be noted that the correct kinematics is relativistic, and the non-relativistic approach is an approximation for lower energies. Despite some of the non-relativistic formulations being simpler at first sight the use of relativistic invariants with e.g. the concept of the *invariant total energy* that includes the rest energy as well as kinetic energy, on the other hand, makes the description more consistent and lucid.

The Q value of nuclear reactions is given by the mass differences between exit- and entrance-channel particles

$$Q = \left[(m_1 + m_2) - \sum_{n_{\text{out}} m_i} \right] c^2. \tag{3.11}$$

Because of the constancy of the total energy Q can also be expressed by the difference between the kinetic energies. The Q value determines the quite different behavior of endothermic ($Q < 0$), elastic ($Q = 0$), or exothermic ($Q > 0$) reactions, especially at low energies. The endothermic reactions start at a threshold energy, at which the energy of the relative motion ("energy in the center-of-mass system" $E_{\text{c.m.}}$) is just equal to Q.

3.2.2 Momentum Conservation

The conservation of momentum is quantum-mechanically equally important as in classical collision processes. Together with energy conservation it determines the kinematics of all decays and nuclear reactions. Certain processes like pair creation are e.g. possible only if a heavier collision partner takes care of the momentum conservation.

3.2.3 Reaction Kinematics

Another important application of both conservation laws is the transformation (non-relativistic: Galilei transformation, relativistic: Lorentz transformation) between different coordinate systems, especially between the laboratory and the center-of-mass (c.m.) systems. The two-body system of the entrance channel of a reaction is reduced to a one-body problem by separation of the relative motion from the motion of the center of mass. For the classical equations of motion as well as for the Schrödinger equation we see that the motion of the center of mass is straight and uniform and is not influenced by the collision dynamics. The collision dynamics, however, depends only on the relative motion, to which corresponds the energy *available* for a reaction, which is always smaller than the total energy or the total kinetic energy in the laboratory system. Theoretical considerations are normally performed in the c.m. system and experimental data must therefore be transformed kinematically into this system (or also vice versa). Full expositions of the kinematics of nuclear (and particle) reactions are part of many books on nuclear physics/nuclear reactions (see e.g. Refs. [MAR70, PER82, SAT90, FRA86, PER82, MAR68] that only when necessary kinematical considerations are shown in this text.

3.2.4 Conservation of Angular Momentum

Angular-momentum conservation is connected with continuous transformations, namely with rotations in 3D space. The generator of an infinitesimal rotation $\delta\phi$ e.g. about the z-axis is

$$R = 1 + \delta\phi \frac{\partial}{\partial\phi}. \tag{3.12}$$

With the z component of the angular momentum operator $\mathbf{J}$

$$J_z = -i\hbar \left(x\frac{\partial}{\partial y} - y\frac{\partial}{\partial x} \right) = -i\hbar \frac{\partial}{\partial\phi} \tag{3.13}$$

one gets:

$$R_{\text{inf}} = 1 + \frac{i}{\hbar} J_z \delta\phi \tag{3.14}$$

or

$$R_{\text{fin}} = \lim_{n \to \infty} \left(1 + \frac{i}{\hbar} J_z \delta\phi \right)^n = \exp\left(\frac{i}{\hbar} J_z \Delta\phi \right) \quad \text{for finite rotations.} \tag{3.15}$$

J_z is conserved ($[J_z, \mathbf{H}] = 0$) if the potential does not explicitly depend on ϕ. Similarly, $\mathbf{J}$ or $\mathbf{J}^2$ are conserved if the potential depends explicitly only on r, but not on θ, ϕ, i.e. like in classical physics for central forces, because the operator of $\mathbf{J}$ acts only on angular variables.

The usual description of nuclear reactions takes care of the conservation of angular momentum by describing the observables in the angular-momentum representation. The solution of the Schrödinger equation is normally done in spherical polar coordinates. The angular-momentum eigenfunctions $Y_\ell^m(\theta, \phi)$ are the angle-dependent parts of the wave function and are separable from the spatial wave-function part. For particles with spin the description is much more complicated because the orbital angular momentum and the channel spins of the incoming and the outgoing channels must each be coupled to the total angular momentum J, which is the only conserved angular-momentum quantity. Depending on the observable, Racah algebra, i.e. 3j- 6j- or 9j-symbols may be used, see also Sect. 22.3 and Chap. 5. The most general formalism (for neutral particles) was published by Welton [WEL63], formulating general (polarization) observables of the exit channel as functions of those of the entrance channel in a partial-wave expansion and showing the clear separation between the dynamics (transition-matrix elements), angular-momentum algebra, and the geometry (rotation functions $D_{MM'}^L(\theta, \phi, \beta)$). This formalism was extended to charged particles by Heiss [HEI72] (for details, see Chaps. 5 and 7).

For spinless particles and central forces the orbital angular momentum ℓ is a conserved quantity, for reactions with particles, whose spin is $\neq 0$ only the total angular momentum J is conserved. For two-particle nuclear reactions we thus have

$$\vec{\ell}_{\text{in}} + \vec{s}_a + \vec{s}_A \to \vec{J}(\text{ Intermediate state}) \to \vec{\ell}_{\text{out}} + \vec{s}_b + \vec{s}_B. \quad (3.16)$$

The vector sum $\vec{S}_{\text{in}} = \vec{s}_a + \vec{s}_A$ or $\vec{S}_{\text{out}} = \vec{s}_b + \vec{s}_B$ are the *entrance or exit channel spins*.

3.2.5 Conservation of Parity

The parity operation P is the spatial reflection of the physical system at the origin and is a discrete transformation:

$$P\vec{r} = -\vec{r}, \quad (3.17)$$

$$Pr = r, \quad P\theta = \pi - \theta, \quad P\phi = \pi + \phi, \quad (3.18)$$

and

$$Pt = t, \quad P\vec{p} = -\vec{p}, \quad P\vec{\ell} = \vec{\ell}. \quad (3.19)$$

It is assigned the multiplicative quantum number "parity" π by

$$P\Psi = \pi\Psi. \quad (3.20)$$

3.2 Conserved Quantities in Nuclear Reactions

Subsequent execution of two parity operations leads back to the original system

$$P^2 \Psi = \Psi, \tag{3.21}$$

thus the eigenvalue of the parity operation is

$$\pi = \pm 1. \tag{3.22}$$

The system behaves under parity conservation either "parity-even" (e) or "parity-odd" (o). The P operation transforms a (unitary) operator appropriately:

$$P U_{e,o} P^{-1} = \pm U_{e,o}. \tag{3.23}$$

For transition-matrix elements thus

$$\langle \alpha | U_{e,o} | \beta \rangle \begin{cases} \neq 0, \\ = 0, \end{cases} \text{if } \alpha \text{ and } \beta \text{ have } \begin{cases} \text{equal} \\ \text{different} \end{cases} \text{parity}. \tag{3.24}$$

Correspondingly for eigenstates with well-defined parity (i.e. if parity conservation holds) the expectation value of a parity-odd operator must vanish. An example are *pseudo-scalars* like the helicity (longitudinal polarization) $\langle \hat{\sigma} \vec{p} \rangle$:

$$\langle \alpha_{\pm} | \hat{\sigma} \vec{p} | \alpha_{\pm} \rangle = 0. \tag{3.25}$$

3.2.6 Nuclear Reactions Under Parity Conservation

Parity appears in nuclear reactions in two instances:

- Every particle has an *intrinsic parity*. If the nucleons are assigned positive parity the parities of all other particles are fixed via connecting nuclear reactions (example: From the reaction $np \to d\pi$—together with isospin conservation—the negative parity of the pions results).
- Reaction theory shows that the parity behavior of the spherical harmonics

$$P Y_\ell^m(\theta, \phi) = (-)^\ell Y_\ell^m(\theta, \phi) \tag{3.26}$$

determines the parity of the relative motion

$$\pi = (-)^\ell. \tag{3.27}$$

Thus, for a nuclear reaction

$$\pi_a \cdot \pi_A \cdot (-)^{\ell_{\text{in}}} = \pi_{\text{intermediate state}} = \pi_b \cdot \pi_B \cdot (-)^{\ell_{\text{out}}}. \tag{3.28}$$

Figure 3.1 shows the effect of space reflection on nuclear reactions. Together with angular-momentum conservation parity conservation restricts the number of

Fig. 3.1 Effect of the parity operation in a nuclear reaction with spins

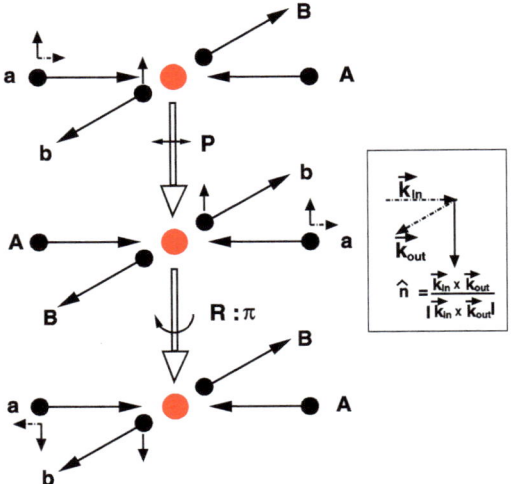

possible angular momentum channels and intermediate states in compound nuclear reactions. For one well-defined intermediate state (resonance, compound state) with fixed parity π the orbital angular momenta in the entrance exit channels thus are either even or odd only, excluding the other possibility even if allowed by angular momentum coupling.

The conservation of parity is considered valid for the *strong interaction* and equally its complete violation in the *weak interaction*. This has numerous consequences. On the one hand the description of nuclear reactions (especially of the polarizations observables) is often substantially simplified because a number of observables vanish under parity conservation. For others there are at least restrictions e.g. on the number of possible angular-momentum states. On the other hand, nuclear reactions offer the possibility to find effects of parity violation by trying to measure "forbidden" observables.

3.2.7 Nuclear Reactions Under Parity Violation

A wave function with "good" parity is either parity-even (e) or parity-odd (o). A wave function of states of non-conserved parity is always "parity-mixed" with a mixing parameter F with

$$\Psi = (1-F)^{1/2}\Psi_e + F^{1/2}\Psi_o. \tag{3.29}$$

An operator has non-vanishing expectation values for such mixed wave functions only if its transformation behavior is parity-odd:

3.2 Conserved Quantities in Nuclear Reactions

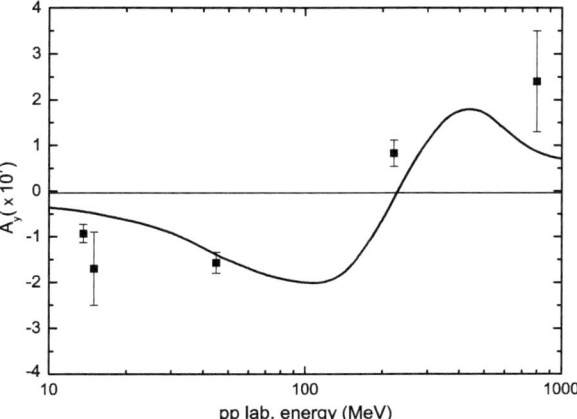

Fig. 3.2 Measured values of the longitudinal analyzing power in pp scattering at 15 MeV (Los Alamos, Bonn), 45 MeV (SIN/PSI), and 800 MeV (ANL) in comparison to predictions of the standard model (DDH: Desplanques, Donoghue, and Holstein [DES80])

$$\langle \Psi | U | \Psi \rangle = (1-F)^{1/2}[\langle \Psi_e | U | \Psi_e \rangle + \langle \Psi_o | U | \Psi_o \rangle] + 2F^{1/2} \cdot \langle \Psi_e | U | \Psi_o \rangle$$
$$= 2F^{1/2} \cdot \langle \Psi_e | U | \Psi_o \rangle. \quad (3.30)$$

A longitudinal polarization or analyzing power is such an observable (this is discussed in detail in Chaps. 5 and 6). More precisely: The axis of quantization $\hat{y}$ of the spins in nuclear reactions with polarized particles is given as usual by

$$\hat{y} \equiv \hat{n} = \frac{\vec{k}_{\text{in}} \times \vec{k}_{\text{out}}}{|\vec{k}_{\text{in}} \times \vec{k}_{\text{out}}|}, \quad (3.31)$$

where $\vec{k}_{\text{in}} \equiv \hat{z}$ and $\vec{k}_{\text{out}}$ designate (with $p = \hbar k$) the momenta of the incident or outgoing particles. It is normal to the scattering plane, spanned by $\hat{z}$ and $\vec{k}_{\text{out}}$, and is the only component invariant under the P operation.

The two other components (along x and z axes) of the polarization or analyzing power in $\hat{x}$ and $\hat{z}$ directions must vanish under parity conservation or can be non-vanishing under parity violation, see also Sect. 6.5.

Thus we have observables P_z or A_z that are sensitive to just such a violation. The longitudinal analyzing power A_z has been investigated especially for pp scattering at several energies. The magnitude of the parity-violating effect caused by the weak nuclear interaction on the cross section is on the order-of-magnitude $1 \cdot 10^{-7}$ of the superimposed effects of the strong interaction. In order to measure it with sufficient statistics an absolute precision of better than $1 \cdot 10^{-8}$ and very small systematic errors are necessary. Figure 3.2 shows the results of the experiments, which prove a clear parity violation in agreement with the standard model.

The study of [DES80] has been updated by a recent survey in [RAM06]. A comprehensive study of the status of hadronic parity violation and future developments of the field (especially in the framework of effective field theory) was published recently, see [HOL09].

3.3 Isospin in Nuclear Reactions

The concept of isospin (and its conservation) dates back to Heisenberg (1932) and was suggested by the similarity of the cross sections (more precisely: of the scattering lengths (see below) derived from the low-energy cross sections) in the pp (after subtraction of the direct Coulomb effects), the np, and the nn interactions. Later additional similarities in other observables e.g. the magnetic form factors of the nucleons were found. Iso-multiplets in mirror nuclei with (assumed) identical nuclear structure differ approximately only in their Coulomb energies. Small remaining differences may be attributed to non-electromagnetic effects due to differences in quark masses (Nolen-Schiffer anomaly). In particle physics iso-multiplets play an important role, also because there the Coulomb-energy differences are small compared to the total energy. For further reading see e.g. Refs. [FOX66, TEM67, AND69], and [WIL69].

3.3.1 Formalism

Formally the isospin can be treated like the spin. One defines for the hadronic interaction isospin operators, which are vectors with the usual commutation relations, ladder operators, isospin coupling, isospin substates etc. The details of the formalism for spin polarization that are in part also applicable here are set forth in Chap. 5. The pertinent symmetry transformations are rotations in isospin space. Isospin conservation means invariance under rotations in the three-dimensional isospin space, i.e.

$$\mathbf{U} = e^{i\theta\vec{n}\mathbf{T}/\hbar} \qquad (3.32)$$

with $\det \mathbf{U} = 1$. Under the strong interaction the length $\sqrt{T(T+1)}$ of the vector $\mathbf{T}$ under rotations remains constant (it is a scalar in T space). In the electromagnetic interaction the T multiplets are split into $2T + 1$ components that correspond to different charge states with

$$Q = T_3 + Y/2, \qquad (3.33)$$

the *Nishijima relation* introduced by M. Gell-Mann, with Y the *hypercharge*; for nuclei $Y = A = B$, the baryon number. Thus $\mathbf{T}$ in the electromagnetic interaction is not a conservation quantity. In the strong interaction the assumption of a conservation of $\mathbf{T}$ is suggestive because of the close agreement in the properties of the members of iso-multiplets. A closer look shows, however, that in reality isospin symmetry is broken to an order-of-magnitude of about 1 %. The third component of T, T_3, because of charge conservation, remains a perfect conservation quantity (it formally has the index 3, not z as with spin, because isospin has no similar meaning in real space). In nuclear physics the nucleon, which is a *two-state system* is assigned the isospin $\mathbf{t} = 1/2$, the proton the "UP" projection $t_3 = +1/2$, and the neutron "DOWN" projection $t_3 = -1/2$, see Fig. 3.3. For composite systems the

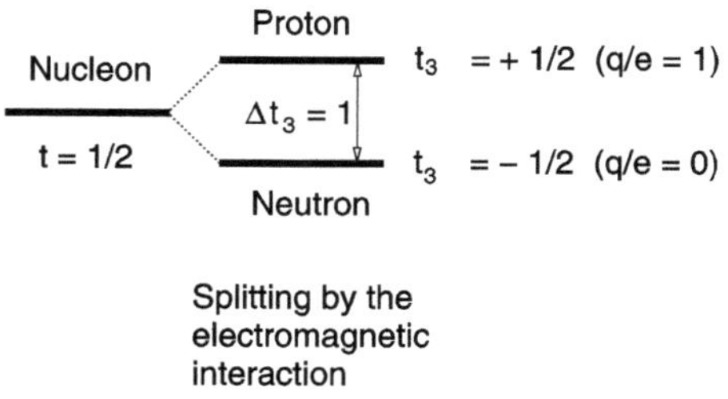

3.3 Isospin in Nuclear Reactions

Fig. 3.3 Breaking of the isospin symmetry by the electromagnetic interaction

Nucleon $t = 1/2$ splits into Proton $t_3 = +1/2$ ($q/e = 1$) and Neutron $t_3 = -1/2$ ($q/e = 0$), with $\Delta t_3 = 1$. Splitting by the electromagnetic interaction.

Table 3.2 Allowed partial-wave states of the two-nucleon systems

Isospin channel spin	$T = 0$		$T = 1$	
	$S = 0$	$S = 1$	$S = 0$	$S = 1$
$L = 0$	–	$^3S_1(d)$	$^1S_0(d^*)$	–
$L = 1$	$^1P_1$	–	–	$^3P_{0,1,2}$
$L = 2$	–	$^3D_{1,2,3}$	$^1D_2$	–

total isospin is the vector sum of the component isospins, for the projections the scalar sums:

$$\mathbf{T} = \sum_{i=1}^{A} \mathbf{t}^{(i)}; \qquad T_3 = \sum_{i=1}^{A} t_3^{(i)}, \qquad (3.34)$$

and thus for nuclei: $T_3 = -1/2(N - Z)$ and $T_{\min} = 0$ for even A, $T_{\min} = 1/2$ for odd A, $T_{\max} = A/2$.

Formally, also under the assumption of isospin conservation the generalization of the *Pauli principle* is possible by including, besides space and spin, also the isospin. A fermion state is then described by a wave function, which is an antisymmetrized product-wave function of space, spin, and isospin wave functions. For the nucleon-nucleon system e.g. every other of the—in principle possible—two-nucleon states is excluded, as shown in Table 3.2.

Under isospin conservation e.g. the transition from the "normal" (bound) triplet deuteron to the (unbound) *singlet deuteron* in the pn final-state interaction of the breakup reaction

$$p + d \rightarrow p + (n + p)_{^1S_0 = d^*} \qquad (3.35)$$

is strictly forbidden (the singlet deuteron d^* is a slightly unbound state of the pn system with antiparallel nucleon spins that appears only in pn scattering, see also Tables 3.2 and 3.3 and Chap. 9). However, experimental indications for this transition occurring have been found [GAI88, NIE92], which points to the isospin not being completely conserved, i.e. its conservation would be weakly violated. In fact, in the electromagnetic interaction the isospin is conserved trivially, which incited the question whether the (weak) observed isospin breaking is caused by influences of the Coulomb force or is a more fundamental feature of the strong interaction. Today the latter is the accepted point of view.

3.3.2 Isospin as Conserved Quantity

The question of isospin conservation (or: violation) reaches far beyond the pure formalism. At first sight the behavior of certain reactions seemed to prove isospin conservation (if Coulomb effects are taken into account). Examples are:

- The branching ratios of some reactions may be to a large extent explained using isospin coupling and applying Clebsch-Gordan (3j) coefficients (see Sect. 22.3). This implies that the transition-matrix element does not depend significantly on the third isospin component:

$$p + d \to {}^3\text{He} + \pi^0 (1) \text{ is to be compared with } p + d \to {}^3\text{H} + \pi^+ (2). \quad (3.36)$$

With the isospin CG coefficients the ratio of the cross sections for the two branches is calculated in agreement with the experiments is

$$\frac{(d\sigma/d\Omega)_1}{(d\sigma/d\Omega)_2} = \left(\frac{\langle \frac{1}{2} \frac{1}{2} 10 | \frac{1}{2} \frac{1}{2} \rangle}{\langle \frac{1}{2} - \frac{1}{2} 11 | \frac{1}{2} \frac{1}{2} \rangle} \right)^2 = \left(\frac{-\sqrt{1/3}}{\sqrt{2/3}} \right)^2 = 1 : 2. \quad (3.37)$$

Other examples are pion-nucleon scattering or reactions such as $pp \to d\pi^+$ and $pn \to d\pi^0$.
- Resonances in isospin-forbidden compound processes show much smaller (by orders-of-magnitude) formation and decay probabilities, expressed by their total widths Γ, than comparable allowed resonance transitions. An example is the series of narrow resonances in the elastic protons scattering from light nuclei such as ^{12}C, ^{24}Mg and others, for which the entrance channel has the isospin $T = 1/2$, the resonances, however, belong to states with isospin $T = 3/2, T = 5/2$:

$$p + {}^{12}\text{C} \to {}^{13}\text{N}^* (E_p = 14.231 \text{ MeV}) \to p + {}^{12}\text{C}. \quad (3.38)$$

This resonance has a width of only 1.1 keV, whereas comparable neighboring resonances have widths of about 100 keV. (This resonance—because of its sharpness—has widely been used for energy calibration of accelerators.) Figure 3.4 shows a measurement of excitation functions across this narrow isospin-forbidden resonance in comparison with a typical nearby much wider isospin-allowed state.
- In 1967 the *Barshay-Temmer Theorem* was formulated [BAR67] that allows the identification of members of isospin multiplets (isobaric analogs): *In a nuclear reaction $a + A \to b + b'$ where b and b' are exactly connected by a rotation in isospace and where a or A have isospin $T = 0$, the differential cross section of the reaction will be exactly symmetric about 90° in the c.m. system, independently of the reaction mechanism.* This excludes reactions, in which such a symmetry is imposed by other mechanisms such as CN reactions (see Chap. 8 and Sect. 11.5), or identical particles (see Sect. 3.4). A typical reaction (among others) for this test has been $d + {}^4\text{He} \leftrightarrow {}^3\text{H} + {}^3\text{He}$.

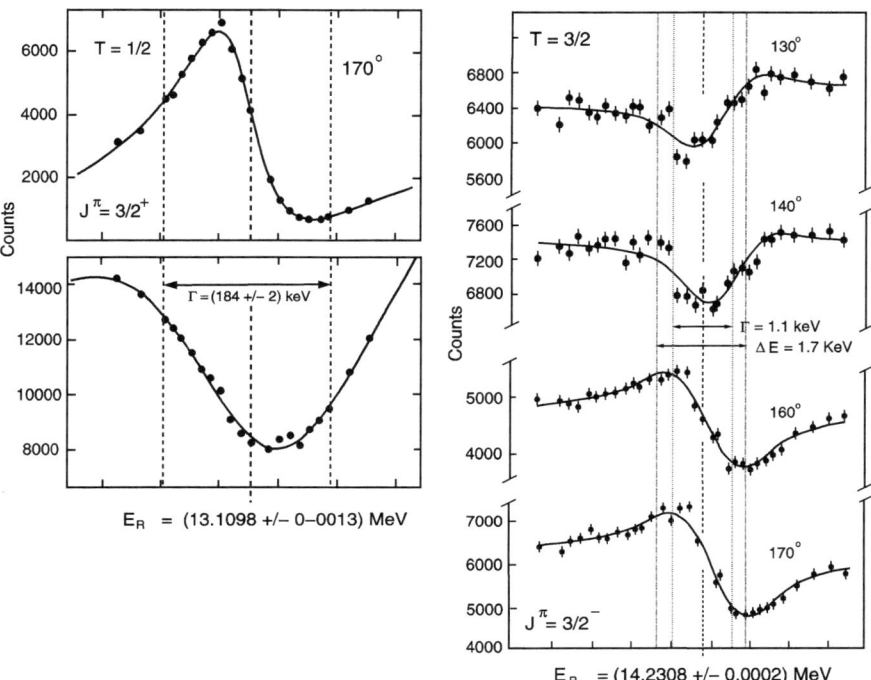

Fig. 3.4 Excitation functions of the ^{12}C + p scattering across the $T = 1/2$ resonance (*left top*: elastic channel, *left bottom*: p_1 inelastic channel) and the $T = 3/2$ resonance (*right*; at different lab. angles). The *solid lines* are Breit-Wigner fits including interference with a background amplitude and an experimental energy spread. This experimental resolution (mainly the energy spread of the incident proton beam from the tandem Van-de-Graaff accelerator) is larger (about 1.7 keV) than the true resonance width (1.1 keV). This resonance has been used to calibrate the tandem energy (the value of E_R is known from spectrograph measurements)

3.3.3 Isospin Breaking

The non-vanishing of the width of the $T = 3/2$ resonance in ^{13}N is an indication and actually a measure of the incomplete isospin conservation (or: slight isospin violation).

Isobaric-Analog resonances (IAR) are another good example for partial isospin conservation and, at the same token, weak isospin breaking. In heavy nuclei no isospin conservation had been expected due to the strength of the Coulomb interaction. The appearance of strong IAR, however, pointed to a relatively good isospin purity of the systems, which is largely conserved by the rapidity of the decay at the high excitation energies, preventing strong mixing of states. The *doorway mechanism*, through which the IAR with isospin $T^> = T_0 + 1/2$ decay into (and mix with) the normal high level-density compound states with $T^< = T_0 - 1/2$ before they decay into the exit channel implies this mixture of states with different isospins. The structure of lighter nuclei also shows examples of isospin conservation (isobar multiplets like ^{3}H and ^{3}He), where, however, after careful Coulomb corrections, a remainder of isospin breaking is left unexplained (*Nolen-Schiffer anomaly*).

Only in 2003 a cross section, small, but finite, of the isospin-forbidden reaction

$$d + d \to {}^4\text{He} + \pi^0 \qquad (3.39)$$

was measured, which constitutes a clear breaking of charge symmetry [STE03, MIL06].

Our present understanding of the nucleon-nucleon interaction—first formulated by Hideki Yukawa in 1935 [YUK35] for a hypothetical exchange particle, first believed to be the muon μ, later identified with the pion π—is that the long- and medium-range parts of the interaction are mediated by the exchange of (virtual) mesons whereas the short-range behavior can probably only be described by the QCD or Effective-Field Theories (EFT) derived from it. Thus, the following non-trivial causes of isospin breaking are mainly:

- The mass difference of π^0 and $\pi^\pm$. On the more microscopic level of the QCD this is caused by the (unexplained) mass difference between the up and down quarks.
- The mixture of the two mesons ρ with $T = 1$ and ω with $T = 0$ in the meson exchange between nucleons as well as the $\eta - \pi^0$ mixing.

Henley and Miller [HEN79] published a detailed classification of the different possibilities of isospin breaking or mixing (Class I to Class IV) shown in Table 3.3. In the nucleon-nucleon system the effects of *charge independence* (equality of the observables of the *np*- and the *pp/nn* systems) and *charge symmetry* (equality of the observables of the *nn*- and of the *pp* systems) or their breaking are the most interesting. The respective scattering lengths show in a very sensitive way (by a "magnifying glass" effect of about 20 % an isospin violation in the potential of about 1 % is detected) the breaking of charge independence while a small charge-symmetry breaking was proved recently in difficult intermediate-energy experiments (TRIUMF, IUCF). In nuclear physics isospin is a weakly broken symmetry. The status of charge-symmetry breaking and its relation to QCD around 2006 is summarized in Ref. [MIL06].

3.4 Exchange Symmetry in Nuclear Reactions of Identical Particles

In the scattering of identical particles a detector at the c.m. angle θ is unable to distinguish whether it registers forward-scattered ejectiles under θ or, under the angle $\pi - \theta$, backward-emitted recoils. This is shown in Fig. 3.5. The formal scattering theory (see below) shows that the angular distributions must be symmetric around $\pi/2$ and therefore must be described by even-order Legendre polynomials. Quantum-mechanically, in addition, it is to be expected that the forward- and backward-scattered particle waves interfere. In this case no classical description of

3.4 Exchange Symmetry in Nuclear Reactions of Identical Particles

Table 3.3 Henley classification of isospin breaking in two-nucleon systems

Class	Form of the iso potential	Effects	Example
I	Isoscalar $V_{ij}^I = A + B\vec{\tau}(i) \cdot \vec{\tau}(j)$	Charge Symmetry (CS) Charge Independence (CI) Rotation Invariance in Iso Space	Valid in the strong int. in first order $nn = np = pp$
II	Iso Tensor $V_{ij}^{II} = C[\tau_3(i)\tau_3(j) - \frac{1}{3}\vec{\tau}(i)\vec{\tau}(j)]$	CS, not CI Invariance under Reflection in the 1–2 plane of iso space	$nn = np$ $np(T=1) = np(T=0)$ $np(T=1) \neq nn/pp$
III	Third Component Iso Vector $V_{ij}^{III} = D[\tau_3(i) + \tau_3(j)]$	no CS, no CI but Symmetry at Interchange $1 \longleftrightarrow 2$	$nn \neq np$ $[V_{ij}^{III}, T^2] \propto [T_3, T^2] = 0$ no "mixing" acts not on np
IV	$V_{ij}^{IV} = E[\tau_3(i) - \tau_3(j)] + F[\vec{\tau}(i) \times \vec{\tau}(j)]$	no CS, no CI Antisymmetry at Exchange $1 \longleftrightarrow 2$	Only for np $[V_{ij}^{IV}, T^2] \neq 0$ "Mixing" Transitions $^3L_j \longleftrightarrow ^1L_j$ $d \longleftrightarrow d^*$ **Origin:** not Coulomb, but $m_d \neq m_u$, $\rho^0 - \omega-$, $\pi^0 - \eta$-Mixing
		Test: e.g., in np scattering: $\Delta P(\theta) \equiv P_n(\theta) - P_p(\pi - \theta) \neq 0?$ $\Delta A(\theta) \equiv A_n(\theta) - A_p(\pi - \theta) \neq 0?$	

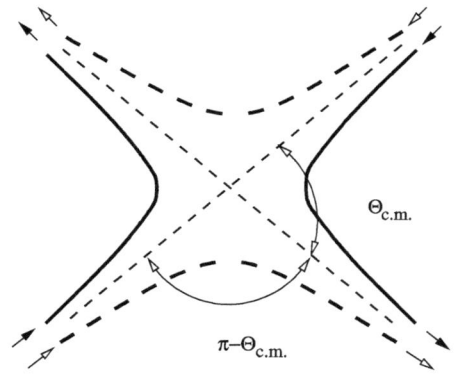

Fig. 3.5 Trajectories of identical particles in the c.m. system

the scattering process is possible. In addition, the details of the interference depend on the spin structure of the interacting particles: identical bosons behave differently from identical fermions, and when the particles have spin $\neq 0$ (i.e. always for fermions) the spin states must be coupled and superimposed in the cross section with their spin multiplicities as weighting factors. The following examples, which can be tested experimentally will explain this.

3.4.1 Identical Bosons with Spin $I = 0$

Here

$$[d\sigma/d\Omega(\theta)]_B = |f_1(\theta) + f_2(\pi - \theta)|^2. \tag{3.40}$$

3.4.2 Identical Fermions with Spin $I = 1/2$

For the fermions the spin singlet cross section

$$[d\sigma/d\Omega(\theta)]_s = |f_1(\theta) - f_2(\pi - \theta)|^2 \tag{3.41}$$

and the triplet cross section

$$[d\sigma/d\Omega(\theta)]_t = |f_1(\theta) + f_2(\pi - \theta)|^2 \tag{3.42}$$

in the total (integrated) cross section must be added *incoherently*, each weighted with their spin multiplicities:

$$[d\sigma/d\Omega(\theta)]_F = \frac{1}{4}|f_1(\theta) + f_2(\pi - \theta)|^2 + \frac{3}{4}|f_1(\theta) - f_2(\pi - \theta)|^2. \tag{3.43}$$

In these two cases the interference has opposite sign, which e.g. at $\theta = \pi/2$ has the consequence that in the case of two bosons there is an interference maximum, for fermions a minimum. Under the special assumption that there is no spin-spin force acting ($f_s = f_t = f$), and with $f(\theta) = f(\pi - \theta)$ one obtains for identical fermions a decrease, for identical bosons an increase each by the factor 2 as compared to the classical cross section.

For pure (Sub-)Coulomb scattering (meaning: Coulomb scattering at energies below the Coulomb barrier) of identical particles the scattering amplitudes can be calculated explicitly (i.e. also summed over partial waves) since we deal with the Rutherford amplitude known from scattering theory, see Sect. 2.3.2:

$$\left(\frac{d\sigma}{d\Omega}\right)_{\text{Coul}} = \left(\frac{Z^2 e^2}{4E_\infty}\right)^2 \left\{ \frac{1}{\sin^4 \frac{\theta}{2}} + \frac{1}{\cos^4 \frac{\theta}{2}} + \frac{2(-1)^{2s} \cos[\eta_s \ln \tan^2 \frac{\theta}{2}]}{(2s+1) \sin^2 \frac{\theta}{2} \cos^2 \frac{\theta}{2}} \right\}. \tag{3.44}$$

In addition to the forward-scattering Rutherford cross section there is a corresponding recoil Rutherford term plus an interference term between both. Figure 3.6 shows this behavior (which is analogous to that of light in Young's double-slit experiment, but additionally shows the influence of spin and statistics).

3.5 Time-Reversal Invariance

The CPT symmetry is considered to be a very deeply founded principle of particle physics (but not, however, exempt from experimental scrutiny). Via the CPT

3.5 Time-Reversal Invariance   55

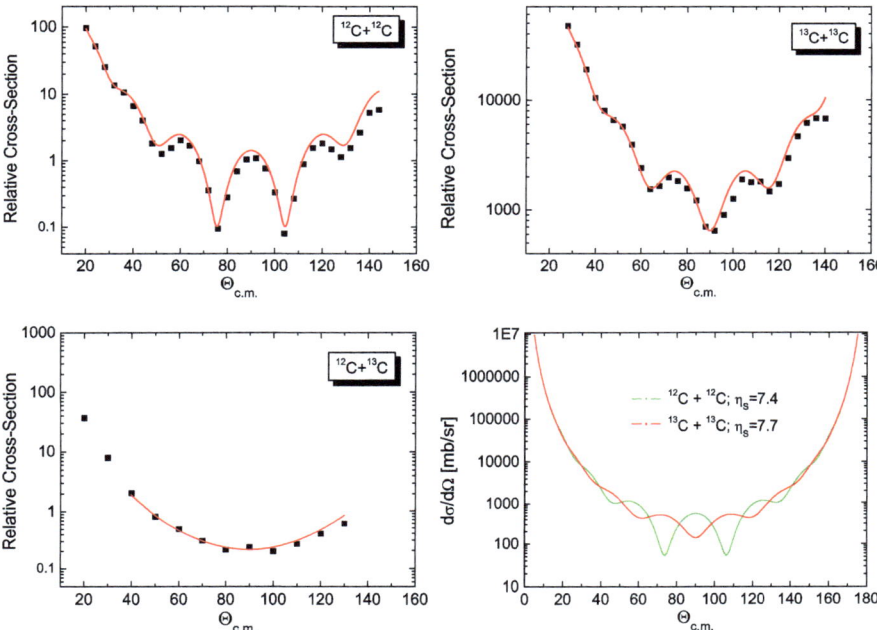

Fig. 3.6 Experimental c.m. angular distributions of Coulomb scattering of two identical bosons (^{12}C) and fermions (^{13}C) as well as of two non-identical particles of nearly equal masses and theoretical cross sections at $E_{\text{lab}} = 7$ MeV. The angular distribution for the non-identical particles is obtained when the spectra of the forward and backward scattered particles cannot be separated by the detector, which is the case for (nearly) equal masses. Otherwise one would obtain a typical Rutherford distribution for each particle separately. The data were measured by students of an advanced lab. course at IKP Cologne in 2003

theorem the proven violation of CP invariance in the weak interaction of the $K^0 - \bar{K}^0$ system made the validity of time-reversal invariance questionable. A direct and independent evidence for time-reversal asymmetry was found only recently [LEE12, BER12] by comparing "in" and "out" channels in the entangled $B^0\bar{B}^0$ system and found to be in agreement with the amount of CP violation thus confirming also the validity of the CPT theorem.

Compared to other invariances the time-reversal invariance is a special case. There is no conserved quantity (and quantum number), because the operator T of time reversal is *anti-linear and unitary = anti-unitary*. It acts as

- $Tt = -t$,
- $T\vec{r} = \vec{r}$,
- $T\vec{p} = -\vec{p}$,
- $T\vec{L} = -\vec{L}$,
- $T\vec{S} = -\vec{S}$.

Fig. 3.7 Effect of the time-reversal operation on nuclear reactions with spins

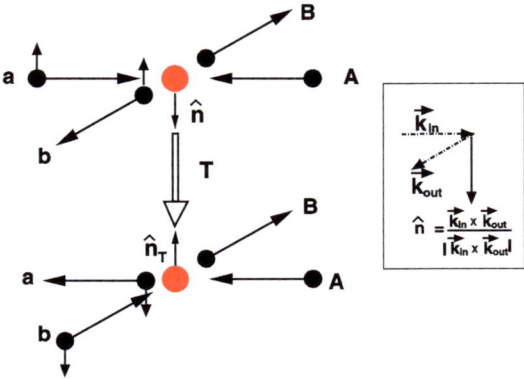

The time-reversal operation does not leave the (time-dependent) Schrödinger equation *form-invariant*:

$$T: -\frac{\partial \Psi}{\partial t} = \frac{\hbar^2}{2\mu}\nabla^2\Psi \rightarrow -\frac{\partial \Psi}{\partial (-t)} = \frac{\hbar^2}{2\mu}\nabla^2\Psi = \frac{\partial \Psi}{\partial t}. \quad (3.45)$$

The form-invariance under the T operation can only be restituted by the additional operation of complex conjugation (operator CC):

$$T = T_0 \cdot CC. \quad (3.46)$$

Thus:

$$T\Psi(t) = \Psi^*(-t). \quad (3.47)$$

The wave function of free particles is transformed into its complex conjugate by the T operation

$$T: e^{i(kr-Et/\hbar)} \rightarrow e^{-i(kr-Et/\hbar)}. \quad (3.48)$$

Applied to motions the operator T produces an *inversion of motion*. In nuclear reactions it interchanges entrance and exit channels, incoming and outgoing momenta as well as spins. Time-reversal invariance is of principal and also practical importance. In principle its validity can be tested also in nuclear reactions. Its validity simplifies the description of nuclear reactions, especially for the polarization observables of particles with spin. Figure 3.7 shows the effect of the time-reversal operation on a nuclear reaction with spins.

3.5.1 Time Reversal, Reciprocity, and Detailed Balance

In the framework of perturbation theory reciprocity allows to find relations between observables of forward (designated by $\rightarrow$) versus backward ($\leftarrow$) reactions. *Fermi's*

3.5 Time-Reversal Invariance

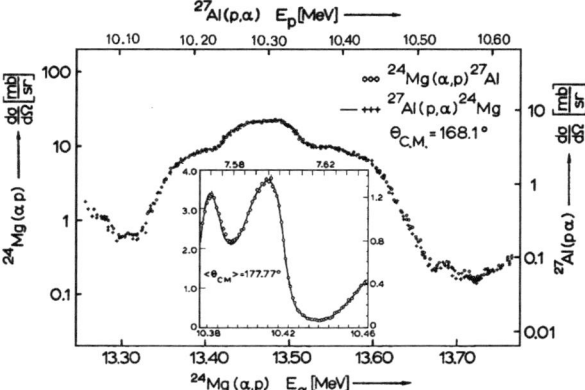

Fig. 3.8 Cross sections of the forward and backward reactions $^{24}\text{Mg}(\alpha, p)^{27}\text{Al}$, measured at the same c.m. energies and angles. The large picture shows the result of the first such experiment to test detailed balance, the *inset* an improved later measurement at some other energies and thus also different reaction mechanism. Adapted from Refs. [VWI68, BLA83]

Golden Rule connects both via the transition matrix element:

$$\left(\frac{d\sigma}{d\Omega}\right)_{\rightarrow} = \frac{2\pi}{\hbar}\left|\langle\Psi_{\text{out}}|H_{\text{if}}|\Psi_{\text{in}}\rangle\right|^2 \rho_{\rightarrow}, \quad (3.49)$$

$$\left(\frac{d\sigma}{d\Omega}\right)_{\leftarrow} = \frac{2\pi}{\hbar}\left|\langle\Psi_{\text{out}}|H_{\text{if}}|\Psi_{\text{in}}\rangle\right|^2 \rho_{\leftarrow}. \quad (3.50)$$

Under time-reversal invariance for spinless particles $H_{\text{if}} = H_{\text{fi}}$, from which follows the

Principle of Detailed Balance *The cross sections of forward and backward reactions are—except for phase-space factors—identical. More precisely*:

$$\frac{(\frac{d\sigma}{d\Omega})_{\rightarrow}}{(\frac{d\sigma}{d\Omega})_{\leftarrow}} = \frac{(2s_a+1)(2s_A+1)}{(2s_b+1)(2s_B+1)} \cdot \frac{k_{\text{in}}}{k_{\text{out}}}. \quad (3.51)$$

This relation is of practical importance because it can—via the reverse reaction—provide observables not directly measurable e.g. because no stable targets exist (such as radioactive targets). The relation can be used to test time-reversal invariance. Its validity was investigated in measurements of several different reactions (e.g. $^{24}\text{Mg}(\alpha, p)^{27}\text{Al}$ [RIC66, VWI68]) and their inverse reactions, and in different energy ranges (where partly different reaction mechanisms dominate the cross sections) by trying to find significant deviations from detailed balance. The results were a confirmation of the non-violation of time-reversal invariance with a precision (i.e. the upper limit of a possible violation) of $\approx 2 \cdot 10^{-3}$ (for a discussion about the determination of the significance limit of such a "null" experiment see e.g. Refs. [HAR86, HAR90, KLE74]). The precision reached, however, was in all cases much too low to show any effects from the time-reversal non-invariance stemming from the known CP violation in conjunction with a valid CPT theorem. Figure 3.8 shows the results of two different detailed-balance experiments.

For reactions with polarized particles time-reversal invariance predicts relations between polarization observables of the forward (designated by $\rightarrow$) and backward ($\leftarrow$) reaction:

$$(A_y)_\rightarrow = (P_y)_\leftarrow. \tag{3.52}$$

The vector analyzing power of the forward reaction with polarized particles in the entrance channel is equal to the vector polarization that is produced with unpolarized particles in the exit channel of the backward reaction. However, the polarization experiments performed to look for time-reversal violations showed less significance than the cross-section measurements.

For elastic scattering, for which forward and backward reactions are the same this means that in a *double-scattering experiment*, in which the same reaction is performed consecutively at the same relative energy and the same c.m. angle the analyzing power (identical to the outgoing polarization) can be determined absolutely (except for the sign) via the relation

$$\left(\frac{d\sigma}{d\Omega}\right)_\text{pol} = \left(\frac{d\sigma}{d\Omega}\right)_\text{unpol} (1 + A_y^2), \tag{3.53}$$

see also Chap. 5.

3.5.2 Other Nuclear Observables of Possible Time Reversal Violation TRV

Only briefly we mention that time-reversal invariance forbids the existence of static nuclear moments: odd electric moments (E1, E3, ...), and even magnetic moments (M0, M2, ...). The standard model predicts TRV by an electric dipole moment of $10^{-33} < d_n < 10^{-31}$ e·cm. A crude estimate from observations would give

$$d_n \approx e \cdot \underbrace{r_\text{rms}}_{10^{-13}\text{ cm}} \cdot \underbrace{F^{PV}}_{10^{-7}} \cdot \underbrace{P^{TRV}}_{10^{-3}} \leq 10^{-23} \text{ e·cm} \tag{3.54}$$

where F^{PV} are the measured degree of parity violation and P^{TRV} the minimum possible TRV from CP non-invariance and with CPT validity.

Novel methods for the determination of d of protons, deuterons, and other light nuclei using storage-ring accelerators and polarization technology have been proposed at BNL Brookhaven and COSY/Jülich, see e.g. Ref. [PRE13]. The basic idea is to have a perturbation by the interaction of a radial electric field with the dipole moment along the rings build up over many revolutions. The injected longitudinally polarized ions would gain a transverse polarization component that could be measured with polarimeters of high precision and high sensitivity. Several designs (purely electric, purely magnetic, and mixed electric/magnetic) are being discussed. The hope is to increase the sensitivity from the present limit of $< 2.9 \cdot 10^{-26}$ e·cm for the neutron and $< 7.9 \cdot 10^{-25}$ e·cm for the proton to $\approx 1 \cdot 10^{-29}$ e·cm. This is still far away from the predictions of the standard model but may be in range of some supersymmetric (SUSY) models ($10^{-29} < d_n < 10^{-24}$ e·cm).

Recently, two nuclides with static octupole (E3) deformation ("pear-shape") have been identified, ^{220}Rn and ^{224}Ra, using γ spectroscopy after Coulomb excitation of radioactive beams in the isotope separator REX-ISOLDE (CERN), and the multi-detector array MINIBALL [GAF13], see also Sect. 17.7 and Fig. 17.11.

3.6 Exercises

3.1. Calculate the lab. threshold energy for the production of antiprotons in the collision of protons on protons. Which exit channels are possible?
3.2. Why—in pp reactions—are strange particles always produced in pairs ("associated production")? Find the lowest-energy example.
3.3. List a number of (parity-non-invariant) pseudo-scalar observables.
3.4. Prove the forbiddenness of static electric dipole moments in nuclei by time-reversal invariance. If $d(n) \neq 0$: Why must parity invariance then be violated too?
3.5. Which reactions are possible/forbidden? Why?

- $p + d \to p + p + n$ at $E_{p,\text{lab}} = 1.0$ MeV
- $p + p \to \bar{p} + p$
- $d + d \to {}^4\text{He} + \gamma$
- $\alpha + {}^{12}\text{C} \to d + {}^{14}\text{N}$ ($J^\pi = 1^+$ and $J^\pi = 0^+$)
- $\gamma + p \to p + p + \bar{p}$.

3.6. Construct a table of the allowed states of the $d + d$ system.
3.7. Show explicitly that the expectation value of a spinor $\vec{\sigma}$ is reversed under the time-reversal operation, whereas the helicity $\langle \hat{p} \cdot \vec{\sigma} \rangle$ is invariant.
3.8. In Fig. 3.8 the cross sections of the reaction $^{27}\text{Al}(p,\alpha)^{24}\text{Mg}$ and its inverse are plotted as functions of their respective lab. energies such that they have the same c.m. energies and c.m. angles. Verify the relations between these energies and angles.

References

[AND69] J.D. Anderson et al., *Conf. on Nuclear Isospin, Asilomar* (Academic Press, San Diego, 1969)
[BAR67] S. Barshay, G.M. Temmer, Phys. Rev. Lett. **12**, 728 (1967)
[BER12] J. Bernabéu, F. Martinez-Vidal, P. Villanueva-Pérez, doi:10.1007/JHEP08(2012)064
[BLA83] E. Blanke, H. Driller, W. Glöckle, H. Genz, A. Richter, G. Schrieder, Phys. Rev. Lett. **51**, 355 (1983)
[DES80] B. Desplanques, J.F. Donoghue, B.R. Holstein, Ann. Phys. (N.Y.) **124**, 449 (1980)
[FOX66] J. Fox, D. Robson (eds.), *Conf. on Isobaric Spin in Nuclear Physics, Tallahassee* (Academic Press, New York, 1966)

[FRA86] H. Frauenfelder, E.M. Henley, *Nuclear and Particle Physics A: Background and Symmetries*, 2nd printing edn. Lect. Notes and Suppl. in Phys. (Benjamin/Cummings, Reading, 1986)
[GAF13] L.P. Gaffney et al., Nature **497**, 911 (2013)
[GAI88] N.O. Gaiser, S.E. Darden, R.C. Luhn, H. Paetz gen. Schieck, S. Sen, Phys. Rev. C **38**, 1119 (1988)
[HAR86] H.L. Harney, A. Richter, H.A. Weidenmüller, Rev. Mod. Phys. **58**, 607 (1986)
[HAR90] H.L. Harney, A. Hüpper, A. Richter, Nucl. Phys. A **518**, 35 (1990)
[HEI72] P. Heiss, Z. Phys. **251**, 159 (1972)
[HEN79] E.M. Henley, G.A. Miller, in *Mesons in Nuclei*, ed. by M. Rho, H.D. Wilkinson (North-Holland, Amsterdam, 1979), p. 416
[HOL09] B.R. Holstein, Eur. Phys. J. A **41**, 279 (2009)
[KLE74] G. Klein, H. Paetz gen. Schieck, Nucl. Phys. A **219**, 422 (1974)
[LEE12] J.P. Lees, et al. (BaBar collaboration), Phys. Rev. Lett. **109**, 211801 (2012)
[MAR68] J.B. Marion, F.C. Young, *Nucl. Reaction Analysis: Graphs and Tables* (North-Holland, Amsterdam, 1968)
[MAR70] P. Marmier, E. Sheldon, *Physics of Nuclei and Particles*, 2 Vols. (Academic Press, New York, 1969/1970)
[MIL06] G.A. Miller, A.K. Opper, E.J. Stephenson, Annu. Rev. Nucl. Part. Sci. **56**, 253 (2006)
[NIE92] P. Niessen, S. Lemaître, K.H. Nyga, G. Rauprich, R. Reckenfelderbäumer, L. Sydow, H. Paetz gen. Schieck, P. Doleschall, Phys. Rev. C **45**, 2570 (1992)
[NOE18] E. Noether, Nachr. Akad. Wiss. Göttingen, Math. Physik. Kl. IIa, Math. Phys. Abt. **1918**, 235 (1918)
[PER82] D.H. Perkins, *Introd. to High Energy Physics* (Addison-Wesley, Reading, 1982)
[PRE13] J. Pretz et al. (for the JEDI collaboration), arXiv:1301.2937v2 (hep-ex) (2013)
[RAM06] M.J. Ramsey-Musolf, S.A. Page, Annu. Rev. Nucl. Part. Sci. **56**, 1 (2006)
[RIC66] A. Richter, A. Bamberger, P. von Brentano, T. Mayer-Kuckuk, W. von Witsch, Z. Naturforsch. **21a**, 1002 (1966)
[SAT90] G.R. Satchler, *Introd. Nucl. Reactions*, 2nd edn. (McMillan, London, 1990)
[STE03] E.J. Stephenson, A.D. Bacher, C.E. Allgower, A. Gårdestig, C.M. Lavelle, G.A. Miller, H. Nann, J. Olmsted, P.V. Pancella, M.A. Pickar, J. Rapaport, T. Rinckel, A. Smith, H.M. Spinka, U. van Kolck, Phys. Rev. Lett. **91**, 142303 (2003)
[TEM67] G.M. Temmer, in *Fundamentals in Nuclear Theory*, ed. by A. de Shalit, C. Villi (IAEA, Vienna, 1967), Chap. 3
[VWI68] W. von Witsch, A. Richter, P. von Brentano, Phys. Rev. **169**, 923 (1968)
[WEL63] T.A. Welton, in *Fast Neutron Physics II*, ed. by J.B. Marion, J.J. Fowler (Interscience, New York, 1963), p. 1317
[WIL69] D.H. Wilkinson (ed.), *Isospin in Nuclear Physics* (North-Holland, Amsterdam, 1969)
[YUK35] H. Yukawa, Proc. Phys. Math. Soc. Jpn. **17**, 48 (1935)

Chapter 4
Cross Sections

4.1 General Appearance of Cross Sections

The coarse behavior of low-energy cross sections, i.e. that without consideration of details of the nuclear interaction, depends on a small number of conditions: whether we deal with neutral or charged particles, and whether the Q-value of the reaction discussed is such that the reaction is elastic ($Q = 0$), endo- ($Q < 0$), or exothermic ($Q > 0$). Especially exothermic reactions with neutrons have no Coulomb barrier to surmount as well as no energetic threshold. Thus, they can be measured down to energies approaching 0. In this case the neutrons have to be produced in a nuclear reactor, and the methods to obtain beams with good geometry and well-defined energies are different from those for charged particles, see Part II of this text.

In this low-energy case one can understand the cross sections without reference to internal degrees of freedom of the nuclear interaction. We assume validity of Fermi's *Golden Rule* for a two-particle reaction $a + A \rightarrow b + B$ and particles with spins I:

$$\frac{d\sigma}{d\Omega} = \frac{(2I_b + 1)(2I_B + 1)}{(2\pi \hbar^2)^2} \mu_i \mu_f \frac{k_f}{k_i} |H'_{if}|^2 \qquad (4.1)$$

where i and f designate the initial (or incident) and the final (or outgoing) channels, the μ_i, μ_f, k_i, and k_f the reduced masses and momenta in these channels. H'_{if} is the transition matrix element of the interaction causing the reaction, treated as a small perturbation. Application of this rule results in the characteristics (for s waves), described below. The general behavior of low-energy cross sections has been discussed in Refs. [MAR70, SAT90], see also, especially in connection with the astrophysical S-factor, Chap. 14 and [ROL88].

4.1.1 Neutral Particles—Elastic Scattering (n, n)

With $Q = 0$; $v_i = v_f$ and $H_{if} = $ const, the (angle-integrated) cross section is constant with energy. One of many examples is shown in Fig. 4.1.

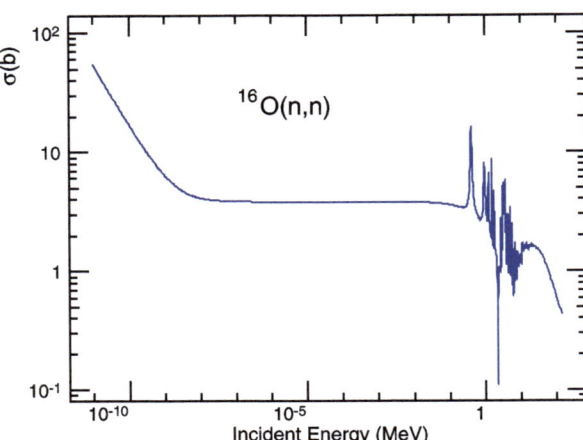

Fig. 4.1 Excitation function of elastic neutron scattering from ^{16}O at low energies. The wide central region is constant with energy. The regions below and above are dominated by neutron diffraction (coherent scattering) and onset of inelastic processes, respectively. Adapted from [NNDC]

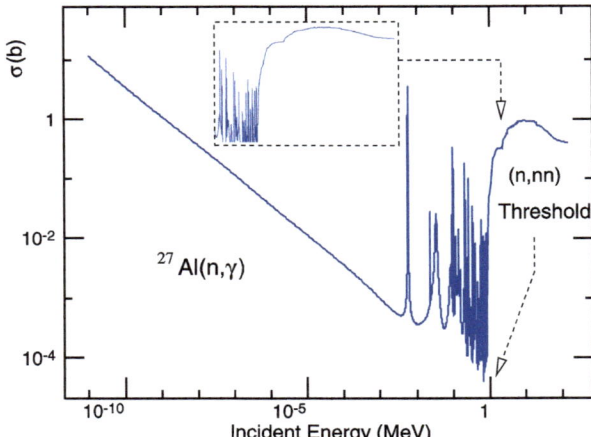

Fig. 4.2 Excitation function of non-elastic neutron reactions (n,γ), (n,n') and $(n,2n)$ from ^{27}Al at low energies. The *onset* of the inelastic reactions leads to an $\approx$ parabolic threshold behavior. The regions below the threshold shows the $E^{-1/2}$ behavior of n capture and resonances. Adapted from [NNDC]

4.1.2 Neutral Particles—Inelastic Neutron Scattering

The reaction (n, n') is *endothermic*, i.e. $Q < 0$. Near the threshold (i.e. in a small energy interval) with $v_i \approx$ const follows:

$$\sigma \propto v_f = \sqrt{2\mu_f(E_i - E_{\text{threshold}})}, \qquad (4.2)$$

which corresponds to a parabolic increase of the cross sections starting from the threshold energy. Figure 4.2 shows an example.

4.1 General Appearance of Cross Sections

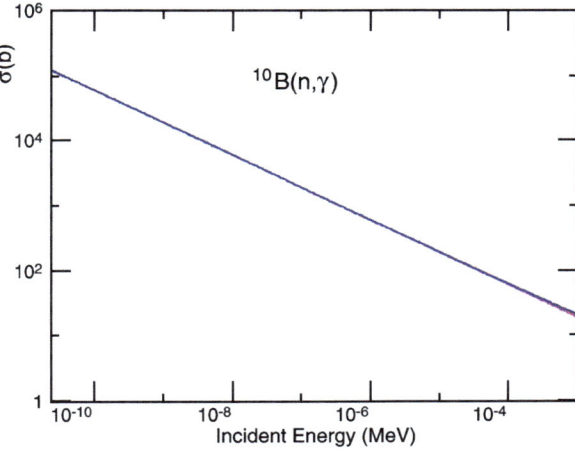

Fig. 4.3 Excitation function of a neutron capture reaction (n, γ) from ^{10}B at low energies. The very high cross section makes this reaction suitable for removing neutrons e.g. in a nuclear reactor. Adapted from [NNDC]

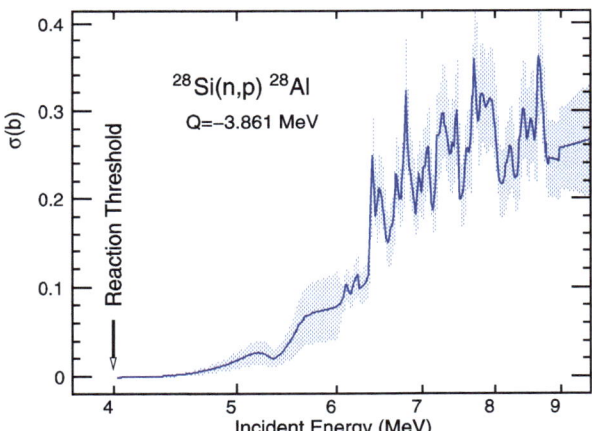

Fig. 4.4 Excitation function of the reaction ^{28}Si$(n, p)^{28}$Al which has a negative Q-value, therefore a threshold, and an exponential increase of σ because of the exit-channel Gamow factor. Adapted from [NNDC]

4.1.3 Neutral Particles—Exothermic Reactions with Thermal Neutrons (n, γ), (n, p), (n, f) etc.

Typically $Q \approx 1$ MeV. Thus, $v_f \approx$ const, and (see Fig. 4.3)

$$\sigma \propto 1/v_i \propto E^{-1/2}. \tag{4.3}$$

4.1.4 Neutral Incident Particles—Endothermic Reactions ($Q < 0$) with Charged Exit Channel

For charged particles in the exit channel (typical reactions are (n, α), (n, p), ...) cross sections are dominated by the *Gamow* factor e^{-G_b}, which is determined by the

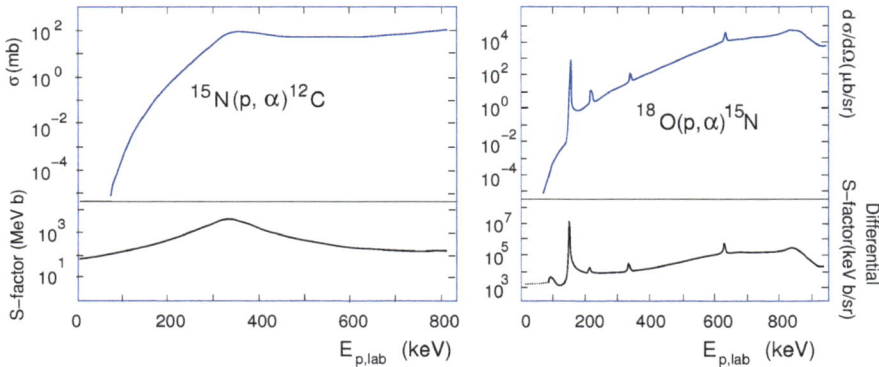

Fig. 4.5 Examples of two reactions in the astrophysical low-energy range showing the enormous influence of the Coulomb barrier on the cross sections increasing over many decades. Nevertheless such studies are important for obtaining reaction rates of reactions of nucleosynthesis and at the same time difficult. The S-factors (one "total", the other a "differential" S-factor) are a good way to divide out the Coulomb behavior showing the essential nuclear features more clearly such as the s-wave behavior that allows extrapolation to zero energy, or resonances. Adapted from [RED82] (*left*) and [LOR79] (*right*)

Coulomb barrier. An energy dependence of the matrix elements H'_{if} enters here as a *penetrability*. We thus obtain an integrated cross section that increases exponentially starting from the threshold (Fig. 4.4).

4.1.5 Charged Particles in the Entrance and Exit Channels

The same applies here with the difference of having two Gamow factors and the cross section going with $\propto e^{-(G_a+G_b)}$. This is similar for exothermic and endothermic reactions (except for the threshold behavior). Figure 4.5 shows two examples of such reactions.

4.1.6 Threshold Effects

When the increasing incident energy passes thresholds of channels opening up successively this shows up as an increase of the cross section at each threshold. A typical example is the (n, f) reaction (f = fission) with increasing number n of emitted neutrons ($n = 0, 1, 2, \ldots$).

If one considers only one channel, e.g. the elastic channel, it can be observed that whenever a new channel opens up a marked decrease of the cross section is visible. This may be explained by the conservation of particle flux

4.1 General Appearance of Cross Sections 65

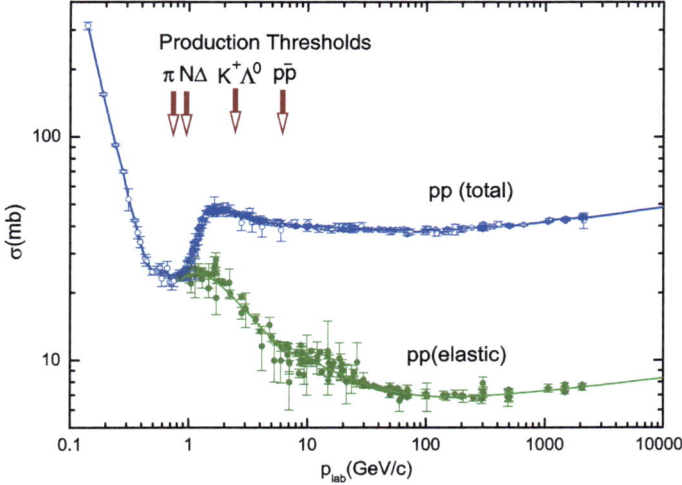

Fig. 4.6 Data of elastic and inelastic scattering of protons from protons (with *error bars*) as functions of the lab. momentum of the incident proton. A few thresholds for the production of real particles such as the pion are shown. The inelasticity begins at the pion production threshold and causes a strong decrease of the purely elastic scattering cross section together with a weak threshold effect (cusp) and an increase of the total cross section. Adapted from [NNDC]

(= probability conservation), which theoretically is a consequence of the unitarity of the scattering matrix. An example is the pp cross section at and around the pion-production threshold.

Occasionally the cross sections around that energy show some resonance-like excursions (*cusps*), which are not resonances in the true sense as discussed here, but can be described with the unitarity of the S-matrix. The shape of these *cusps* has nothing to do with the Breit-Wigner shape of regular resonances. Figure 4.6 shows these effects at the example of pp scattering.

4.1.7 Other Phenomena

The features of s-wave cross sections discussed so far were determined solely by the phase space (a constant matrix element was assumed) or showed a simple transmission behavior (Coulomb penetrability). Similarly the penetrability for partial waves with $\ell > 0$ could be parametrized. In nuclear astrophysics this basic behavior is often "divided out" by discussing, instead of the cross section, the astrophysical S function or S factor, see Chap. 14.

Thus, any "special" energy dependence of the matrix elements becomes more clearly visible. Especially this is true for *resonances* that are eigenstates of the Hamilton operator of compound systems and measure the lifetime of these states by their widths. Examples are the low-energy neutron resonances of uranium isotopes,

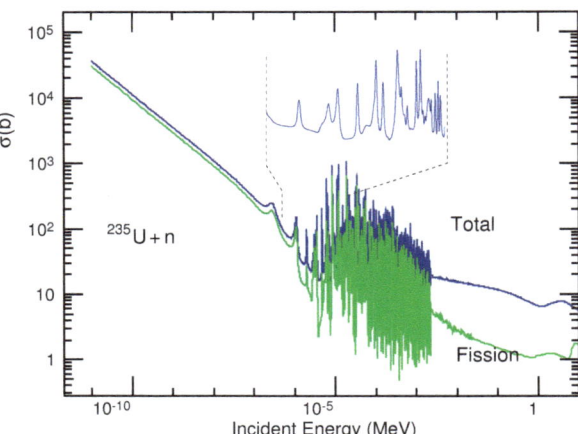

Fig. 4.7 Integrated cross section of the interaction of neutrons with ^{235}U. This isotope undergoes fission by thermal neutrons. The total cross section is a sum of capture and fission reactions. The number of single compound-nucleus resonances increases exponentially with increasing energy where they finally overlap and appear averaged. Adapted from [NNDC]

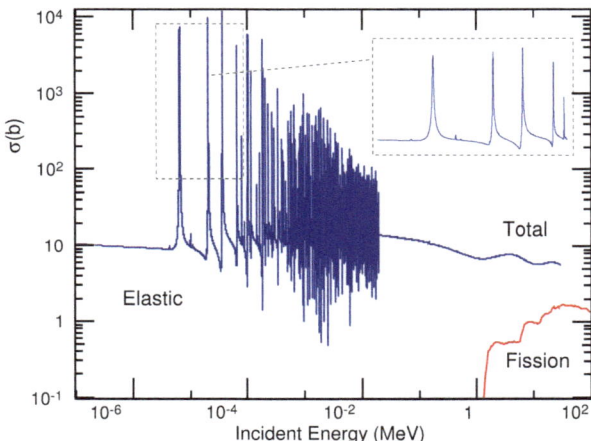

Fig. 4.8 Integrated cross section of the interaction of neutrons with ^{238}U. This isotope undergoes fission only by fast neutrons for which there is a threshold energy. The total cross section at lower energies is purely elastic scattering, at higher energies a sum of elastic and fission reactions. Again the density of the number of single C.N. resonances increases with energy. Adapted from [NNDC]

which play such an important role in nuclear reactors, see Fig. 4.8 and charged-particle resonances with large cross sections playing a role in nucleosynthesis. An example is the reaction ^{18}O$(p, \alpha_0)^{15}$N [LOR79, ROL88], see Fig. 4.5. These neutron resonances show different features for the two most important target isotopes ^{235}U and ^{238}U: ^{235}U is fissionable by thermal neutrons, thus shows the low-energy behavior typical for capture reactions with $1/v$, see Fig. 4.7. ^{238}U needs fast neutrons for fission, i.e. the fission cross section is $\neq 0$ only above a threshold of about 1.5 MeV, below which elastic scattering dominates the cross section (constant with energy). In addition, both reactions show strong resonances in an intermediate energy range, which corresponds to a strong energy dependence of the matrix elements of the Hamiltonian due to excitation of compound-nuclear states.

Fig. 4.9 Scattering quantum-mechanically

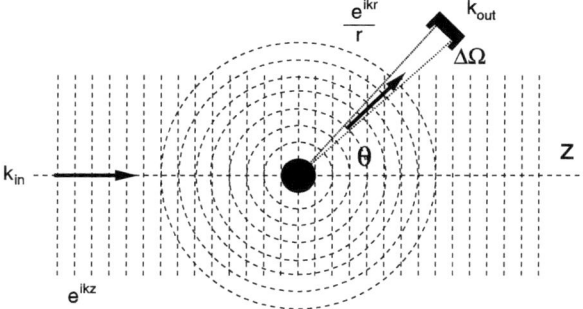

4.2 Formal Description of Nuclear Reactions

The formal description of nuclear reactions aims at a clean separation of the dynamics from geometry- and angular-momentum dependences of the observables. In the study and exploration of nuclear forces it is the dynamical part that is of special interest. In order to make the comparison between theoretical predictions and experimental data easy and meaningful a suitable interface between both has to be defined. This "matching" could be on the level of the observables directly or of the matrix elements derived from some additional data reduction (e.g. via a phase-shift analysis). The latter is advantageous because the wealth of different data (especially polarization observables) is reduced to the common essential dynamical quantities, after separation of non-dynamical parameters (kinematics, geometry, and angular-momentum algebra, see Chap. 5).

4.2.1 Wave Function and Scattering Amplitude

The scattering is described by stationary wave functions for asymptotic states. The incident wave is prepared as a plane wave at $z \to -\infty, t \to -\infty$. The scattered wave is an outgoing spherical wave, which is asymptotically described at $r \to \infty, t \to \infty$. This situation that also corresponds to the usual experimental setup is shown in Fig. 4.9. The modification of the incident wave by the scattering process is described by the scattering amplitude $f(\theta, \phi)$. The total wave function

$$\Psi \to e^{ik_{in}z} + f(\theta, \phi)\frac{e^{ik_{out}r}}{r} \qquad (4.4)$$

is the solution of either a Schrödinger equation with boundary conditions or a corresponding integral equation (the Lippmann-Schwinger equation, which is normally formulated as to contain the boundary conditions automatically). These conditions are taken into account by imposing them on the wave function. If there is a scattering potential $\neq 0$ the wave function in the external region i.e. that of the free particles is matched at a suitable point, often at the nuclear surface $r = R$ or the "edge" of the potential, to the wave function in the nuclear interior i.e. that influenced by the potential. We require:

- The wave functions and their derivatives must be continuous at $r = R$.
- For bound states $\Psi \to 0$ for $r \to \infty$.
- For scattering states Ψ approaches the asymptotic solution for a free particle with $r \to \infty$.
- We require $\Psi \to 0$ for $r \to 0$ in order to avoid the singularity at the origin.

4.2.2 Scattering Amplitude and Cross Section

We start from the general definition of the cross section

$$d\sigma = \frac{\vec{j}_{\text{out}} d\vec{A}}{|\vec{j}_{\text{in}}|} \tag{4.5}$$

and use a classical continuity equation for the particle flux expressed by $\vec{j}$:

$$\vec{j} = \rho \cdot \vec{v}. \tag{4.6}$$

Quantum mechanically

$$\vec{j} = \frac{i\hbar}{2\mu} \left[\Psi^* \vec{\nabla} \Psi - \Psi \vec{\nabla} \Psi^* \right] \quad \text{and} \quad \rho = \Psi \Psi^*, \tag{4.7}$$

which provides a connection to the wave functions that may be solutions of the Schrödinger equation. With

$$\Psi_{\text{in}} \equiv \Phi = a e^{i k_{\text{in}} z} \quad \text{and} \quad \Psi_{\text{out}} = a f(\theta, \phi) \frac{e^{i k_{\text{out}} r}}{r} \tag{4.8}$$

we obtain

$$\vec{j}_{\text{in}} = \hbar/2\mu |a|^2 2\vec{k}_{\text{in}} = |a|^2 \vec{v}_{\text{in}} \tag{4.9}$$

$$\vec{j}_{\text{out}} = \hbar/2\mu |a|^2 |f(\theta, \phi)|^2 \frac{k_{\text{out}}}{r^2} = |a|^2 \vec{v}_{\text{out}} \frac{|f(\theta, \phi)|^2}{r^2}. \tag{4.10}$$

With the outgoing flux $\vec{j}_{\text{out}} d\vec{A} = \vec{v}_{\text{out}} |a|^2 |f(\theta, \phi)|^2 dA/r^2$ through the area $dA = r^2 d\Omega$ we get

$$d\sigma = \frac{\vec{j}_{\text{out}} d\vec{A}}{|\vec{j}_{\text{in}}|} = \frac{v_{\text{out}} |a|^2 |f(\theta, \phi)|^2 d\Omega}{v_{\text{in}} |a|^2}, \tag{4.11}$$

and

$$\frac{d\sigma}{d\Omega} = \frac{k_{\text{out}}}{k_{\text{in}}} |f(\theta, \phi)|^2 = \frac{k_{\text{out}}}{k_{\text{in}}} f \cdot f^*, \tag{4.12}$$

and

$$\frac{d\sigma}{d\Omega} = |f(\theta,\phi)|^2 \quad \text{for elastic scattering.} \qquad (4.13)$$

Please note (for more detail see Chap. 5):

- For particles with spin the scattering amplitude has to be replaced by a matrix describing the transitions between the different spin-substates of the entrance and exit channels (*M* matrix). The complex square, the absolute value of $f \cdot f^*$, is replaced by taking the trace of this matrix: $d\sigma/d\Omega \propto \text{Tr}(MM^\dagger)$.
- In addition to the (unpolarized) cross section there are many other (*polarization*) observables such as polarization, analyzing power, polarization-transfer coefficients, spin-correlation coefficients etc. For many more detailed investigations in nuclear physics the study of polarization observables has been and still is important, in some cases indispensable. More detailed description of polarization effects in nuclear reactions and their measurement may be found in Refs. [HGS12, NUR13].

4.2.3 Schrödinger Equation

For the application to nuclear reactions stationary solutions are required, i.e. the time-independent Schrödinger equation is to be used. Corresponding to the geometry of the scattering problem the use of spherical polar coordinates is useful. For a central potential $V(\vec{r})$ the equation reads

$$-\frac{\hbar^2}{2\mu}\left[\frac{1}{r^2}\frac{\partial}{\partial r}\left(r^2\frac{\partial}{\partial r}\right) + \frac{1}{r^2\sin\theta}\frac{\partial}{\partial \theta}\left(\sin\theta\frac{\partial}{\partial \theta}\right) + \frac{1}{r^2\sin^2\theta}\frac{\partial^2}{\partial \phi^2}\right]\Psi(\vec{r})$$
$$+ V(\vec{r})\Psi(\vec{r}) = E\Psi(\vec{r}), \qquad (4.14)$$

where $\Psi(\vec{r})$ is an abbreviation for $\Psi^{(+)}_{\vec{k}_i}(\vec{r})$, which corresponds to the stationary scattering wave function. The angle-dependent part of the Hamilton operator may be expressed by the angular-momentum operator

$$L^2 = L_x^2 + L_y^2 + L_z^2 = -\hbar^2\left[\frac{1}{\sin\theta}\frac{\partial}{\partial \theta}\left(\sin\theta\frac{\partial}{\partial \theta}\right) + \frac{1}{\sin^2\theta}\frac{\partial^2}{\partial \phi^2}\right]. \qquad (4.15)$$

Eigenvalues and eigenfunctions of L^2 and L_z are given by the eigenvalue equations:

$$L^2 Y_{\ell m}(\theta,\phi) = \ell(\ell+1)\hbar^2 Y_{\ell m}(\theta,\phi), \qquad (4.16)$$

$$L_z Y_{\ell m}(\theta,\phi) = m\hbar Y_{\ell m}(\theta,\phi). \qquad (4.17)$$

The well-known commutation relations apply here. With these the Hamiltonian may be written

$$H = -\frac{\hbar^2}{2\mu}\left[\frac{1}{r^2}\frac{\partial}{\partial r}\left(r^2\frac{\partial}{\partial r}\right) - \frac{L^2}{\hbar^2 r^2}\right] + V(r), \qquad (4.18)$$

i.e. the substitution leads to the appearance of a term with L^2, which may be added to the potential term, providing the *centrifugal potential*. Because of $[H, L^2]$ and $[H, L_z] = 0$ one searches for the common eigenfunctions to H, L^2 and L_z for a product Ansatz, which corresponds to a separation of radial and angular parts of the wave function and is at the same time a partial-wave expansion

$$\Psi(\vec{r}) = \sum_{\ell=0}^{\infty}\sum_{m=-\ell}^{\ell} C_{\ell m}(k) R_{\ell m}(kr) Y_{\ell m}(\theta, \phi). \qquad (4.19)$$

The separated radial equation (for $m = 0$ we have azimuthal independence of the scattering problem) is a *Bessel* differential equation

$$\left[\frac{d^2}{dr^2} + k^2 - \frac{\ell(\ell+1)}{r^2} - U(r)\right] u_\ell(k, r) = 0, \qquad (4.20)$$

with the substitutions $u_\ell(k, r) = r R_\ell(k, r)$ and $U(r) = 2\mu V(r)/\hbar^2$.

The solutions of this equation for free particles ($U = 0$) are

-
$$j_\ell(kr) \to_{r\to\infty} \frac{\sin(kr - \ell\pi/2)}{kr} \equiv \frac{e^{i[kr-\ell\pi/2]} - e^{-i[kr-\ell\pi/2]}}{2ikr}. \qquad (4.21)$$

These Spherical Bessel Functions are regular for $r \to 0$.

-
$$n_\ell(kr) \to_{r\to\infty} \frac{\cos(kr - \ell\pi/2)}{kr} \equiv \frac{e^{i[kr-\ell\pi/2]} - e^{-i[kr-\ell\pi/2]}}{-2kr}. \qquad (4.22)$$

These Spherical Neumann Functions are irregular for $r \to 0$. Figure 4.10 shows the few lowest-order Spherical Bessel and Neumann Functions (also: *Bessel Functions of the second kind* or *Weber Functions*).

- Solutions are also the linear combinations

$$h_\ell^{(1,2)}(kr) = j_\ell(kr) \pm i n_\ell(kr) \qquad (4.23)$$

(Hankel Functions).

For bound states, because of the behavior at the origin, only the solution $j_\ell(kr)$ can be used.

The scattering from a potential shifts the phase of the scattering wave by δ_ℓ: The wave numbers in free space ($k_{\text{free}} = \sqrt{2\mu E/\hbar^2}$) and inside the potential region ($k_{\text{pot}} = \sqrt{2\mu(E-V)/\hbar^2}$) are different. After passage of a wave through a potential "layer" of thickness d the phase of the transmitted wave has changed relative to that

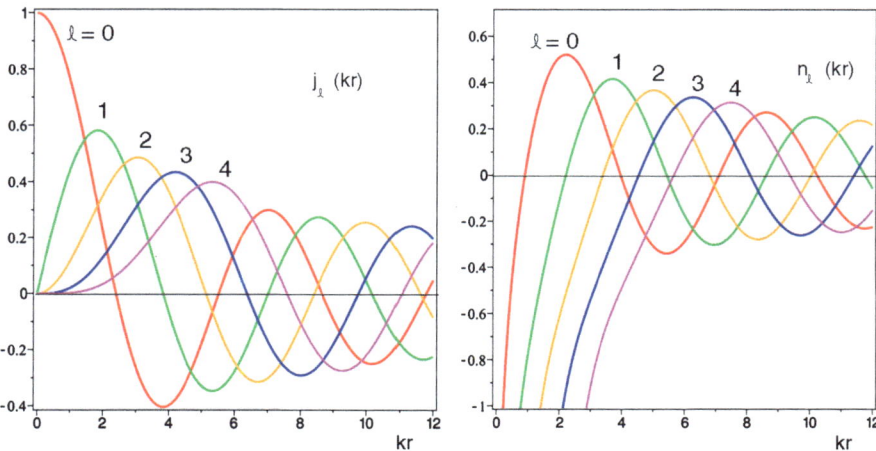

Fig. 4.10 The behavior of the lowest-order Spherical Bessel and Neumann Functions $j_\ell(kr)$ and $n_\ell(kr)$ as functions of $x = kr$

of the free wave by $\delta = (k_{\text{pot}} - k_{\text{free}})d$. In classical optics terms like *phase shift*, *refractive index* and *optical path length* have a similar origin and meaning.

In order to also expand the total wave function into partial waves a similar expansion of the plane incident wave e^{ikz} is needed. This is given by the mathematical identity (Rayleigh)

$$e^{ikz} = \sum_{\ell=0}^{\infty} (2\ell+1) i^\ell j_\ell(kr) P_\ell(\cos\theta) \qquad (4.24)$$

where the $j_\ell(kr)$ are the Bessel Functions of the first kind and the $P_\ell(\cos\theta)$ the Legendre Polynomials. The lowest-order Legendre Polynomials are depicted in Fig. 4.11. Basically, this expansion is the angular-momentum representation of the wave function.

Thus the total wave function becomes

$$\Psi_{tot} \to \propto \sum_{\ell=0}^{\infty} (2\ell+1) i^\ell \frac{e^{i[kr-\ell\pi/2]} - e^{-i[kr-\ell\pi/2]}}{2ikr} P_\ell(\cos\theta) + f(\theta,\phi) \frac{e^{ikr}}{r}. \qquad (4.25)$$

On the other hand this total wave function must also satisfy a simple partial-wave expansion—but with a phase shift caused by the potential.

$$\delta = (k_{\text{pot}} - k_{\text{free}}) \cdot d \qquad (4.26)$$

with

$$k_{\text{free}} = \sqrt{\frac{2\mu E}{\hbar^2}} \quad \text{and} \quad k_{\text{pot}} = \sqrt{\frac{2\mu(E-V)}{\hbar^2}}. \qquad (4.27)$$

Fig. 4.11 The four lowest-order Legendre Polynomials $P_L(\cos\theta)$, which determine the angular distributions of unpolarized cross sections, see also Chaps. 5 and 7. Only the *left half* of the picture is relevant for angular distributions, θ being a polar angle

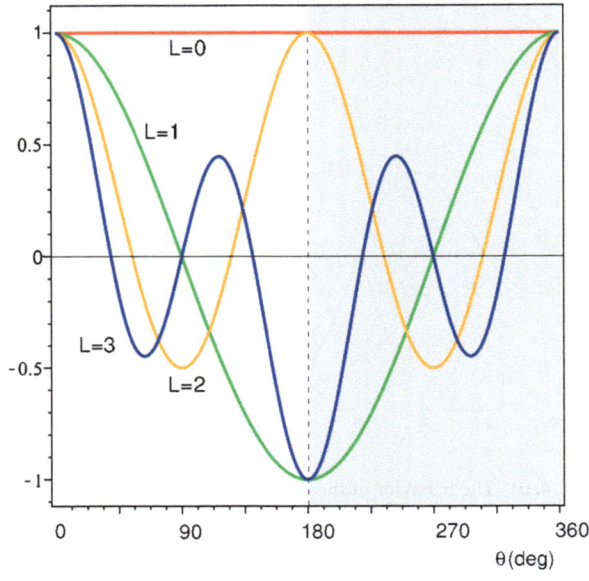

Fig. 4.12 Illustration (in one dimension) of the action of a potential region on the wavelength (or wave number k) of a wave, causing a phase shift δ relative to the free wave. The sign of the phase shift δ depends on whether the potential is attractive ($\delta < 0$) or repulsive ($\delta > 0$)

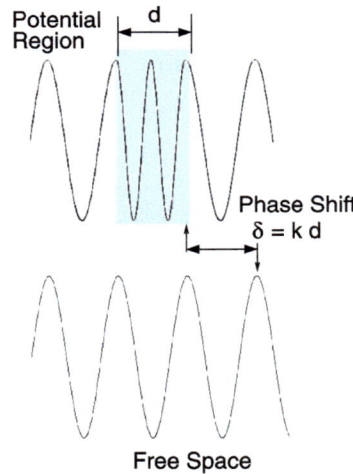

The simplified Fig. 4.12 illustrates the connection between the potential V and the phase shift of a wave.

$$\Psi_{tot} \to \sum_{\ell=0}^{\infty} C_\ell \frac{1}{2ikr}\left[e^{i(kr-\ell\pi/2+\delta_\ell)} - e^{-i(kr-\ell\pi/2+\delta_\ell)}\right]P_\ell(\cos\theta). \quad (4.28)$$

4.2 Formal Description of Nuclear Reactions

By comparing coefficients in the incoming and the outgoing waves in both expansions one obtains on the one hand a normalization:

$$C_\ell(k) = (2\ell + 1)i^\ell e^{i\delta_\ell}, \tag{4.29}$$

and on the other an expression for $f(\theta)$

$$\sum_\ell (2\ell+1) i^\ell e^{i\delta_\ell} \frac{1}{2ikr} P_\ell(\cos\theta) e^{i(kr - \ell\pi/2 + \delta_\ell)}$$

$$= \sum_\ell (2\ell+1) i^\ell \frac{1}{2ikr} e^{i(kr - \ell\pi/2)} P_\ell(\cos\theta) + f(\theta,\phi) \frac{e^{ikr}}{r}, \tag{4.30}$$

from which follows

$$f(\theta) = \frac{i}{2k} \sum_{\ell=0}^{\infty} (2\ell+1)\left(1 - e^{2i\delta_\ell}\right) P_\ell(\cos\theta)$$

$$= \frac{1}{k} \sum_{\ell=0}^{\infty} (2\ell+1) \sin\delta_\ell e^{i\delta_\ell} P_\ell(\cos\theta). \tag{4.31}$$

The quantity $\eta_\ell = \exp(2i\delta_\ell)$ is the *scattering function* and identical with the simplest form of the general scattering matrix $S_\ell(k)$. This function contains the dynamics of the interaction and—via the scattering amplitude (or more generally: via the transfer (T) matrix or, for particles with spin, the M matrix)—determines the observables like $d\sigma/d\Omega$ and polarization components.

4.2.4 The Optical Theorem

Following from the definition of the scattering amplitude f, as derived above, and the (integrated) cross section $\sigma_{\text{int,el}}$ for purely elastic scattering is an interesting relation between both, the *Optical Theorem*. The imaginary part of the scattering amplitude

$$f(\theta = 0°) = \frac{1}{k} \sum_{\ell=0}^{\infty} (2\ell+1) e^{i\delta_\ell} \sin\delta_\ell \tag{4.32}$$

$$= \frac{1}{2ik} \sum_{\ell=0}^{\infty} (2\ell+1)[\eta_\ell - 1] \tag{4.33}$$

is

$$\mathrm{Im}[f(\theta = 0°)] = \frac{1}{k} \sum_{\ell=0}^{\infty} (2\ell+1) \sin^2\delta_\ell. \tag{4.34}$$

When comparing this with

$$\sigma_{\text{int,el}} = \frac{4\pi}{k^2} \sum_{\ell=0}^{\infty} (2\ell+1) \sin^2 \delta_\ell \qquad (4.35)$$

$$\boxed{\sigma_{\text{int,el}} = \frac{4\pi}{k} \cdot \text{Im}\bigl[f(\theta=0°)\bigr]} \qquad (4.36)$$

for elastic scattering results. In the case of contributions from non-elastic channels (absorption, see Sect. 8.2) this is

$$\sigma_{\text{tot}} \equiv \sigma_{\text{int,el}} + \sigma_{\text{abs}} = \frac{4\pi}{k} \text{Im}\bigl[f_\alpha(\theta=0°)\bigr] \qquad (4.37)$$

with f_α the scattering amplitude of the elastic channel. The optical theorem connects a global quantity, the total cross section, with the forward scattering amplitude. It arises from the conservation of probability flux (or, equivalently, the unitarity of the S-matrix) requiring a destructive interference between the incident and the scattered waves in the forward direction, something like a "shadow" of the incident beam, cast by the target, and removing particles from it in proportion of the total cross section, see e.g. Ref. [JOA83].

4.3 Remark and Exercise

4.1 For charged particles the solutions of the Schrödinger equation with the Coulomb potential only, see Eq. (2.25), are the regular and irregular *Coulomb Functions*

$$F_\ell \longrightarrow \sin(kr - \ell\pi/2 - \eta_S \ln 2kr + \sigma_\ell), \qquad (4.38)$$

$$G_\ell \longrightarrow \cos(kr - \ell\pi/2 - \eta_S \ln 2kr + \sigma_\ell) \qquad (4.39)$$

with the Coulomb phase shifts $\sigma_\ell = arg\,\Gamma(\ell+1+i\eta_S)$. The Coulomb potential is of long range and increasingly so towards lower energies, often requiring a large number of partial waves before, at a certain large "screening" radius, the series can be truncated. For short-range potentials such as the hadronic interaction between nuclei often the series can be truncated after a few partial waves, especially at very low energies with s-waves acting only. If both types of potentials are acting (which is the normal case for charged particles) this different behavior suggests treating them separately by adding the corresponding scattering amplitudes. This leads to the sum of the two cross sections (for Rutherford scattering one of them is just the closed-form Rutherford cross section), but in addition there is an interference term, which needs summing over the many partial waves of the long-range Coulomb amplitude f_C

$$d\sigma/d\Omega = |f(\theta)|^2 = |f_C(\theta)|^2 + |\hat{f}(\theta)|^2 + 2\text{Re}\bigl[f_C^*(\theta) \cdot \hat{f}(\theta)\bigr]. \qquad (4.40)$$

However, in this case the short-range amplitude $\hat{f}$ is different from that without any Coulomb potential

$$\hat{f}(\theta) = \frac{i}{2k} \sum (2\ell + 1)e^{2i\sigma_\ell}\left(1 - e^{2i\delta_\ell}\right) P_\ell(\cos\theta), \qquad (4.41)$$

(compare Eq. (4.31). For details see Ref. [JOA83]).

Show that for the case of neutral projectiles like neutrons the Coulomb functions in the asymptotic limit become the spherical Bessel functions $j_\ell(kr)$ and $n_\ell(kr)$ (now often $y_\ell(kr)$).

References

[HGS12] H. Paetz gen. Schieck, *Nuclear Physics with Polarized Particles*. Lecture Notes in Physics, vol. 842 (Springer, Heidelberg, 2012)
[JOA83] C. Joachain, *Quantum Collision Theory*, 3rd edn. (North-Holland, Amsterdam, 1983)
[LOR79] H. Lorenz-Wirzba, P. Schmalbrock, H.P. Trautvetter, M. Wiescher, C. Rolfs, Nucl. Phys. A **313**, 346 (1979)
[MAR70] P. Marmier, E. Sheldon, *Physics of Nuclei and Particles, Vol. I* (Academic Press, New York, 1970). Chap. 11.2 ff.
[NNDC] Natl. Nucl. Data Center, EANDC, Nucl. Reactions, http://www.nndc.gov
[NUR13] S.B. Nurushev, M.F. Runtso, M.N. Strikhanov, *Introd. to Polarization Physics*. Lecture Notes in Physics, vol. 859 (Springer, Heidelberg, 2013)
[RED82] A. Redder, H.W. Becker, H. Lorenz-Wirzba, C. Rolfs, P. Schmalbrock, H.P. Trautvetter, Z. Phys. A **305**, 325 (1982)
[ROL88] C. Rolfs, W.S. Rodney, *Cauldrons in the Cosmos* (University of Chicago Press, Chicago, 1988). Chap. 4 ff.
[SAT90] G.R. Satchler, *Introd. Nucl. Reactions*, 2nd edn. (McMillan, London, 1990). Chap. 3.7 ff.

Chapter 5
Polarization in Nuclear Reactions—Formalism

When performing the usual (unpolarized) cross section measurements this implies—for particles with spin $\neq 0$—that in the description and experimentally there is summing over initial-channel and averaging over final-channel spin states. At the same time this means that a certain amount of information is being discarded that could be exploited if the nuclear reaction between spin substates could be studied. How this is achieved is the content of this chapter, in which the formalism necessary is discussed first. In many cases the polarization observables provide additional and more subtle information, in others the desired information can only be gained by using polarization, e.g. for detecting and measuring symmetry violations. Some typical results obtained by using polarized particles or by measuring polarization observables will be discussed in the pertinent following sections of this text. Two recent references have treated this subject in more detail. One, Ref. [NUR13] deals with polarization phenomena in high-energy physics and emphasizes Soviet and Russian developments and thus complements nicely Ref. [HGS12] which describes low-energy, non-relativistic polarization phenomena. Other references, especially concerning the polarization formalism, are Refs. [FIC71, EBE74, CON94]. Two early publications aimed at achieving an introductory, but basic understanding of spin polarization are Refs. [BAR67, DAR67].

5.1 Polarization Formalism

Quantum mechanics deals with statistical statements about the result of measurements on an ensemble of states (particles, beams, targets). In other words: by giving an expectation value of operators it provides probabilities (better: probability amplitudes) for the result of a measurement on an ensemble. Here two limiting cases can be distinguished. One is the case that our knowledge about the system is complete e.g. when all members of an ensemble are in the same spin state. This state will then be characterized completely by a state vector (ket). A special case is the spin state of a *single* particle, which is always completely (spin-)polarized.

We call this a pure state. In general, our knowledge of a system is incomplete and can only be described by superposition of such pure states weighted with the probability of their occurrence in this superposition. Such a state is called a *mixed state*. In the wavefunction of such a state probabilities enter in twofold ways: one is the quantum-mechanical description of a state by its probability amplitude or its square, its quantum-mechanical probability, the other the classical probability, with which each quantum-mechanical state is mixed into the ensemble. The appropriate and also practical description of such states is by using the density operator or density matrix.

5.2 Expectation Value and Average of Observables in Measurements

Carrying out a number of measurements of an observable **A** on a (generally) mixed ensemble results in an expectation value, which is the statistical ensemble average of the quantum-mechanical expectation values $\langle \Psi^{(i)}|A|\Psi^{(i)}\rangle$ with respect to the pure states $|\Psi^{(i)}\rangle$ present (or considered) in the ensemble. These should be expandable in an eigenstate basis $|u_n\rangle$ (i.e. $A|u_n\rangle = a_n|u_n\rangle$) with $\langle u_n|u_m\rangle = \delta u_{nm}$

$$\langle \mathbf{A}\rangle = \sum_i p_i \langle \Psi^{(i)}|A|\Psi^{(i)}\rangle = \sum_i \sum_n p_i |\langle u_n|\Psi^{(i)}\rangle|^2 a_n \qquad (5.1)$$

with

$$|\Psi^{(i)}\rangle = \sum_n \langle u_n|\Psi^{(i)}\rangle \langle |u_n\rangle = \sum_n c_n^{(i)} |u_n\rangle. \qquad (5.2)$$

The two types of probability are: once as $|\langle u_n|\Psi^{(i)}\rangle|^2$, which is the probability to find the state $|\Psi^{(i)}\rangle$ in an eigenstate $|u_n\rangle$ of A (with eigenvalue a_n) in the measurement, but also as the probability p_i of finding the ensemble in a quantum mechanical state characterized by $|\Psi^{(i)}\rangle$. The number of terms in the n-sums is N while i depends on the composition of the statistical ensemble. In this representation the properties of the ensemble and of the observable **A** factorize.

5.3 Density Operator, Density Matrix

With the foregoing relations the density operator ρ can be defined as

$$\rho = \sum_i p_i |\Psi^{(i)}\rangle\langle \Psi^{(i)}|. \qquad (5.3)$$

Its matrix elements are

$$\langle \rho_{n,m} = b_m|\rho|b_n\rangle = \sum_i p_i \langle b_m|\Psi^{(i)}\rangle\langle \Psi^{(i)}|b_n\rangle. \qquad (5.4)$$

5.3 Density Operator, Density Matrix

With this definition the ensemble average of an operator (an observable) **A** can be written comfortably

$$\langle \mathbf{A} \rangle = \sum_n \sum_m \langle b_m | \rho | b_n \rangle \langle b_n | A | b_m \rangle = \text{Tr}(\rho \mathbf{A}). \tag{5.5}$$

Since the trace of a matrix is independent of its different representations $\langle \mathbf{A} \rangle$ can be evaluated in any suitable basis.

5.3.1 General Properties of ρ

The density matrix (see e.g. [FAN57, FEY72]) is especially suited for a description of an arbitrary (pure or mixed) polarization state (differently from a wave function). With it, averages and expectation values as well as statistical distributions of measurable quantities can be described.

- Every quantized statistical mixture is described exactly and as completely as possible by its density operator.
- Pure and mixed states are being treated in identical ways and operator techniques can be used consistently for the description.
- A wave function $|\Psi\rangle$ can be determined only up to a phase factor $e^{i\phi}$ (which plays no role for the observables). The density matrix ρ, however, is identical for $|\Psi\rangle$ and $|\Psi\rangle e^{i\phi}$.

Other properties of ρ are:

- The trace of ρ is 1

$$\text{Tr}(\rho) = \sum_i \sum_n p_i \langle b_n | \Psi^{(i)} \rangle \langle \Psi^{(i)} | b_n \rangle = \sum_i p_i \langle \Psi^{(i)} | \Psi^{(i)} \rangle = 1. \tag{5.6}$$

(This follows also from $\langle \mathbf{A} \rangle = \text{Tr}(\rho \mathbf{A})$ with $\mathbf{A} = \mathbf{E}$ ($\mathbf{E}$ = unit matrix) and $\sum_k p_k = 1$.)
- For **A** to have real expectation values ρ must be hermitean:

$$\rho = \rho^\dagger : \rho_{ik} = \rho_{ki}*. \tag{5.7}$$

- ρ is positively definite (i.e. all diagonal elements are ≥ 0):

$$\langle b_n | \rho | b_n \rangle = \sum_i p_i \langle b_n | \Psi^{(i)} \rangle \langle \Psi^{(i)} | b_n \rangle = \sum_i p_i |\langle b_n | \Psi^{(i)} \rangle|^2 \geq 0. \tag{5.8}$$

5.3.2 Density Matrix of the General Mixed State

The density matrix is equally useful for pure and mixed states. The trace $\text{Tr}(\rho^2) \leq 1$ is a measure of the degree of mixing of an ensemble (i.e. its deviation from the pure state)

$$\mathrm{Tr}(\rho^2) = \sum_n \sum_i \sum_k p_i p_k \langle n|\Psi^{(i)}\rangle\langle\Psi^{(i)}|\Psi^{(k)}\rangle\langle\Psi^{(k)}|n\rangle$$

$$= \sum_i \sum_k p_i p_k |\langle\Psi^{(i)}|\Psi^{(k)}\rangle|^2$$

$$= \sum_k (p_k)^2 \le \left\{\left(\sum_k p_k\right)^2 = [\mathrm{Tr}(\rho)]^2 = 1\right\}, \qquad (5.9)$$

where the Schwarz inequality and the following symbolic relations have been used:

$$\mathrm{Tr}\sum = \sum \mathrm{Tr}, \qquad (5.10)$$

$$\mathrm{Tr}(ABC) = \mathrm{Tr}(BCA) = \mathrm{Tr}(CAB) \quad \text{and} \qquad (5.11)$$

$$\sum |n\rangle\langle n| = 1. \qquad (5.12)$$

While the density matrix

$$\rho = \begin{pmatrix} 1 & 0 & 0 \\ 0 & 0 & 0 \\ 0 & 0 & 0 \end{pmatrix} \qquad (5.13)$$

represents a pure state (one of complete polarization in the $+x$ direction), the density matrices

$$\rho = 1/2 \begin{pmatrix} 1 & 0 & 0 \\ 0 & 1 & 0 \\ 0 & 0 & 0 \end{pmatrix} \quad \text{or} \quad \rho = 1/3 \begin{pmatrix} 1 & 0 & 0 \\ 0 & 1 & 0 \\ 0 & 0 & 1 \end{pmatrix} \qquad (5.14)$$

are those of mixed (partially polarized) states. The second one—with a uniform occupation of all spin substates—characterizes a completely unpolarized (or maximally mixed) state. All spin substates belonging to spin S have the same weight $p_i = 1/(2S+1)$ and

$$\rho = \sum_m |m\rangle \frac{1}{2S+1}\langle m| \quad \text{and} \quad \mathrm{Tr}(\rho^2) = \frac{1}{2S+1} < 1. \qquad (5.15)$$

5.3.3 Examples for Density Matrices

Spin $S = 1/2$ As outlined above a pure state is characterized completely by its wave function

$$|\Psi\rangle = a|\chi_{+1/2}\rangle + b|\chi_{-1/2}\rangle \equiv a|\uparrow\rangle + b|\downarrow\rangle = a\begin{pmatrix}1\\0\end{pmatrix} + b\begin{pmatrix}0\\1\end{pmatrix}. \qquad (5.16)$$

5.3 Density Operator, Density Matrix

Its density matrix is then

$$\rho = \begin{pmatrix} |a|^2 & ab^* \\ ba^* & |b|^2 \end{pmatrix} \quad \text{with } |a|^2 + |b|^2 = 1 \qquad (5.17)$$

and the relation

$$\text{Tr}(\rho) = \text{Tr}(\rho^2) = 1 \qquad (5.18)$$

holds.

A completely unpolarized beam with all spin-substates equally occupied has the density matrix

$$\rho = 1/2 \begin{pmatrix} 1 & 0 \\ 0 & 1 \end{pmatrix} = 1/2 \left[\begin{pmatrix} 1 & 0 \\ 0 & 0 \end{pmatrix} + \begin{pmatrix} 0 & 0 \\ 0 & 1 \end{pmatrix} \right]. \qquad (5.19)$$

(This corresponds to a superposition of pure spin states with equal weights of $1/2$.)

A general beam, partially polarized in an arbitrary direction relative to the z axis, is described by a density matrix consisting of two contributions: one contribution p ($0 \leq p \leq 1$) that is completely polarized, and the other $(1-p)$ that is completely unpolarized

$$\begin{aligned} \rho &= (1-p)\frac{1}{2} \begin{pmatrix} 1 & 0 \\ 0 & 1 \end{pmatrix} + p \begin{pmatrix} \rho_{++} & \rho_{+-} \\ \rho_{-+} & \rho_{--} \end{pmatrix} \\ &= \begin{pmatrix} \frac{1-p}{2} + p\cos^2\frac{\beta}{2} & \frac{p}{2}\sin\beta e^{-i\phi} \\ \frac{p}{2}\sin\beta e^{i\phi} & \frac{1-p}{2} + p\sin^2\frac{\beta}{2} \end{pmatrix}. \end{aligned} \qquad (5.20)$$

Upon diagonalization (e.g. by a rotation by $-\beta$, a *unitary transformation*, for details see Chap. 5.4) this becomes

$$\rho = 1/2 \begin{pmatrix} 1+p & 0 \\ 0 & 1-p \end{pmatrix}. \qquad (5.21)$$

In this case the state can be interpreted as a superposition of the two pure states defined with respect to the quantization axis with the contributions (occupation numbers) $|a|^2$ and $|b|^2$ (and N_+ and N_-, respectively).

$$\rho = N_+ \begin{pmatrix} 1 & 0 \\ 0 & 0 \end{pmatrix} + N_- \begin{pmatrix} 0 & 0 \\ 0 & 1 \end{pmatrix} = \begin{pmatrix} N_+ & 0 \\ 0 & N_- \end{pmatrix}. \qquad (5.22)$$

By equating Eqs. (5.21) and (5.22) we see that

$$p = \frac{N_+ - N_-}{N_+ + N_-} \qquad (5.23)$$

has the usual form of a polarization (it is in fact the modulus of the vector polarization of a spin-$1/2$ system). It is therefore suggestive to introduce the general definition.

Definition of "Polarization" Spin polarization is the expectation value of a spin operator. Its components are expectation values of the corresponding spin-operator components with equal transformation properties (scalar, vector, tensor, and higher ranks).

If we take $\vec{S}$ to be a spin operator (with vector character such as for the spin-1/2 case) in the diagonal representation with respect to the z direction as quantization axis of ρ one obtains for the z component[1]

$$\begin{aligned} p_z = \langle S_z \rangle &= \frac{\mathrm{Tr}(\rho S_z)}{\mathrm{Tr}(\rho)} \\ &= \frac{1}{\mathrm{Tr}(\rho)} \mathrm{Tr}\left[\begin{pmatrix} N_+ & 0 \\ 0 & N_- \end{pmatrix}\begin{pmatrix} 1 & 0 \\ 0 & -1 \end{pmatrix}\right] \\ &= \frac{1}{\mathrm{Tr}(\rho)} \mathrm{Tr}\begin{pmatrix} N_+ & 0 \\ 0 & -N_- \end{pmatrix} \\ &= \frac{N_+ - N_-}{N_+ + N_-}, \end{aligned} \qquad (5.24)$$

which agrees with the usual (naïve) definition of a polarization.

Spin $S = 1$ In the diagonal representation also a naïve definition of the (vector) polarization can be introduced, analogous to the spin-1/2 case

$$p_z = \frac{N_+ - N_-}{N_+ + N_0 + N_-}. \qquad (5.25)$$

However, since in this definition no statement about the occupation of the state $|\chi_0\rangle$ has been made, it is evident that at least one additional independent quantity has to be defined in order to be able to describe the spin-1 situation completely. This quantity is called tensor polarization (sometimes also "alignment" for spin-1 particles such as photons as different from vector "polarization") and it is defined as the (normalized) difference between the sum of N_+ and N_- and N_0

$$p_{zz} = \frac{N_+ + N_- - 2N_0}{N_+ + N_0 + N_-}. \qquad (5.26)$$

For a pure state, similar to the spin-1/2 case (for the up, parallel, and down states also indices $+$, 0, and -1 or $\uparrow$, $\rightarrow$, or $\downarrow$ are in use)

$$|\Psi\rangle = a|\chi_1\rangle + b|\chi_0\rangle + c|\chi_{-1}\rangle \equiv a|1\rangle + b|0\rangle + c|-1\rangle = \begin{pmatrix} a \\ b \\ c \end{pmatrix} \qquad (5.27)$$

---

[1] For the following discussion a description in Cartesian coordinates is assumed. For spin-1 particles the indices z or zz with z along an arbitrary axis are explicitly written to distinguish between vector and tensor polarization.

5.3 Density Operator, Density Matrix

and

$$\rho = \begin{pmatrix} |a|^2 & ab^* & ac^* \\ ba^* & |b|^2 & bc^* \\ ca^* & cb^* & |c|^2 \end{pmatrix}. \tag{5.28}$$

For mixed states the ensemble average over the pure states that constitute the ensemble has to be taken.

For Spin $S = 1$ the state vectors and density matrices are

$$|\uparrow\rangle = \begin{pmatrix} 1 \\ 0 \\ 0 \end{pmatrix} \quad |\rightarrow\rangle = \begin{pmatrix} 0 \\ 1 \\ 0 \end{pmatrix} \quad |\downarrow\rangle = \begin{pmatrix} 0 \\ 0 \\ 1 \end{pmatrix}$$

$$\rho_{+1} = \begin{pmatrix} 1 & 0 & 0 \\ 0 & 0 & 0 \\ 0 & 0 & 0 \end{pmatrix} \quad \rho_0 = \begin{pmatrix} 0 & 0 & 0 \\ 0 & 1 & 0 \\ 0 & 0 & 0 \end{pmatrix} \quad \rho_{-1} = \begin{pmatrix} 0 & 0 & 0 \\ 0 & 0 & 0 \\ 0 & 0 & 1 \end{pmatrix} \tag{5.29}$$

p_z	1	0	-1
p_{zz}	1	-2	1

Mixed states contain the pure states with their respective statistical weights. A completely unpolarized beam is described by

$$\rho = \frac{1}{3} \begin{pmatrix} 1 & 0 & 0 \\ 0 & 1 & 0 \\ 0 & 0 & 1 \end{pmatrix}. \tag{5.30}$$

For a beam with complete positive "alignment" the $+1$ and -1 states are completely occupied and the density matrix is

$$\rho = \frac{1}{2} \begin{pmatrix} 1 & 0 & 0 \\ 0 & 0 & 0 \\ 0 & 0 & 1 \end{pmatrix} \tag{5.31}$$

with $p_z = 0$, $p_{zz} = 1$. The situation of maximal negative alignment is obtained when only the 0 state is occupied:

$$\rho = \begin{pmatrix} 0 & 0 & 0 \\ 0 & 1 & 0 \\ 0 & 0 & 0 \end{pmatrix} \tag{5.32}$$

with $p_z = 0$, $p_{zz} = -2$. A case occurring e.g. in the Lambshift polarized-ion source is a mixed state where the $+1$ state has the (relative) occupation number of 2/3, the 0 state that of 1/3

$$\rho = \frac{1}{3} \begin{pmatrix} 2 & \sqrt{2} & 0 \\ \sqrt{2} & 1 & 0 \\ 0 & 0 & 0 \end{pmatrix} \tag{5.33}$$

and $p_z = 2/3$, $p_{zz} = 0$. A further case realized there is

$$\rho = \frac{1}{3}\begin{pmatrix} \frac{1}{2} & 0 & 0 \\ 0 & 2 & 0 \\ 0 & 0 & \frac{1}{2} \end{pmatrix} \quad (5.34)$$

and $p_z = 0$, $p_{zz} = -1$.

Rotation of a Pure $S = 1$ State The rotation of a spin-1/2 state has been indicated above, see Sect. 5.3.3. The general formalism for rotations will be described below in Chap. 5.4. The rotation of ρ by rotation functions according to

$$\rho' = \mathbf{D}\rho\mathbf{D}^\dagger \quad (5.35)$$

is described there in detail. For

$$\rho = \begin{pmatrix} 1 & 0 & 0 \\ 0 & 0 & 0 \\ 0 & 0 & 0 \end{pmatrix}$$

under a rotation by β, ϕ the rotated density matrix is

$$\rho' = \frac{1}{4}\begin{pmatrix} (1+\cos\beta)^2 & \sqrt{2}(1+\cos\beta)\sin\beta e^{i\phi} & \sin^2\beta e^{2i\phi} \\ \sqrt{2}\sin\beta(1+\cos\beta)e^{-i\phi} & 2\sin^2\beta & \sqrt{2}\sin\beta(1-\cos\beta)e^{i\phi} \\ \sin^2\beta e^{-2i\phi} & \sqrt{2}(1-\cos\beta)\sin\beta e^{-i\phi} & (1-\cos\beta)^2 \end{pmatrix} \quad (5.36)$$

with the rotation matrix (see Sect. 5.4) for $S = 1$

$$\mathbf{D} = \frac{1}{2}\begin{pmatrix} (1+\cos\beta)e^{i\phi} & -\sqrt{2}\sin\beta e^{i\phi} & (1-\cos\beta)e^{i\phi} \\ \sqrt{2}\sin\beta & 2\cos\beta & -\sqrt{2}\sin\beta \\ (1-\cos\beta)e^{-i\phi} & \sqrt{2}\sin\beta e^{-i\phi} & (1+\cos\beta)e^{-i\phi} \end{pmatrix} \quad (5.37)$$

ρ' has the trace of 1 and represents—which is not easily recognizable from its external form—also a pure state.

5.3.4 Complete Description of Spin Systems

The number of parameters for a complete description of a system with spin S depends on the value of S. For example for $S = 1$: besides the intensity (or number of particles) being a scalar quantity or a tensor of rank 0, the vector polarization rank 1) needs three, the tensor polarization (a tensor of rank 2) needs nine components minus one for a normalization condition (making it traceless). Altogether these are 11 parameters (12 with the intensity). The polarized beam e.g. from an ion source has

5.3 Density Operator, Density Matrix

rotational symmetry around the z axis, the direction of the beam. Therefore the tensor will be symmetric, i.e. $p_{xx} = p_{yy}$, $p_{xy} = p_{yx}$ etc., which reduces the number of parameters to 8 (9). For the polarization of particles produced in a nuclear reaction this reduction does not hold. But symmetries like parity conservation help reduce the number of observables. A special case are again pure states. Such a state is described by $4S = 2(2S+1) - 2$ real parameters ($2S+1$ complex numbers minus one normalization condition minus one common phase). Normally for mixed states the number of parameters is $4S(S+1)$, i.e. $2(2S+1)^2$ real numbers minus $(2S+1)^2$ hermiticity conditions minus one normalization: $2(2S+1)^2 - (2S+1)^2 - 1 = 4S(S+1)$. In numbers this is shown in Table 5.1:

Table 5.1 Number of parameters necessary for a description of spin states

	Pure state in z direction	Pure state in arbitrary direction β, ϕ relative to x, y, z	Mixed state general
$S = 1/2$	1	2	3
	p_z	e.g. β, ϕ	e.g. p_x, p_y, p_z
$S = 1$	2	4	8
	e.g. p_z, p_{zz}	e.g. p_z, p_{zz} β, ϕ	e.g. p_x, p_y, p_z $p_{yy}, p_{zz}, p_{xz}, p_{xy}, p_{yz}$
$S = 3/2$	3	6	15

5.3.5 Expansions of the Density Matrix, Spin Tensor Moments

The density matrix directly is not well suited to describe observables. It is rather used to calculate expectation values of operators. The intensity of the incoming or outgoing particles in a nuclear reaction, and also the number of target particles are proportional to the trace of the relevant density matrix ρ. Since these quantities transform as scalars they are, up to a normalization, equal to $1 = \text{Tr}(\rho E)$. It is customary to normalize the incident density matrix exactly to 1 ($\text{Tr}(\rho)_{\text{in}} = 1$) which automatically leads to $\text{Tr}(\rho_{\text{fin}}) \neq 1$. The expansion of the density matrix into basis systems with well-defined properties (e.g. under transformations like rotations) provides a description of observables with corresponding behavior. Another requirement of the definition is certainly the correspondence of the so-defined quantities with older naïvely defined quantities. Rank-1 polarization must behave like a vector with a maximum value of 1, rank-2 polarization like a rank-2 tensor etc.

Therefore the prescription is to expand the density matrix into a complete set of **orthogonal** basis matrices with the desired properties and with the expansion coefficients being the new parameter set

$$\rho = \sum_j \lambda_j U_j^\dagger. \tag{5.38}$$

For the U only orthogonality is required: $\mathrm{Tr}(U_i U_k^\dagger) = \delta_{ik}(2S+1)$, but not necessarily hermiticity since in general the λ_j are complex. Because ρ has $(2S+1)^2$ complex elements $(2S+1)^2$ basis matrices are needed (e.g. four for $S=1/2$). The expansion runs from $j=1$ to $(2S+1)^2$. Every tensor of rank k has $2k+1$ components, therefore

$$(2S+1)^2 = \sum_{k=0}^{k_{\max}} (2k+1) = \frac{1}{2}(k_{\max}+1)(2k_{\max}+2) = (k_{\max}+1)^2. \quad (5.39)$$

Thus, the maximum rank of spin tensors necessary for a complete description of a spin system is $k_{\max} = 2S$.

For the interpretation of the expansion coefficients λ_j: multiply both sides with U_i, form the trace, use orthogonality and the definition of the expectation value of an operator

$$\mathrm{Tr}(\rho U_i) = \mathrm{Tr}(U_i \rho) = \mathrm{Tr}\left(\sum_{j=1}^{(2S+1)^2} \lambda_j U_i U_j^\dagger\right) = \sum_j \lambda_j \mathrm{Tr}(U_i U_j^\dagger)$$
$$= \sum_j \lambda_j (2S+1)\delta_{ij} = (2S+1)\lambda_i \equiv \langle U_i \rangle \quad (5.40)$$

where the average is to be taken over the ensemble (e.g. the beam). By comparison one obtains

$$\lambda_i = \frac{1}{2S+1} \langle U_i \rangle_{\mathrm{beam}} \quad (5.41)$$

and

$$\rho = \frac{1}{2S+1} \sum_{j=1}^{(2S+1)^2} \langle U_j \rangle_{\mathrm{beam}} U_j^\dagger. \quad (5.42)$$

Like in other expansions (electromagnetic, mass moments …) the coefficients $\langle U_j \rangle$ are called **moments** of the expansion, here: **(spin) tensor moments**. This concept was introduced by Fano [FAN53].

The most important of such expansions are those into *Cartesian* and into *spherical tensors*. The latter behave under rotations like Spherical Harmonics $Y_\ell^m(\theta, \phi)$. It is useful to choose irreducible representations for the U_j. All transformations, e.g. rotations are then linear and different ranks of submatrices will not be mixed by the transformation, but only transform within their rank (e.g. components of the tensor polarization do not produce a vector polarization).

5.3 Density Operator, Density Matrix

Expansion of ρ in a Cartesian Basis for Spin $S = 1/2$

$$|\chi\rangle = \begin{pmatrix} a \\ b \end{pmatrix}$$

$$\rho = \begin{pmatrix} \overline{|a|^2} & \overline{ab^*} \\ \overline{ba^*} & \overline{|b|^2} \end{pmatrix}. \tag{5.43}$$

As basis matrices U_i, the unit matrix, and the three Pauli matrices are chosen: $\vec{\sigma}$

$$U_1 = \mathbf{E} = \begin{pmatrix} 1 & 0 \\ 0 & 1 \end{pmatrix}$$

$$U_2 = \sigma_x = \begin{pmatrix} 0 & 1 \\ 1 & 0 \end{pmatrix}$$

$$U_3 = \sigma_y = \begin{pmatrix} 0 & -i \\ i & 0 \end{pmatrix} \tag{5.44}$$

$$U_4 = \sigma_z = \begin{pmatrix} 1 & 0 \\ 0 & -1 \end{pmatrix}.$$

This leads to

$$\rho = \frac{1}{2}(1 + \langle \sigma \rangle \sigma) = \frac{1}{2}(1 + \vec{p}\vec{\sigma})$$

$$= 1/2 \begin{pmatrix} 1 + p_z & p_x - ip_y \\ p_x + ip_y & 1 - p_z \end{pmatrix}. \tag{5.45}$$

By comparing coefficients we find

$$p_x = 2\operatorname{Re}(\overline{ba^*}); \qquad p_y = 2\operatorname{Im}(\overline{ba^*}); \qquad p_z = \overline{|a|^2} - \overline{|b|^2}. \tag{5.46}$$

The result for p_z corresponds to the naïve definition of a polarization!

Spin $S = 1$ Here one needs $(2S + 1)^2 = 9$ basis matrices of rank 2. They are obtained applying the direct product of tensors with rank 1 (vectors). Because of their rotation properties the three Cartesian Pauli matrices for $S = 1$ and the unit matrix are chosen, from which one can form 13 3×3 matrices of rank 2. Beginning with

$$\mathbf{E} = \begin{pmatrix} 1 & 0 & 0 \\ 0 & 1 & 0 \\ 0 & 0 & 1 \end{pmatrix} \qquad S_x = \frac{1}{\sqrt{2}} \begin{pmatrix} 0 & 1 & 0 \\ 1 & 0 & 1 \\ 0 & 1 & 0 \end{pmatrix}$$

$$S_y = \frac{1}{\sqrt{2}} \begin{pmatrix} 0 & -i & 0 \\ i & 0 & -i \\ 0 & i & 0 \end{pmatrix} \qquad S_z = \begin{pmatrix} 1 & 0 & 0 \\ 0 & 0 & 0 \\ 0 & 0 & -1 \end{pmatrix} \tag{5.47}$$

one forms

$$S_x^2 = \frac{1}{2}\begin{pmatrix} 1 & 0 & 1 \\ 0 & 2 & 0 \\ 1 & 0 & 1 \end{pmatrix} \quad S_y^2 = \frac{1}{2}\begin{pmatrix} 1 & 0 & -1 \\ 0 & 2 & 0 \\ -1 & 0 & 1 \end{pmatrix}$$

$$S_z^2 = \begin{pmatrix} 1 & 0 & 0 \\ 0 & 0 & 0 \\ 0 & 0 & 1 \end{pmatrix} \quad S_x S_y = \frac{1}{2}\begin{pmatrix} i & 0 & -i \\ 0 & 0 & 0 \\ i & 0 & -i \end{pmatrix} \tag{5.48}$$

$$S_x S_z = \frac{1}{\sqrt{2}}\begin{pmatrix} 0 & 0 & 0 \\ 1 & 0 & -1 \\ 0 & 0 & 0 \end{pmatrix} \quad S_y S_z = \frac{1}{\sqrt{2}}\begin{pmatrix} 0 & 0 & 0 \\ i & 0 & i \\ 0 & 0 & 0 \end{pmatrix}$$

and similarly for $S_y S_x$, $S_z S_x$ and $S_z S_y$. The S_i are hermitean: $S_i S_j = (S_j S_i)^\dagger$, i.e. the $S_i S_j$ are connected with the $S_j S_i$ via commutation relations. The antisymmetric combinations

$$S_i S_j - S_j S_i = i S_k \tag{5.49}$$

and the symmetric combinations

$$S_i S_j + S_j S_i = (S_i S_j)^\dagger + (S_j S_i)^\dagger = (S_i S_j + S_j S_i)^\dagger \tag{5.50}$$

are automatically hermitean. In a decomposition

$$S_i S_j = \frac{1}{2}(S_i S_j + S_j S_i) + \frac{1}{2}(S_i S_j - S_j S_i) \tag{5.51}$$

into a symmetric and an antisymmetric part the latter provides nothing new because of the commutation relations. Therefore only the symmetric part is kept to form the new combinations

$$S_{ij}' = \frac{1}{2}(S_i S_j + S_j S_i) = 1/2\big(S_i S_j + (S_i S_j)^\dagger\big). \tag{5.52}$$

Of these there are exactly six. Five are needed, which allows the introduction of a physical condition into the final definition of the S_{ij}: The polarization of the unpolarized ensemble ought to be zero: $(\text{Tr}(S_j) = \text{Tr}(S_i S_j') = 0)$. Since this is not fulfilled for the S_{ij}' with $i = j$, a new definition is used:

$$S_{ij} =: 3 S_{ij}' - 2\delta_{ij}\mathbf{E} = 3/2\,(S_i S_j + S_j S_i) - 2\delta_{ij}\mathbf{E}, \tag{5.53}$$

leading to: $S_{xx} + S_{yy} + S_{zz} = 0$, $\text{Tr}(S_{ij}) = \text{Tr}(S_i) = 0$. All "polarizations" of the unpolarized ensemble are then zero. Of the three diagonal elements S_{ii} only two are independent. Also, since no two S_{ii} are orthogonal, one uses two orthogonal linear combinations of the S_{ii} instead. Several combinations are possible, e.g. $(S_{xx} - S_{yy}; S_{zz})$ or $(S_{zz} - S_{xx}; S_{yy})$. Here $S_{xx} + S_{yy}$ and $S_{xx} - S_{yy}$ will be chosen. With $\text{Tr}(S_i^2) = 2$, $\text{Tr}(S_{ij}^2) = 9/2$ and $\text{Tr}(S_{xx} + S_{yy})^2 = 6$, $\text{Tr}(S_{xx} - S_{yy})^2 = 18$

5.3 Density Operator, Density Matrix

the Cartesian expansion basis is

$$U_1 = \mathbf{E};$$
$$U_2 = \sqrt{3/2}S_x; \quad U_3 = \sqrt{3/2}S_y; \quad U_4 = \sqrt{3/2}S_z;$$
$$U_5 = \sqrt{2/3}S_{xy}; \quad U_6 = \sqrt{2/3}S_{xz}; \quad U_7 = \sqrt{2/3}S_{yz}; \quad (5.54)$$
$$U_8 = \sqrt{1/2}(S_{xx} + S_{yy});$$
$$U_9 = \sqrt{1/6}(S_{xx} - S_{yy}).$$

With $\langle S_{xx}\rangle + \langle S_{yy}\rangle + \langle S_{zz}\rangle = 0$ and $S_{xx} + S_{yy} + S_{zz} = 0$ one obtains

$$\rho = \frac{1}{3}\sum_{k=1}^{(2S+1)^2}\langle U_k\rangle U_k^\dagger$$
$$= \frac{1}{3}\left(\mathbf{E} + \frac{3}{2}\sum_i \langle S_i\rangle S_i + \frac{1}{3}\sum_{ij}\langle S_{ij}\rangle S_{ij}\right) \quad (5.55)$$

(for a different derivation see [OHL72]). The sum over ij is meant such that for $i = j$ one, for $i \neq j$ two terms appear). Written out:

$$\rho = \frac{1}{3}\begin{pmatrix} 1 + \frac{3}{2}p_z + \frac{1}{2}p_{zz} & \frac{1}{\sqrt{2}}[\frac{3}{2}(p_x - ip_y) + (p_{xz} - ip_{yz})] & \frac{1}{2}(p_{xx} - p_{yy}) - ip_{xy} \\ \frac{1}{\sqrt{2}}[\frac{3}{2}(p_x + ip_y) + (p_{xz} + ip_{yz})] & 1 - p_{zz} & \frac{1}{\sqrt{2}}[\frac{3}{2}(p_x - ip_y) - (p_{xz} - ip_{yz})] \\ \frac{1}{2}(p_{xx} - p_{yy}) + ip_{xy} & \frac{1}{\sqrt{2}}[\frac{3}{2}(p_x + ip_y) - (p_{xz} + ip_{yz})] & 1 - \frac{3}{2}p_z + \frac{1}{2}p_{zz} \end{pmatrix}. \quad (5.56)$$

The polarization parameters of rank k ($k = 0, 1, 2$) form tensors of rank k. For $k = 0$ it is a scalar proportional to some intensity, for $k = 1$ it is a vector with three components (the vector polarization) and for $k = 2$ it is a tensor of rank 2, represented by a symmetric traceless $(k+1) \times (k+1)$ (thus 3×3) matrix of the form

$$\begin{pmatrix} p_{xx} & p_{xy} & p_{xz} \\ p_{xy} & p_{yy} & p_{zy} \\ p_{xz} & p_{zy} & p_{zz} \end{pmatrix} \quad (5.57)$$

with 5 independent parameters. Taken together the spin-1 system has 9 parameters.

The relation to the components a, b, c of an ($S = 1$) spinor is

$$p_x = \sqrt{2}\,\text{Re}\,\overline{(ba^* + cb^*)}; \qquad p_y = \sqrt{2}\,\text{Im}\,\overline{(ba^* + cb^*)};$$

$$p_z = \overline{|a|^2} - \overline{|c|^2};$$

$$p_{xz} = \frac{1}{\sqrt{2}}\,\text{Re}\,\overline{(ba^* - cb^*)}; \qquad p_{yz} = \frac{1}{\sqrt{2}}\,\text{Im}\,\overline{(ba^* - cb^*)}; \qquad (5.58)$$

$$p_{xy} = \text{Im}\,\overline{(ca^*)};$$

$$p_{zz} = -(p_{xx} + p_{yy}) = 1 - 3\overline{|b|^2} = \overline{|a|^2} + \overline{|c|^2} - 2\overline{|b|^2}$$

in agreement with the naïve definitions of p_z and p_{zz}.

Limiting Values of the Polarization Components Because of $\text{Tr}(\rho) = \overline{|a|^2} + \overline{|b|^2} + \overline{|c|^2} = 1$

$$\overline{|a|^2} + \overline{|c|^2} \le 1 \qquad (5.59)$$

holds and thus

$$\left|\overline{|a|^2} - \overline{|c|^2}\right| \le \overline{|a|^2} + \overline{|c|^2} \le 1. \qquad (5.60)$$

From this follows

$$|p_z| \le 1 \quad \text{or} \quad -1 \le p_z \le +1 \qquad (5.61)$$

and, because the choice of axes is arbitrary,

$$|p_x| \le 1 \quad \text{and} \quad |p_y| \le 1. \qquad (5.62)$$

In the same way it can be shown that

$$|p_{xz}| \le 3/2; \qquad |p_{yz}| \le 3/2; \qquad |p_{xy}| \le 3/2; \qquad |p_{xx} - p_{yy}| \le 3 \qquad (5.63)$$

and therefore

$$-2 \le p_{zz} \le +1. \qquad (5.64)$$

The components of the vector and tensor polarization are not independent of each other but are related via the occupation numbers (or the occupation probabilities) N_i of the three substates of the spin-1 system. In the diagonal representation ($\overline{|a|^2} = N_+$; $\overline{|b|^2} = N_0$; $\overline{|c|^2} = N_-$) the normalization with $N_+ + N_0 + N_- = 1$ holds together with the "positivity condition" $N_i \ge 0$. This is sufficient to determine the values of the vector and tensor polarization as functions of the occupation numbers (Fig. 5.1).

Expansion into Spherical Tensors The representation of any vector can be Cartesian or spherical. Since tensors (tensor operators) of arbitrary rank can be generated by the combination (the direct or external product) from the components of a given

5.3 Density Operator, Density Matrix

Fig. 5.1 Range of values of the polarization components of a spin-1 system as functions of the (relative) occupation numbers of a spin-1 system with additional conditions $N_+ + N_0 + N_- = 1$ and $N_i \geq 0$

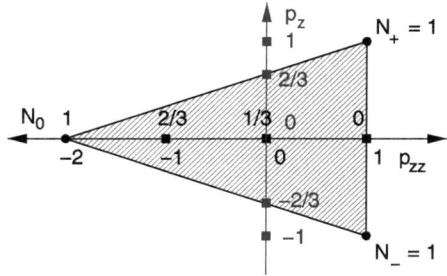

set of vectors (vector operators) the spherical tensors τ_{kq} (e.g. of rank $k = 2$ or higher) can be constructed from spin operators $\vec{S}$ in their spherical representation τ_{kq} (for $k = 1$). The resulting product operators in general are **reducible**, i.e. their transformation properties e.g. under rotations will not be simple. By forming the direct product of two vectors in three dimensions, a (Cartesian) tensor with nine elements will result, which behave differently under rotations, namely like a scalar (the trace of this tensor), like a vector (whose components are the three cross products of the original vector components), and like a tensor of rank 2.

Reducible tensors can be decomposed into objects (tensors) which still behave differently but *independently* e.g. under rotations. More formally: A reducible Cartesian tensor T_{ij} of rank 2 can be generated from two vectors **U** and **V** using the prescription $\mathbf{T} = \mathbf{U} \otimes \mathbf{V}$, i. e. $T_{ij} \equiv U_i V_j$. However, it can be decomposed ("reduced") in the following way:

$$U_i V_j = 1/3\, \mathbf{UV} \delta_{ij} + 1/2\,(U_i V_j - U_j V_i) + \left[1/2\,(U_i V_j + U_j V_i) - 1/3\, \mathbf{UV} \delta_{ij}\right]. \tag{5.65}$$

Thus, the tensor is decomposed into a scalar (tensor of rank 0), an antisymmetric tensor (vector or cross product), which transforms like a vector (tensor of rank 1), and a 3×3 traceless, symmetric tensor of rank 2. Symbolically:

$$3 \otimes 3 = 1 \oplus 3 \oplus 5. \tag{5.66}$$

This, however, is just a decomposition of the reducible Cartesian tensor into spherical components (with rotation properties like the spherical harmonics of ranks 0, 1 and 2). The matrices describing the tensor can then be decomposed into submatrices along the main diagonal, which transform only linearly according to their rank and without influencing the other submatrices. This "pure" behavior under rotations is displayed only by the special linear combinations of the tensor components defined above, but not by these alone. Under rotations of systems represented by reducible tensors all components would have to be transformed in common and according to the usual transformation rules for tensors—namely non-linearly in the rotation functions.

Especially for higher spins and generally because of these transformation properties the use of irreducible **spherical tensors** is preferred for the description of polarization. Generalizing this to arbitrary spin systems: They transform under rotations **linearly in the rotation functions** like the angular-momentum eigenfunctions, the spherical harmonics $Y_\ell^m(\theta, \phi)$.

These tensors are generated by a special combination of spin vector operators (which are irreducible per se!), by applying the principles of vector coupling of angular momentum operators using Clebsch-Gordan coefficients where operators of higher rank with identical transformation properties as those of the constituents are generated:

$$\tau_{KQ} = \sum_{q,q'}(kk'qq'|KQ)\tau_{kq}\tau_{k'q'}. \tag{5.67}$$

The density matrices of higher-spin systems can be expanded into a complete set of such basis matrices—just like for spin-1/2 systems. We define as a "spherical basis" the irreducible tensors of rank K in a spherical coordinate representation (short: "spherical tensors"). They have to fulfill at least one condition: τ_{KQ} has to be transformed like the spherical harmonic of rank K and component Q Y_K^Q under spatial rotations. The spherical tensors, like the spherical harmonics, are irreducible, i.e. under transformations (rotations) they will always be transformed into tensors of equal rank (about rotations see Sect. 5.4) with

$$\tau_{KQ'} = \sum_Q D_{Q'Q}^K(\alpha, \beta, \gamma)\tau_{KQ}. \tag{5.68}$$

Example for the Construction of a Set of Spherical Tensors for $S = 1$ We choose as basis

$$\mathbf{E} = \begin{pmatrix} 1 & 0 & 0 \\ 0 & 1 & 0 \\ 0 & 0 & 1 \end{pmatrix}$$

$$S_1 = -\frac{1}{\sqrt{2}}(S_x + iS_y) = \begin{pmatrix} 0 & -1 & 0 \\ 0 & 0 & -1 \\ 0 & 0 & 0 \end{pmatrix}$$

$$S_{-1} = \frac{1}{\sqrt{2}}(S_x - iS_y) = \begin{pmatrix} 0 & 0 & 0 \\ 1 & 0 & 0 \\ 0 & 1 & 0 \end{pmatrix} \tag{5.69}$$

$$S_0 \equiv S_z = \begin{pmatrix} 1 & 0 & 0 \\ 0 & 0 & 0 \\ 0 & 0 & -1 \end{pmatrix}.$$

These basis matrices are non-hermitean: $S_{\pm 1}^\dagger = -S_{\mp 1}$ and

$$\text{Tr}(S_1 S_1^\dagger) = \text{Tr}(S_{-1} S_{-1}^\dagger) = \text{Tr}(S_0 S_0^\dagger) = 2. \tag{5.70}$$

5.3 Density Operator, Density Matrix

From these one basis tensor of rank 0, three of rank 1, and five of rank 2 are constructed in analogy to the procedure with spherical harmonics (which are by definition spherical tensors), for which

$$Y_K^Q = \sum_{q_1 q_2} \langle k_1 k_2 q_1 q_2 | K Q \rangle Y_{k_1}^{q_1} Y_{k_2}^{q_2} \tag{5.71}$$

holds, e.g. for $K = 2$

$$Y_2^Q = \sum_{q=-1}^{+1} \langle 11 q Q - q | 2 Q \rangle Y_1^q Y_1^{Q-q}. \tag{5.72}$$

Similarly tensor operators of still higher rank can be constructed by coupling of tensor operators of lower rank ("contraction"). As spin operators for $S = 1$ the four operators defined above $\mathbf{E}$, S_0 and $S_{\pm 1}$ can be used to construct the missing operators of rank 2 S_{2Q}

$$S_{2Q} = \sum_{m=-1}^{1} \langle 11 m Q - m | 2 Q \rangle S_m S_{Q-m}. \tag{5.73}$$

Thus, e.g. (with $Q = 0$)

$$S_{20} = \sum_m \langle 11 m Q - m | 20 \rangle S_m S_{Q-m} = 1/\sqrt{6}(S_1 S_{-1} + S_{-1} S_1) + \sqrt{2/3} S_0^2$$

$$= 1/\sqrt{6}(S_1 S_1^\dagger + S_{-1} S_{-1}^\dagger) + \sqrt{2/3} S_0^2$$

$$= -1/\sqrt{6}(S^2 - S_0^2) + \sqrt{2/3} S_0^2 = 1/\sqrt{6}(-2 + 3 S_0^2) \tag{5.74}$$

(since S^2 is diagonal with $S(S+1) = 2$). This procedure can be continued. All one has to know are the Clebsch-Gordan vector coupling coefficients. In this way for $S = 1$ the following irreducible spherical tensors (normalized according to Lakin [LAK55], i.e. following the Madison convention) are obtained

$$\tau_{00} = \mathbf{E} = \begin{pmatrix} 1 & 0 & 0 \\ 0 & 1 & 0 \\ 0 & 0 & 1 \end{pmatrix}$$

$$\tau_{11} = \sqrt{3/2} S_{+1} = \sqrt{3/2} \begin{pmatrix} 0 & -1 & 0 \\ 0 & 0 & -1 \\ 0 & 0 & 0 \end{pmatrix}$$

$$\tau_{1-1} = \sqrt{3/2} S_{-1} = \sqrt{3/2} \begin{pmatrix} 0 & 0 & 0 \\ 1 & 0 & 0 \\ 0 & 1 & 0 \end{pmatrix}$$

$$\tau_{10} = \sqrt{3/2}\, S_0 = \sqrt{3/2} \begin{pmatrix} 1 & 0 & 0 \\ 0 & 0 & 0 \\ 0 & 0 & -1 \end{pmatrix} \tag{5.75}$$

$$\tau_{22} = \sqrt{3}\, S_{22} = \sqrt{3}\, S_{+1}^2 = \sqrt{3} \begin{pmatrix} 0 & 0 & 1 \\ 0 & 0 & 0 \\ 0 & 0 & 0 \end{pmatrix}$$

$$\tau_{21} = \sqrt{3}\, S_{21} = \sqrt{3/2}(S_0 S_1 + S_1 S_0) = \sqrt{3/2} \begin{pmatrix} 0 & -1 & 0 \\ 0 & 0 & 1 \\ 0 & 0 & 0 \end{pmatrix}$$

$$\tau_{20} = \sqrt{3}\, S_{20} = 1/\sqrt{2}(3 S_0^2 - 2) = 1/\sqrt{2} \begin{pmatrix} 1 & 0 & 0 \\ 0 & -2 & 0 \\ 0 & 0 & 1 \end{pmatrix}.$$

The missing components are calculated with

$$\tau_{K-Q} = (-)^Q \tau_{KQ}^\dagger. \tag{5.76}$$

For $S = \frac{3}{2}$ spherical tensors are constructed similarly

$$S_{3Q} = \sum_{m=-1}^{1} \langle 21 m\, Q-m | 3 Q \rangle S_{2m} S_{1,Q-m}. \tag{5.77}$$

Spin Tensor Moments The quantities that specify the polarization in the spherical representation are the expectation values of these tensor operators, the so-called **spin tensor moments** $t_{KQ} = \langle \tau_{KQ} \rangle$. The hermiticity of ρ entails the hermiticity of the t_{KQ}: $t_{K-Q} = (-)^Q t_{KQ}^*$. Therefore the density matrix can be expressed in terms of tensor moments.

Spherical Tensors, Density Matrix, and Tensor Moments for Spin $S = 1/2$
The irreducible basis (as linear combinations of Pauli operators the spin tensors are in this case automatically *irreducible*)

$$\begin{aligned}
\tau_{00} &= \begin{pmatrix} 1 & 0 \\ 0 & 1 \end{pmatrix} \\
\tau_{10} &= \begin{pmatrix} 1 & 0 \\ 0 & -1 \end{pmatrix} = \sigma_z \\
\tau_{11} &= -\sqrt{2} \begin{pmatrix} 0 & 1 \\ 0 & 0 \end{pmatrix} = -1/\sqrt{2}\,(\sigma_x + i\sigma_y) \\
\tau_{1-1} &= \sqrt{2} \begin{pmatrix} 0 & 0 \\ 1 & 0 \end{pmatrix} = 1/\sqrt{2}\,(\sigma_x - i\sigma_y).
\end{aligned} \tag{5.78}$$

5.3 Density Operator, Density Matrix

The expansion of the density matrix is

$$\rho = 1/2 \sum_{KQ} \langle \tau_{KQ} \rangle \tau_{KQ}^\dagger \tag{5.79}$$

and

$$\text{Tr}(\tau_{KQ} \tau_{K'Q'}^\dagger) = 2\delta_{KQ,K'Q'} \tag{5.80}$$

and therefore

$$\rho = \frac{1}{2}\begin{pmatrix} 1+t_{10} & \sqrt{2}t_{1-1} \\ -\sqrt{2}t_{11} & 1-t_{10} \end{pmatrix} = \frac{1}{2}\begin{pmatrix} 1+p_z & p_x - ip_y \\ p_x + ip_y & 1-p_z \end{pmatrix}. \tag{5.81}$$

By comparison

$$t_{00} \text{ is proportional to an intensity}$$

$$t_{10} = p_z \tag{5.82}$$

$$t_{1\pm 1} = \mp 1/\sqrt{2}(p_x \pm ip_y).$$

Density Matrix and Tensor Moments for Spin $S = 1$

$$\rho = \frac{1}{3}\begin{pmatrix} 1+\sqrt{\frac{3}{2}}t_{10}+\frac{1}{\sqrt{2}}t_{20} & \sqrt{\frac{3}{2}}(t_{1-1}+t_{2-1}) & \sqrt{3}\,t_{2-2} \\ -\sqrt{\frac{3}{2}}(t_{11}+t_{21}) & 1-\sqrt{2}\,t_{20} & \sqrt{\frac{3}{2}}(t_{1-1}-t_{2-1}) \\ \sqrt{3}\,t_{22} & -\sqrt{\frac{3}{2}}(t_{11}-t_{21}) & 1-\sqrt{\frac{3}{2}}t_{10}+\frac{1}{\sqrt{2}}t_{20} \end{pmatrix}. \tag{5.83}$$

By comparison with Eqs. (5.28) and (5.56) the connection between the t_{kq}, the wave-function amplitudes, and the Cartesion tensor components is obtained Tensor moments of **rank $k = 0$** are proportional to an intensity (one scalar, invariant under rotations)

$$t_{00} = 1 = \overline{|a|^2} + \overline{|b|^2} + \overline{|c|^2}. \tag{5.84}$$

Tensor moments of **rank $k = 1$** describe the vector polarization (three components, transformation properties of a vector)

$$t_{1\pm 1} = -\sqrt{3/2}\,\overline{(ba^* + cb^*)} = \mp\sqrt{3/2}\,(p_x \pm ip_y)$$
$$t_{10} = \sqrt{3/2}\,\overline{(|a|^2 - |c|^2)} = \sqrt{3/2}\,p_z. \tag{5.85}$$

Tensor moments of **rank $k = 2$** describe the tensor polarization (eight independent components, transformation properties of a second-rank tensor)

$$t_{20} = 1/\sqrt{2}\,\overline{(|a|^2 + |c|^2 - 2|b|^2)} = 1/\sqrt{2}\,(1 - 3\overline{|b|^2}) = 1/\sqrt{2}\,p_{zz}$$

$$t_{2\pm1} = \sqrt{3/2}\,\overline{(cb^* - ba^*)} = \mp 1/\sqrt{3}\,(p_{xz} \pm ip_{yz}) \tag{5.86}$$
$$t_{2\pm2} = \sqrt{3}\,\overline{ca^*} = 1/(2\sqrt{3})\,(p_{xx} - p_{yy} \pm 2ip_{xy}).$$

Inversely:

$$\begin{aligned}
p_x &= -1/\sqrt{3}\,(t_{11} - t_{1-1}) \\
p_y &= i/\sqrt{3}\,(t_{11} + t_{1-1}) \\
p_z &= \sqrt{2/3}\,t_{10} \\
p_{xx} &= \sqrt{3}/2\,(t_{22} + t_{2-2}) - 1/\sqrt{2}\,t_{20} \\
p_{yy} &= -\sqrt{3}/2\,(t_{22} + t_{2-2}) - 1/\sqrt{2}\,t_{20} \\
p_{zz} &= \sqrt{2}\,t_{20} \\
p_{xy} &= p_{yx} = -i\sqrt{3}/2\,(t_{22} - t_{2-2}) \\
p_{xz} &= p_{zx} = -\sqrt{3}/2\,(t_{21} - t_{2-1}) \\
p_{yz} &= p_{zy} = i\sqrt{3}/2\,(t_{21} + t_{2-1}).
\end{aligned} \tag{5.87}$$

For analyzing powers of nuclear reactions the same relations apply (with uppercase variables T_{kq} and A_i or A_{ik}). In analogy to other expansions into momenta in physics (Mass distributions → moments of inertia, charge/current distributions → Eℓ and Mℓ multipole moments, etc.) the expansion coefficients of spin tensor-moment expansions have geometrical interpretations. Visualizations have been attempted in Refs. [DAR71, BAR67, DAR67, HGS12]. The vector polarization (tensor of rank 1) behaves like a classical vector and is fully described by its three components. It points into a distinct *direction*, which is characterized by a sign—different from the tensor polarization that has only an orientation ("*alignment*") of an anisotropic spin distribution with respect to an *axis*.

Polarization of Particles with Higher Spin The fact that particles with $S = 3/2$ (^{7}Li, ^{23}Na) have been polarized (see Sect. 16.6.2) requires an appropriate description of tensor moments. This case has been discussed in Ref. [DAR71], so will not be detailed here. It is clear that tensor moments up to rank three have to be considered. They must be constructed from spin operators for $S = 3/2$, which—like in the spin-1/2 and spin-3/2 case (see Eqs. (5.44), (5.78), and (5.69))—can be derived from contractions of lower-spin operators and taking into account commutation relations and transformation properties of spherical tensors as well as a normalization condition. Note that in the following equations spherical tensor moments are thus expressed by spherical combinations of the spin-1 Cartesian tensors

$$t_{30} = \frac{1}{6\sqrt{5}}\langle 20S_z^3 - 41 S_z\rangle$$

$$t_{3\pm 1} = \mp \frac{1}{24\sqrt{15}}\langle(60S_z^2 - 51)(S_x \pm i S_y) + (S_x \pm i S_y)(60S_z^2 - 51)\rangle$$
(5.88)

$$t_{3\pm 2} = \frac{1}{\sqrt{6}}\langle S_z(S_x \pm i S_y)^2 + (S_x \pm i S_y)^2 S_z\rangle$$

$$t_{3\pm 3} = \mp \frac{1}{3}\langle(S_x \pm i S_y)^3\rangle.$$

Although spins > 1 have not been considered in the Madison convention its application, e.g. concerning coordinate systems is straightforward. It is evident that for higher spins the Cartesian notation becomes impractical and spherical tensor moments should be used (the term "efficiency tensor (moment)" in [DAR71] should be replaced by "analyzing power").

5.4 Rotations, Angular Dependence of the Tensor Moments

5.4.1 Generalities

It is important to be able to describe polarization observables in rotated coordinate systems. Typical applications are: Spin precession in magnetic fields, deflection of polarized-particle beams by optical elements, nuclear reactions, double scattering and polarization transfer. Especially in nuclear reactions it is necessary to describe observables in different coordinate systems rotated against each other such as for spin-transfer experiments.

We start with the rotation of the density matrix ρ.

A reminder: In quantum mechanics finite rotations of a system are described by rotation operators that are integrals over operators for infinitesimal rotations (which are linear!). E.g. the rotation operator for the finite rotation by an angle α around the z axis has the form (the components of S are given in units of $\hbar$ (i.e. $\hbar \equiv 1$))

$$\mathbf{D}(\alpha) = e^{-i\alpha S_z}.$$
(5.89)

5.4.2 The Description of Rotations by Rotation Operators

The most general rotation is composed from three rotations by the Euler angles (Attention! Several conventions exist; her we follow Condon/Shortley [CON67], Rose [ROS57] and Brink/Satchler [BRI71]): a sequence of right-handed rotations about the z, then the new y, and then about the then new z axis. This is equivalent to a sequence of rotations about the respective old z, y, and x axes. In addition, one has to distinguish between "active" rotations of the system in a fixed coordinate system and "passive" rotations of the coordinate system. A very useful discussion of these conventions, tables of terms (Table 3.1) and of the definitions and the phase factors of rotation functions used by different authors (Tables 4.1 and 4.2) can be found in [CHA98]). The definition of rotation operators used here is

$$\mathbf{D}(\alpha\beta\gamma) = e^{-i\gamma S_z'} e^{-i\beta S_y''} e^{-i\alpha S_z}$$
$$= e^{-i\alpha S_z'} e^{-i\beta S_y} e^{-i\gamma S_z} = e^{-i\beta S_y} e^{-i(\alpha+\gamma)S_z}. \quad (5.90)$$

They have the matrix elements

$$\langle IM'|\mathbf{D}(\alpha\beta\gamma)|IM\rangle = D^I_{MM'}(\alpha\beta\gamma) \quad (5.91)$$

and

$$\langle IM'|\mathbf{D}^\dagger(\alpha\beta\gamma)|IM\rangle = D^I_{MM'}{}^*(\alpha\beta\gamma). \quad (5.92)$$

D is unitary, i.e.

$$\mathbf{D}^\dagger(\alpha\beta\gamma) = \mathbf{D}^{-1}(\alpha\beta\gamma) = \mathbf{D}(-\alpha, -\beta, -\gamma), \quad (5.93)$$

i.e.

$$D^I_{MM'}{}^*(\alpha\beta\gamma) = D^I_{M'M}(-\gamma, -\beta, -\alpha) \quad (5.94)$$

and

$$\sum_{M'} D^I_{M'N}{}^* D^I_{M'M} = \delta_{MN}. \quad (5.95)$$

For the product (two rotations in sequence)

$$D^L_{MM'} = \sum_{m_1 m_1' m_2 m_2'} D^{j_1}_{m_1 m_1'} D^{j_2}_{m_2 m_2'} \langle j_1 j_2 m_1 m_2 | LM \rangle \langle j_1 j_2 m_1' m_2' | LM' \rangle. \quad (5.96)$$

(The symbol $\langle j_1 j_2 m_1 m_2 | LM \rangle$ denotes the Clebsch-Gordan (vector coupling) coefficients, see Sect. 22.3.)

When choosing, as usual, the eigenfunctions of S_z as basis vectors, $\mathbf{D}$ simplifies to

$$D^I_{MN}(\alpha\beta\gamma) = e^{-i(\alpha M + \gamma N)} \langle IM|e^{-i\beta S_y}|IN\rangle. \quad (5.97)$$

The first factor is a phase factor, the second a matrix element of the **reduced rotation functions** $d^I_{MN}(\beta)$. Condon/Shortley [CON67] define the d^k_{mn} as

5.4 Rotations, Angular Dependence of the Tensor Moments

$$d_{mn}^k = \sum_t (-)^t \frac{[(k+m)!(k-m)!(k+n)!(k-n)!]^{1/2}}{(k+m-t)!(k-n-t)!t!(t+n-m)!}$$

$$\times \left(\cos\frac{\beta}{2}\right)^{2k+m-n-2t} \left(\sin\frac{\beta}{2}\right)^{2t+n-m}. \tag{5.98}$$

The sum goes over all t leading to non-negative factorials. For M or $N = 0$ **D** becomes a spherical harmonic via

$$D_{M0}^I(\beta\alpha) = \left(\frac{4\pi}{2I+1}\right)^{1/2} Y_I^M(\beta, \alpha) = \left(\frac{4\pi}{2I+1}\right)^{1/2} Y_I^{-M}(\beta, \alpha) \tag{5.99}$$

and

$$D_{00}^I(\beta) = P_I(\cos\beta) \quad \text{(Legendre polynomial)}. \tag{5.100}$$

Example $S = 1/2$, $S_y = 1/2\sigma_y$:

$$e^{-i\beta S_y} = e^{-i\beta\sigma_y/2} = \cos\frac{\beta}{2} - i\sigma_y \sin\frac{\beta}{2}. \tag{5.101}$$

Thus, the **reduced rotation function** results for a rotation about the y axis

$$d_{mm'}^{1/2} = \begin{pmatrix} \cos\frac{\beta}{2} & -\sin\frac{\beta}{2} \\ \sin\frac{\beta}{2} & \cos\frac{\beta}{2} \end{pmatrix}. \tag{5.102}$$

5.4.3 Rotation of the Density Matrix and of the Tensor Moments

This allows e.g. a description of the density matrix in a rotated system or—equivalently—of a rotated density matrix in the old coordinate system. As an example we rotate the density matrix, which is diagonal with respect to the z axis

$$\rho' = d_{mm'}^{1/2}(\beta) \frac{1}{2}\begin{pmatrix} 1+p & 0 \\ 0 & 1-p \end{pmatrix} d_{mm'}^{1/2\dagger}(\beta)$$

$$= d_{mm'}^{1/2}(\beta) \frac{1}{2}\begin{pmatrix} 1+p & 0 \\ 0 & 1-p \end{pmatrix} d_{mm'}^{1/2\,T}(\beta). \tag{5.103}$$

The polarization direction is completely determined by the two parameters β, ϕ. β is the polar angle (relative to the z axis), ϕ the azimuthal angle = angle of the $(\vec{S}, z)$ plane relative to an arbitrarily defined x axis (where x, y, z form a righthanded system). In the rotation function **D** therefore only two angular parameters are physically relevant

$$D_{MM'}^I(\alpha, -\tilde{\beta}_{\text{Euler}}, -\gamma - \pi/2) \equiv D_{MM'}^I(0, \beta_{\text{polar}}, \phi_{\text{azimut}}). \tag{5.104}$$

Under rotations tensor moments transform more simply than the density matrix. The spherical tensors have been defined such as to transform under rotations like the spherical harmonics

$$t_{kq} = \sum_{q'} D^k_{q'q}(\alpha\beta\tilde{\gamma}) t_{kq'} = \hat{t}_{k0} D^k_{0q}(\beta,\phi). \tag{5.105}$$

Here $\hat{t}_{k0}$ signifies the maximum component, for which ρ also is diagonal, i.e. if the z axis is the quantization axis. Thus the single components for spin $S = 1$ ($S = 1/2$ is trivial) are

$$t_{10} = \hat{t}_{10} \cos\beta$$

$$t_{1\pm 1} = \mp i\hat{t}_{10} \frac{1}{\sqrt{2}} \sin\beta e^{i\phi}$$

$$t_{20} = \hat{t}_{20} \frac{1}{2}(3\cos^2\beta - 1) \tag{5.106}$$

$$t_{21} = -i\hat{t}_{20}\sqrt{\frac{3}{2}} \sin\beta \cos\beta e^{i\phi}$$

$$t_{22} = -\hat{t}_{20}\sqrt{\frac{3}{8}} \sin^2\beta e^{2i\phi}.$$

The connection between Cartesian and spherical tensors is of practical importance because in the literature both descriptions are widely used. For the Cartesian tensors the corresponding transformations result from the connection with the spherical tensors. E.g. for p_{zz}: $p_{zz} = p^*_{zz} P_2(\cos\beta)$, where p^* and p^*_{zz} are the *Cartesian maximum components* of the vector and tensor polarization for S in the direction z. Thus,

$$p_x = p^* \cdot P^1_1(\cos\beta) \cos\phi = p^* \cdot \sin\beta \cos\phi$$

$$p_y = p^* \cdot P^1_1(\cos\beta) \sin\phi = p^* \cdot \sin\beta \sin\phi$$

$$p_z = p^* \cdot P_1(\cos\beta) = p^* \cdot \cos\beta$$

$$p_{zz} = p^*_{zz} \cdot P_2(\cos\beta) = p^*_{zz} \cdot \frac{1}{2}(3\cos^2\beta - 1)$$

$$p_{xx} = p^*_{zz} \cdot \frac{1}{2}[P^2_2(\cos\beta)\cos^2\phi - 1]$$

$$p_{yy} = p^*_{zz} \cdot \frac{1}{2}[P^2_2(\cos\beta)\sin^2\phi - 1] \tag{5.107}$$

$$p_{xx} - p_{yy} = p^*_{zz} \cdot \frac{1}{2} P^2_2(\cos\beta)\cos 2\phi = p^*_{zz} \cdot \left(-\frac{3}{2}\right)\sin 2\beta \cos 2\phi$$

$$p_{xy} = p^*_{zz} \cdot P^2_2(\cos\beta)\sin 2\phi$$

$$p_{yz} = p^*_{zz} \cdot \frac{1}{2} P^1_2(\cos\beta)\sin\phi$$

$$p_{xz} = p^*_{zz} \cdot \frac{1}{2} P^1_2(\cos\beta)\cos\phi = p^*_{zz} \cdot \left(-\frac{3}{4}\right)\sin 2\beta \cos\phi.$$

5.4 Rotations, Angular Dependence of the Tensor Moments

Experimentally interesting special cases are

- $\beta = 0$: only $p_z, p_{zz} \neq 0$
- $\beta = 54.7$: $p_{zz} = 0$
- $\beta = 90$ and e.g. $\phi = 90°$:

$$p_{zz} = -\frac{1}{2}p_{zz}^*$$

$$p_{xx} - p_{yy} = \frac{3}{2}p_{zz}^*$$

$$p_y = p^* \quad \text{and}$$

$$p_x = p_z = p_{xz} = p_{yz} = p_{xy} = 0.$$

5.4.4 Practical Realization of Rotations

In practice the rotation of the quantization axis is achieved using the Larmor precession of the spins in suitable magnetic fields. Only components perpendicular to the spin vector are affected. For charged particles the deflection of a beam is coupled to the spin precession (e.g. in dipole and quadrupole magnets etc.). The component parallel to the magnetic field remains unchanged. In Wien filters the magnetic deflection is compensated by an electric field perpendicular to the magnetic field and the velocity vector of the particles. For homogeneous fields the Wien filter is "straight-looking", when (in MKSA units) the velocity of the particles is $v = E/B$. With a Wien filter rotatable about the beam axis any spin direction in space can be realized (see Sect. 16.8).

5.4.5 Coordinate Systems

Especially for reactions in which polarization observables are measured the appropriate definition of coordinate systems is very important. In "unpolarized" reactions incoming and outgoing particle momenta define one plane, the *scattering plane*. A polarization vector adds another direction (or plane) thus introducing an azimuthal dependence of observables. Two polarizations (as in transfer reactions or spin correlations) add more such preferential directions. In nuclear reactions the outgoing beam in general will be rotated against the incident beam by the scattering angles θ, ϕ. The tensor moments in the entrance and exit channels can be described in different ways. Very common is the description in the helicity formalism, in which each particle is described with respect to its direction of motion $\vec{k}_{in}$ and $\vec{k}_{out}$, respectively, as its positive z axis. The positive y axis follows the convention

$$\hat{y} = \frac{\vec{k}_{in} \times \vec{k}_{out}}{|\vec{k}_{in} \times \vec{k}_{out}|} \qquad (5.108)$$

while the x axis is determined by requiring a right-handed system. In going from the entrance to the exit channel there is the choice of describing the rotation about angles defined either in the laboratory or in the relative (c.m.) system. This convention is the *Madison Convention* [MAK71] and implies the *Basel Convention* [BAK61], see Sect. 6.4.

5.5 Exercises

5.1 (a) What kind of states $\binom{a}{b} = a\binom{1}{0} + b\binom{0}{1}$ are represented by the density matrices

$$\rho = \frac{1}{2}\begin{pmatrix} 0 & 0 \\ 0 & 1 \end{pmatrix}, \quad \frac{1}{2}\begin{pmatrix} 1 & 1 \\ 1 & 1 \end{pmatrix}, \quad \frac{1}{2}\begin{pmatrix} 1 & -1 \\ -1 & 1 \end{pmatrix}, \quad \text{and} \quad \frac{1}{2}\begin{pmatrix} 1 & 0 \\ 0 & 1 \end{pmatrix}? \tag{5.109}$$

(b) Show that the last density matrix can describe different mixed states, differing by phase factors, but with unique density matrix, and thus describing the same physics.

(c) An example of a partially polarized beam is a 75 %/25 % mixture (an "incoherent superposition") of two pure ensembles with probabilities $p(S_{z,\uparrow}) = 0.75$ and $p(S_{x,\uparrow}) = 0.25$. Using Exercise 5.1(a) show that

$$\rho = \frac{1}{8}\begin{pmatrix} 7 & 1 \\ 1 & 1 \end{pmatrix}. \tag{5.110}$$

Give the spin-expectation values ($\equiv$ "ensemble averages") $\langle S_x \rangle$, $\langle S_y \rangle$, and $\langle S_z \rangle$. Show that this ensemble can be composed in different ways (from pure states).

5.2 Why is it sensible to give absolute or relative errors with cross section results whereas for polarization observables relative errors may not be helpful?

5.3 By which angle does the polarization vector of a proton (deuteron) beam polarized in the x direction precess in the x–z plane in a 90° analyzing magnet with the magnetic field in the y direction? Does this precession depend on the beam energy? What is the polarization direction relative to the new z beam direction after the magnet?

5.4 Design a Wien filter (crossed transverse E and B fields) such that the polarization vector of a proton (deuteron) beam, initially transversally polarized in the E-field direction precesses by 90°, becoming longitudinally polarized?

5.5 How do electric fields influence the spin polarization? Could you design an electric-field configuration that transforms the longitudinal polarization of a proton beam into an exactly transverse one? (Assume that we use a field with constant field strength along the bending path.) What electric field strength would be necessary for a 15 MeV proton beam and a bending-path length of 2 m? Why do electric deflectors become impractical at higher energies?

References

[BAK61] Basel Convention in [BAS61] (1961)
[BAR67] H.H. Barschall, Am. J. Phys. **35**, 119 (1967)
[BAS61] P. Huber, K.P. Meyer (eds.), *Proc. Int. Symp. on Polarization Phenomena of Nucleons, Basel, 1960*. Helv. Phys. Acta Suppl., vol. 6 (Birkhäuser, Basel, 1961)
[BRI71] D.M. Brink, G.R. Satchler, *Angular Momentum* (Clarendon Press, Oxford, 1971)
[CON94] H.E. Conzett, Rep. Prog. Phys. **57**, 1 (1994)
[CON67] E.U. Condon, G.H. Shortley, *The Theory of Atomic Spectra* (Cambridge University Press, Cambridge, 1967)
[CHA98] M. Chaichian, R. Hagedorn, *Symmetries in Quantum Mechanics—From Angular Momentum to Supersymmetry*. Graduate Student Series in Physics (Inst. of Phys. Publ., Bristol and Philadelphia, 1998)
[DAR67] S.E. Darden, Am. J. Phys. **35**, 727 (1967)
[DAR71] S.E. Darden, in [MAD71] (1971), p. 39
[EBE74] D. Fick (ed.), *Proc. Meeting on Polarization Nucl. Phys.*, Ebermannstadt, 1973. Lecture Notes in Phys., vol. 30 (Springer, Berlin, 1974)
[FAN53] U. Fano, Phys. Rev. **90**, 577 (1953)
[FAN57] U. Fano, Rev. Mod. Phys. **29**, 74 (1957)
[FEY72] R.P. Feynman, *Statistical Mechanics*, vol. 39 (Benjamin, Reading, 1972)
[FIC71] D. Fick, *Einführung in die Kernphysik mit polarisierten Teilchen* (Teubner, Leipzig, 1971)
[HGS12] H. Paetz gen. Schieck, *Nuclear Physics with Polarized Particles*. Lecture Notes in Physics, vol. 842 (Springer, Berlin, 2012)
[LAK55] W. Lakin, Phys. Rev. **98**, 139 (1958)
[MAD71] H.H. Barschall, W. Haeberli (eds.), *Proc. 3rd Int. Symp. on Polarization Phenomena in Nucl. Reactions*, Madison 1970 (University of Wisconsin Press, Madison, 1971)
[MAK71] Madison Convention in [MAD71], xxv (1971)
[NUR13] S.B. Nurushev, M.F. Runtso, M.N. Strikhanov, *Introd. to Polarization Physics*. Lecture Notes in Physics, vol. 859 (Springer, Heidelberg, 2013)
[OHL72] G.G. Ohlsen, Rep. Prog. Phys. **35**, 717 (1972)
[ROS57] M.E. Rose, *Elementary Theory of Angular Momentum* (Wiley, New York, 1957)

Chapter 6
Nuclear Reactions of Particles with Spin

6.1 General

As for nuclear reactions of spinless particles (this is the only and the simplest case normally treated in lectures and textbooks) scattering amplitudes between entrance and exit states are a useful tool for the description of two-particle nuclear reactions, $a + A \longrightarrow b + B$ (or $A(a,b)B$). These states, however, have to be considered as spin substates in the entrance and exit channels, which leads to the following complications:

- A scattering amplitude will now become a scattering matrix, usually called M matrix.
- Depending on which quantum numbers are conserved (which depends on the symmetry properties of the physical system) angular-momentum coupling has to be performed, e.g. the spins of the incident particles $\mathbf{s}_a, \mathbf{s}_A$ are coupled to the entrance channel spin $\mathbf{S}$, this in turn is coupled to the entrance-channel orbital angular momentum $\mathbf{l}$ to yield the (conserved) total angular momentum $\mathbf{J}$, and analogously for the exit channel ($'$)

$$\mathbf{S}(\mathbf{s}_a + \mathbf{s}_A) + \mathbf{l} \longrightarrow \mathbf{J} \longrightarrow \mathbf{S}'(\mathbf{s}_b + \mathbf{s}_B) + \mathbf{l}'. \tag{6.1}$$

It is useful to apply the formalism of Racah algebra (6j, 9j symbols, or Wigner's W or Z coefficients), in which the summations about magnetic quantum numbers [WEL63] have been performed already and that have relatively simple symmetries and rules.

- In two-particle reactions the entrance and the exit channels may contain up to two particles with spin. The description of the spin state of the entrance and exit channel takes place in a spin state, the dimension of which is the direct product of the spin-space dimensions of each of the two particles: $(2s_a + 1)(2s_A + 1)$ and $(2s_b + 1)(2s_B + 1)$. The corresponding density matrices are also products of the two sub-density matrices and similarly for their expansions into Cartesian or spherical spin tensors. Since in the entrance channel normally there exist no correlations between the spin states of the beam and target the density matrix for the entrance channel can be factorized as well as the corresponding spin tensor. This, however, never applies for the exit channel.

6.2 The M Matrix

The generalization of the scattering amplitude is the *transfer or M matrix*. It is the matrix that transforms the entrance channel density matrix into that of the exit channel. Formally,

$$\rho_{\text{fin}} = \mathbf{M}\rho_{\text{in}}\mathbf{M}^\dagger. \tag{6.2}$$

The density matrices ρ_{in} and ρ_{fin} are the direct-product density matrices of the two particles of the entrance and the exit channels, respectively,

$$\rho_{\text{in}} = \rho_a(\mathbf{s}_a) \otimes \rho_A(\mathbf{s}_A) \quad \text{and} \quad \rho_{\text{fin}} = \rho_b(\mathbf{s}_b) \otimes \rho_B(\mathbf{s}_B). \tag{6.3}$$

This allows in principle a complete description of a nuclear reaction, more precisely of its observables that have been defined as expectation values of certain (spin) operators.

(a) Besides the integrated (total) cross section the simplest observable is the unpolarized differential cross section, for which the beam and target are unpolarized and no polarization, but only intensities are measured in the exit channel. It is defined as being proportional to the normalized expectation value of the unit operator ("intensity")

$$W = \text{Tr}(\rho_{\text{fin}}\mathbf{E}) = \text{Tr}(M\rho_{\text{in}}M^\dagger\mathbf{E}). \tag{6.4}$$

With

$$\rho_{\text{in}} = \rho_a(\mathbf{s}_a) \otimes \rho_A(\mathbf{s}_A) = \frac{1}{2s_a+1}\mathbf{E}(\mathbf{s}_a) \otimes \frac{1}{2s_A+1}\mathbf{E}(\mathbf{s}_A) \tag{6.5}$$

—this is a $(2s_a+1)(2s_A+1) \times (2s_a+1)(2s_A+1)$ matrix—follows

$$\rho_{\text{fin}} = \frac{1}{(2s_a+1)(2s_A+1)}\mathbf{M}\mathbf{M}^\dagger. \tag{6.6}$$

This results in the cross section in the usual sense—the "unpolarized" cross section, if one applies its "usual" definition ("outgoing particle current into the solid-angle element $d\Omega$ at the angle θ, divided by the incident particle current density") and the correct phase-space factors (density of final states, Fermi's "Golden Rule")

$$\left(\frac{d\sigma}{d\Omega}\right)_0 = \frac{1}{(2s_a+1)(2s_A+1)}\frac{k_{\text{fin}}}{k_{\text{in}}}\text{Tr}(\mathbf{M}\mathbf{M}^\dagger). \tag{6.7}$$

(b) When the incident beam is polarized, the target unpolarized, no polarization in the exit channel is measured and an expansion of the density matrix into Cartesian or spherical tensors with their special transformation properties is used, then ρ_{in} may be written e.g. in Cartesian coordinates

$$\rho_{\text{in}} = \frac{1}{(2s_a+1)(2s_A+1)} \sum_i \overbrace{\langle \sigma_i \rangle}^{p_i} (\sigma_i \otimes \mathbf{E}(s_A)) \qquad (6.8)$$

and the cross section

$$\frac{d\sigma}{d\Omega} = \frac{1}{(2s_a+1)(2s_A+1)} \frac{k_{\text{fin}}}{k_{\text{in}}} \sum_i p_i \text{Tr}(\mathbf{M}\sigma_i \mathbf{M}^\dagger)$$

$$= \frac{1}{(2s_a+1)(2s_A+1)} \frac{k_{\text{fin}}}{k_{\text{in}}} \left[\text{Tr}(\mathbf{M}\mathbf{M}^\dagger) + \sum_i p_i \text{Tr}(\mathbf{M}\sigma_i \mathbf{M}^\dagger) \right]$$

$$= \left(\frac{d\sigma}{d\Omega}\right)_0 \left[1 + \sum_i p_i \frac{\text{Tr}(\mathbf{M}\sigma_i \mathbf{M}^\dagger)}{\text{Tr}(\mathbf{M}\mathbf{M}^\dagger)} \right]. \qquad (6.9)$$

The $A_i = \frac{\text{Tr}(\mathbf{M}\sigma_i \mathbf{M}^\dagger)}{\text{Tr}(\mathbf{M}\mathbf{M}^\dagger)}$ are *analyzing powers*, which measure how the reaction is influenced by each single component of the beam polarization p_i. The index i has a general meaning: for $s_a = 1/2$ the beam can at most be vector polarized (rank 1) and the range of i is 1 to 3 or x, y, z, respectively. For spin-1 particles i signifies the components of the vector polarization and of the rank-2 tensor polarization (in the Cartesian case i signifies the combinations x, y, z and (jk) with $j, k = x, y, z$). Therefore these are the components of the vector and tensor analyzing powers.

(c) When the polarization of an outgoing particle (e.g. the ejectile b) is measured while the incident beam is unpolarized it is defined as

$$\vec{p}' = \frac{\text{Tr}(\sigma \rho_{\text{fin}})}{\text{Tr}(\rho_{\text{fin}})} = \frac{\text{Tr}(\sigma \mathbf{M} \rho_{\text{in}} \mathbf{M}^\dagger)}{\text{Tr}(\rho_{\text{fin}})}$$

$$= \frac{(2s_a+1)(2s_b+1)}{(2s_A+1)(2s_B+1)} \frac{\text{Tr}(\mathbf{M}\mathbf{M}^\dagger \sigma)}{\text{Tr}(\mathbf{M}\mathbf{M}^\dagger)} \qquad (6.10)$$

where the transformation properties of $\vec{p}'$ are those of the components of σ for the spin system considered. If, e.g. the σ are the spin-1/2 Pauli operators $\vec{p}$ will be a vector with the Cartesian components p_x, p_y, and p_z.

(d) More complicated cases are those where e.g. both particles in the entrance channel are polarized or the polarizations of both exit-channel particles (ejectile and residual nucleus) are measured in coincidence ("spin correlations") or those, in which the influence of the polarization(s) in the entrance channel on the polarization(s) in the exit channel is measured ("polarization transfer coefficients", "triple-scattering parameters".[1]

For a consistent description especially of more complicated spin states the spherical representation is best suited. The two-particle spin tensors are defined as

- for the entrance channel: $\tau_{kqKQ} = \tau_{kq}\tau_{KQ}$, since in this case polarizations of projectile and target are uncorrelated
- for the exit channel: $\tau_{k'q'K'Q'}$.

Lower-case indices are for projectile or ejectile, resp., upper-case ones for the target or the recoil nucleus. The most general polarization observable therefore depends on the polarization state (or on this state being measured!) of maximally four particles, i.e. in spherical notation on eight indices:

$$\tau_{kqKQ}^{k'q'K'Q'}. \tag{6.11}$$

6.3 Types of Polarization Observables

Besides the most general notation (i.e. the spherical one with kq for the projectile, KQ for the target, $k'q'$ for the ejectile and $K'Q'$ for the recoil (residual) nucleus) the simplified Saclay description in the helicity coordinate system is being used: X_{pqik} with X defining the observable (e.g. $= A$ for analyzing power, C for correlations etc.) [BYS78].

The $pqik$ designate the ejectile, the recoil nucleus, the projectile (the beam) and the target, respectively; their values are: s ("sideways"), n ("normal"), l ("longitudinal") for polarized particles and 0 for unpolarized ones. Thus the following types of observables can be classified (besides the character and name of the observable the nomenclature for the reaction, a typical Cartesian example and the spherical definition is listed)

- **Zero-spin observable**
  Unpolarized cross section X_{0000} or I_{0000}:

$$A(a,b)B \qquad (d\sigma/d\Omega)_0 \propto \mathrm{Tr}(\mathbf{M}\mathbf{M}^\dagger) \tag{6.12}$$

- **One-spin observables**
  - Projectile analyzing power, e.g. A_{00i0} or A_y:

$$A(\vec{a},b)B \qquad T_{kq} \propto \frac{\mathrm{Tr}(\mathbf{M}\tau_{kq}\mathbf{M}^\dagger)}{\mathrm{Tr}(\mathbf{M}\mathbf{M}^\dagger)} \tag{6.13}$$

---

[1]The term "triple scattering" stems from the beginning of polarization experiments when for the production of polarized beams a primary nuclear reaction was needed, the second "scattering" was the one to be investigated, and the third reaction served as an analyzer for the polarization of the outgoing particles. With the advent of polarized-ion sources only "double scattering" was required.

6.3 Types of Polarization Observables

- Target analyzing power, e.g. A_{000k} or A_y:

$$\vec{A}(a,b)B \qquad T_{KQ} \propto \frac{\text{Tr}(\mathbf{M}\tau_{KQ}\mathbf{M}^\dagger)}{\text{Tr}(\mathbf{M}\mathbf{M}^\dagger)} \qquad (6.14)$$

- Outgoing polarization, e.g. p_{p000} or $p_{y'}$:

$$A(a,\vec{b})B \qquad t_{k'q'} \propto \frac{\text{Tr}(\mathbf{M}\mathbf{M}^\dagger \tau_{k'q'})}{\text{Tr}(\mathbf{M}\mathbf{M}^\dagger)} \qquad (6.15)$$

- Similarly for the recoil nuclei, e.g. p_{0q00}:

$$A(a,b)\vec{B} \qquad t_{K'Q'} \propto \frac{\text{Tr}(\mathbf{M}\mathbf{M}^\dagger \tau_{K'Q'})}{\text{Tr}(\mathbf{M}\mathbf{M}^\dagger)} \qquad (6.16)$$

- **Two-spin observables**
  - Polarization transfer coefficients (in the case of the NN interaction they were also designated as "Wolfenstein parameters" $A(=K_z^{x'})$, $A'(=K_z^{z'})$, $R(=K_x^{x'})$, $R'(=K_x^{z'})$, $D(=K_y^{y'})$). Notation: D_{p0i0}, e.g. Cartesian $K_y^{y'}$, or spherical: e.g. $t_{kq}^{k'q'}$. Meaning: transfer of the projectile polarization to that of the ejectile

$$A(\vec{a},\vec{b})B \qquad t_{kq}^{k'q'} \propto \frac{\text{Tr}(\tau_{k'q'}\mathbf{M}\tau_{kq}\mathbf{M}^\dagger)}{\text{Tr}(\mathbf{M}\mathbf{M}^\dagger)}. \qquad (6.17)$$

  - Spin correlation coefficients: The correlation may be considered either in the entrance or the exit channel (in the first case the polarizations of beam and target are of course uncorrelated, in the latter case the are correlated by the nuclear reaction and therefore must be measured in coincidence). The spin tensors are direct products of the tensors of both particles and act e.g. for the entrance channel in a $(2s_a+1)(2s_A+1)$ dimensional spin space. Notation: In the case of the NN interaction e.g. A_{00nn}, Cartesian A_{mn} with $m,n=x,y,z$ (not to be mixed up with the tensor analyzing power A_{ik}) means an entrance-channel correlation of the beam and target polarizations in y ($=$ normal) direction, or $C_{ik,\ell m}$; more generally the spherical notation $t_{kqKQ} = t_{kqKQ}^{0000}$ and $t^{k'q'K'Q'} = t_{0000}^{k'q'K'Q'}$, respectively, may be used. For entrance-channel correlations:

$$\vec{A}(\vec{a},b)B \qquad t_{kq}t_{KQ} \propto \frac{\text{Tr}(\mathbf{M}\tau_{kqKQ}\mathbf{M}^\dagger)}{\text{Tr}(\mathbf{M}\mathbf{M}^\dagger)} \qquad (6.18)$$

  with factorized input tensor moments and similarly (with correlated outgoing tensor moments) for the exit channel.

- **Three- and four-spin observables**: Analogous expressions hold for "generalized analyzing powers" $t_{qkQK}^{q'k'Q'K'}$ using more general spherical (or Cartesian) tensors $\tau_{kqKQ}^{k'q'K'Q'}$.

Fig. 6.1 Coordinate system for the description of the incident polarization as well as construction of the y axis with x along $\vec{y} \times \vec{k}_{in}$, y along $\vec{k}_{in} \times \vec{k}_{out}$, and z along $\vec{k}_{in}$

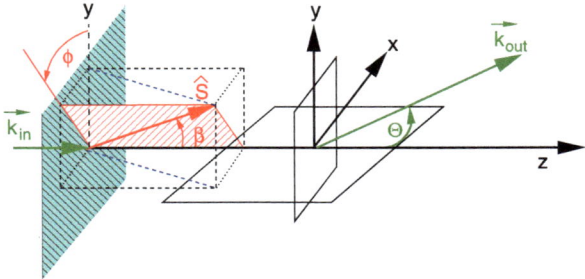

6.4 Coordinate Systems

The notation of polarization tensors (tensor moments) and suitable coordinate systems have been strongly recommended in two international conventions:

- The *Basel Convention* was issued in 1960 [BAK61] and determines that in nuclear reactions with spin-1/2 particles the polarization should be counted positive in the direction $\vec{k}_{in} \times \vec{k}_{out}$. Assuming a positive analyzing power this positive polarization yields a positive left-right asymmetry (L-R).
- The *Madison Convention* [MAK71] and [DAR71] refers to spin-1 particles. A right-handed coordinate system is assumed with the z direction being the momentum direction of the incident or outgoing particles, and where the y direction is along $\vec{k}_{in} \times \vec{k}_{out}$. Cartesian and spherical spin tensors and the corresponding tensor moments are allowed and the components of the polarization are given by p_i, p_{ij} (Cartesian) or t_{kq} (spherical), respectively, those of the analyzing powers A_i, A_{ij}; $i, j = x, y, z$ (Cartesian) or T_{kq}, respectively. The extension to higher spins is straightforward leading to an increasing number of indices for the Cartesian notation.

One consequence of this convention is that when using more than one detector each detector obtains its own coordinate system with $\hat{y}$ perpendicular to the respective scattering plane. Figure 6.1 shows the situation for one system with polarized particles.

6.4.1 Coordinate Systems for Analyzing Powers

Although the (spin) observables to be measured depend only on the polar angle θ, the cross sections including these observables generally exhibit a dependence on the azimuthal angle ϕ. This dependence enters via the need to introduce coordinate systems, in which the detector positions as well as the polarization direction have to be described. While for the vector polarization the ensuing *azimuthal complexity* is a simple $\sin\phi$ or $\cos\phi$ dependence, e.g. for spin 1/2 with the Madison convention

6.4 Coordinate Systems

and parity conservation

$$\frac{d\sigma}{d\Omega}(\theta,\phi) = \left(\frac{d\sigma}{d\Omega}(\theta)\right)_0 \left[1 + \frac{1}{2}p_y A_y(\theta)\right]$$

$$= \left(\frac{d\sigma}{d\Omega}\right)_0 \left[1 + \frac{1}{2}p_Z A_y(\theta)\sin\beta\cos\phi\right]. \quad (6.19)$$

For higher spins and other types of observables the description is more complex. As an example the case of polarized spin-1 projectiles such as deuterons on unpolarized targets will be discussed. Writing out the expansion Eq. (6.9) in Cartesian notation and considering parity conservation, as discussed below, we obtain (the cross sections and analyzing powers depend only on θ and the energy)

$$\frac{d\sigma}{d\Omega} = \left(\frac{d\sigma}{d\Omega}\right)_0 \left[1 + \frac{3}{2}p_y A_y + \frac{1}{2}p_{zz} A_{zz} + \frac{2}{3}p_{xz} A_{xz}\right.$$

$$\left. + \frac{1}{6}(p_{xx} - p_{yy})(A_{xx} - A_{yy})\right]. \quad (6.20)$$

After introducing the β and ϕ dependences of the quantization (spin-symmetry) axis (of the Madison convention [MAK71], see also Fig. 6.1) and parity conservation explicitly we obtain

$$\frac{d\sigma}{d\Omega} = \left(\frac{d\sigma}{d\Omega}\right)_0 \left\{1 + \frac{3}{2}p_Z A_y \sin\beta\cos\phi \right.$$

$$+ p_{ZZ}\left[\frac{1}{4}A_{zz}(3\cos^2\beta - 1) - A_{xz}\sin\beta\cos\beta\sin\phi\right.$$

$$\left.\left. - \frac{1}{4}(A_{xx} - A_{yy})\sin^2\beta\cos 2\phi\right]\right\}, \quad (6.21)$$

where p_Z and p_{ZZ} (often: p_z^* or p^* and p_{zz}^*, also $\hat{p}_Z$ and $\hat{p}_{ZZ}$, see also Eqs. (6.23) for spin correlations) are the coordinate-independent maximum values along the quantization (symmetry) axis of the polarization, e.g. of a beam coming from a polarized-ion source or of a polarized target. We see that the maximum azimuthal complexity is $\propto \cos 2\phi$, which has to be taken into account for the placement of detectors. An arrangement of four detectors $\Delta\phi = 90°$ apart at one polar angle θ is advantageous, because by taking differences and sums of count ratios of the four detectors each of the four analyzing powers can be determined nearly independently of all others [PET67].

6.4.2 Coordinate Systems for Polarization Transfer

For the determination of polarization-transfer coefficients the polarization of the ejectiles from a primary reaction, induced with polarized particles, has to be measured ("double scattering"). This is done using a (calibrated) analyzer reaction. For this again the Madison convention is used, i.e. the direction of motion of the outgoing particles (along $\vec{k}_{\text{out}}$ in Fig. 6.1) is the new z' axis for the second scattering. However, this axis can be defined along the c.m. or the lab. direction. Then the coordinate system for the analyzer reaction may be defined as before as a right-handed system with

$$\vec{x}' \text{ along } \vec{y}' \times \vec{z}', \qquad \vec{y}' \text{ along } \vec{z} \times \vec{k}'_{\text{out}}, \quad \text{and} \quad \vec{z}' = \vec{k}'_{\text{out}}. \tag{6.22}$$

More detailed discussions on polarization transfer can be found in [OHL72, OHL73, SPE83, SYD93, SYD98].

6.4.3 Coordinate Systems for Spin Correlations

In this case two—in principle independent—polarizations have to be considered, because they can be prepared independently with arbitrary spin directions. With the above prescription for both, two coordinate systems can be defined, which will be rotated azimuthally against each other around the z axis, i.e. they are connected by one azimuthal angle Φ. The azimuthal angles describing the detector positions in both systems are therefore connected by trigonometric relations containing Φ.

As mentioned above the observables (such as spin-correlation coefficients) depend only on the polar angle θ whereas the (polarized) cross section generally acquires an azimuthal dependence via the introduction of coordinate systems. These are in principle arbitrary but we follow Ohlsen ([OHL72]) and the Madison Convention ([MAK71]). In Ref. [OHL72] the case of the azimuthal dependence of spin-correlation cross sections is explained for spin-1/2 on spin-1 systems (see also Ref. [PRZ06]). For the spin-1 on spin-1 case this is explained in [HGS10]. Figure 6.2 shows a possible coordinate system, in which the separately defined coordinates of the projectile and target polarizations may be combined into a single azimuthal dependence of the correlation cross section.

Generally, it is advisable to choose such a system that the description of a real experiment is as simple and intuitive as possible. In this respect the Cartesian description is more intuitive than the spherical one. The description of polarization components from polarized sources and polarized targets is best imagined in a space-fixed coordinate system, in which the direction of the polarization vectors are described by two sets of polar and azimuthal angles (β_b, ϕ_b) for the incident beam polarization and (β_t, ϕ_t) for the target. The orientation of the tensor polarization is fixed to that of the polarization vector. As coordinate system here a set of axes x, y, and z is chosen, where z is identical with the incoming beam direction (along $\vec{k}_{\text{in}}$), y may be vertically upward, and x, y, and z form a right-handed screw.

On the other hand, we need a scattering-frame system where a Y axis is defined by the direction of $\vec{k}_{\text{in}} \times \vec{k}_{\text{out}}$, the Z axis coincides with z, and with the X-axis again forming a right-handed system together with $Z = z$. It is clear that this system is

6.4 Coordinate Systems

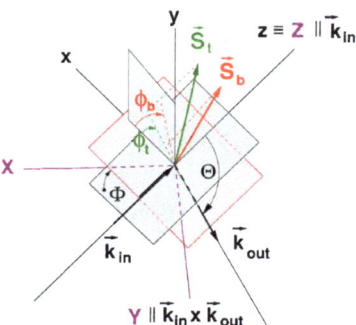

Fig. 6.2 Coordinate systems for spin-1 on spin-1 polarization correlation experiments. The polarization symmetry axes S_b for beam (b) and S_t for target (t) polarizations are defined in the space-fixed coordinate system x, y, and z. The detector(s) are positioned at polar angles θ with respect to the $z = Z$ axis and at angles ϕ as measured clockwise from the x axis along z. Relative to the spin directions the azimuthal angles are $\phi_b - \Phi$ and $\phi_t - \Phi$. The polar angles of the polarizations β_b and β_t are not shown

different for each detector, the position of which must be characterized by a polar angle θ and some azimuthal angle. We demand that for the parts of the cross section with only one particle type (beam or target) being polarized (leading to *analyzing powers*) we have the usual description of the azimuthal dependence on ϕ (with a maximum azimuthal complexity of $\cos 2\phi$), see Sect. 6.4.1. Figure 6.2 shows the relations between the two polarization-symmetry axes and the projectile-helicity frame. The polarization components in the scattering frame of the incident beam are

$$p_X = \hat{p}_Z \sin\beta_b \cos(\phi_b - \Phi)$$
$$p_Y = \hat{p}_Z \sin\beta_b \sin(\phi_b - \Phi)$$
$$p_Z = \hat{p}_Z \cos\beta_b$$
$$p_{XY} = \frac{3}{4}\hat{p}_{ZZ} \sin^2\beta_b \sin 2(\phi_b - \Phi)$$
$$p_{YZ} = \frac{3}{2}\hat{p}_{ZZ} \sin\beta_b \cos\beta_b \sin 2(\phi_b - \Phi) \quad (6.23)$$
$$p_{XZ} = \frac{3}{2}\hat{p}_{ZZ} \sin\beta_b \cos\beta_b \cos(\phi_b - \Phi)$$
$$p_{XX} - p_{YY} = \frac{3}{2}\hat{p}_{ZZ} \sin^2\beta_b \cos 2(\phi_b - \Phi)$$
$$p_{ZZ} = \frac{1}{2}\hat{p}_{ZZ}(3\cos^2\beta_b - 1)$$

and similarly for the target polarization

$$q_X = \hat{q}_Z \sin\beta_t \cos(\phi_t - \Phi)$$

$$q_Y = \hat{q}_Z \sin\beta_t \sin(\phi_t - \Phi)$$

$$q_Z = \hat{q}_Z \cos\beta_t$$

$$q_{XY} = \frac{3}{4}\hat{q}_{ZZ} \sin^2\beta_t \sin 2(\phi_t - \Phi)$$

$$q_{YZ} = \frac{3}{2}\hat{q}_{ZZ} \sin\beta_t \cos\beta_t \sin 2(\phi_t - \Phi) \quad (6.24)$$

$$q_{XZ} = \frac{3}{2}\hat{q}_{ZZ} \sin\beta_t \cos\beta_b \cos(\phi_t - \Phi)$$

$$q_{XX} - q_{YY} = \frac{3}{2}\hat{q}_{ZZ} \sin^2\beta_t \cos 2(\phi_t - \Phi)$$

$$q_{ZZ} = \frac{1}{2}\hat{q}_{ZZ}(3\cos^2\beta_t - 1).$$

The quantities $\hat{p}_i, \hat{p}_{jk}, \hat{q}_i, \hat{q}_{jk}$ are the (coordinate-system independent) vector and tensor polarizations of beam and target as given by the occupation numbers of the hyperfine Zeeman states in a rotationally-symmetric frame along the z axis.

In the spin-correlation cross section terms beam and target polarizations p and q appear as products. Therefore typical azimuthal dependences and complexities arise from combinations such as (and similarly for sin terms)

$$\propto \cos(\phi_b - \Phi) \cdot \cos(\phi_t - \Phi)$$
$$\propto \cos 2(\phi_b - \Phi) \cdot \cos(\phi_t - \Phi) \quad (6.25)$$
$$\propto \cos 2(\phi_b - \Phi) \cdot \cos 2(\phi_t - \Phi).$$

By using trigonometric relations it can be seen that these terms lead to azimuthal dependences $\propto [\cos M\Phi]_{M=0,1,\ldots}$ and therefore to a maximum "complexity" of

$$\cos 4\Phi \quad \text{and also} \quad \sin 4\Phi \quad (6.26)$$

e.g. for the correlation coefficient $C_{xy,xy}$. For this coefficient the product $\sin 2(\phi_b - \Phi) \cdot \sin 2(\phi_t - \Phi)$ may be transformed into

$$\frac{1}{2}(\cos 4\Phi + 1) \sin 2\phi_b \sin 2\phi_t$$
$$-\frac{1}{2}(\cos 4\Phi - 1) \cos 2\phi_b \cos 2\phi_t$$
$$-\frac{1}{2}\sin 4\Phi (\cos 2\phi_b \sin 2\phi_t + \cos 2\phi_t \sin 2\phi_b). \quad (6.27)$$

This complexity has to be met by a sufficiently fine-grained detector arrangement. It is clear that substantial simplifications arise with the choice of special polarization directions. If e.g. in an experiment both polarization vectors point in the x direction then with $\phi_b = \phi_t = 0$ only a simple ϕ dependence

$$\propto (1 - \cos 4\Phi) \qquad (6.28)$$

of the cross section results for this correlation coefficient. Of course the different ϕ dependences have to be established for all coefficients. In Chap. 19.2.1 the complete set of terms entering the general cross section of spin-1 on spin-1 correlations is given. For identical particles some terms are redundant.

6.5 Structure of the M Matrix and Number of "Necessary" Experiments

A nuclear reaction may be considered "completely measured", when all elements of the M matrix have been determined uniquely. How many independent polarization experiments will be necessary to achieve this goal depends on the spin structure of the reaction and can be derived theoretically. A number of investigations into this question has been published, see e.g. Refs. [HOF66, FIC67, KOE68, SIM74] and the conclusions were that in the cases cited no measurement of e.g. four-spin observables was necessary. An example for an experimental program, by which the "complete" determination of all M-matrix elements was attempted was the NN program at the Saclay SATURNE facility. At present the complete set of NN data is maintained by CNS DAC [GWU11] e.g. for use with the phase-shift analysis program SAID [SAID13]. Several other phase-shift analyses of the NN systems are maintained, e.g. at Nijmegen [NNO13].

Because "complete" data sets do not exist for most reactions another approach has been taken that consists in a least-squares fitting procedure of the incomplete set of data of certain reactions. Examples are few-body reactions such as the two DD reactions at very low energies by a multi-channel R-matrix or single-channel T-matrix approach (see e.g. Refs. [HGS10, HGS12].)

In practice—e.g. due to experimental uncertainties—some redundancy is used, i.e. more observables than minimally required will have to be measured. In addition not all M-matrix elements are independent since symmetries impose restrictions and create relations between them:

- Rotational symmetry (conservation of angular momentum):
  The outcome of a measurement of a nuclear reaction is independent of the orientation of the coordinate system or of the orientation of an experiment in a given coordinate system.
- Mirror symmetry (Parity conservation—not for the weak interaction): The outcome of a measurement of a nuclear reaction is the same as that of a reaction reflected at the origin.
- Time-reversal symmetry: Since the time-reversal operator is "anti-unitary" there is no conserved quantity here, but relations exist between the observables of the forward and the backward reactions as reversal of motion with suitable application to the spins of the particles involved).

In principle the behavior of polarization observables under symmetry transformations can be derived by considering each of its components with respect to a given (Cartesian) coordinate system as a product of the corresponding tensor and the unit vector in the coordinate direction. Therefore not only the behavior of the spin tensor itself but also that of the coordinates have to be considered. In the helicity coordinate system the coordinates behave differently under the following two transformations by the parity operator P and time-reversal operator T

$$\begin{array}{cccc} & x & y & z \\ \text{Parity } \mathbf{P}: & -x & y & -z \\ \text{Time reversal } \mathbf{T}: & x & y & z \end{array} \quad (6.29)$$

i.e. only $\hat{y} = \frac{\vec{k}_{in} \times \vec{k}_{fin}}{|\vec{k}_{in} \times \vec{k}_{fin}|}$ is invariant under the parity transformation whereas all components of the Cartesian spin operator $\mathbf{S}$ are invariant. (This follows from the fact that $\mathbf{S}^2$ commutes with $\mathbf{P}$ and from the commutation relations $\mathbf{S} \times \mathbf{S} = i\mathbf{S}$.) Therefore only S_y will be $\mathbf{P}$ invariant. This means that in a nuclear reaction with a parity-conserving interaction only the polarization component $p_{y'} \neq 0$ can be produced in the exit channel or the only component of an analyzing power $\neq 0$ will be A_y. On the other hand the measurement of components such as A_x or $A_z \neq 0$ is very suitable in the search for a parity violation.

From the parity behavior of S not only the parity behavior of higher-rank observables (for spins $> 1/2$) but also of the corresponding tensor moments can be obtained. Parity conservation imposes the condition on analyzing tensor moments (analyzing powers):

$$T_{k-q} = (-1)^{k+q} T_{kq}. \quad (6.30)$$

Like for spin-1/2 systems due to parity conservation the number of observables for larger spins will also be reduced. For $S = 1$ only one vector analyzing power A_y and the three tensor analyzing powers A_{zz}, A_{xz} and $A_{xx} - A_{yy}$ or iT_{11}, T_{20}, T_{21} and T_{22}, resp., can be $\neq 0$. Third-rank Cartesian analyzing powers thus

$$A_{xxx} = A_{zzz} = A_{zzx} = A_{xxz} = A_{yyz} = A_{xyy} = 0. \quad (6.31)$$

As a rule: the sum of the number of indices x and z must be even for the observable to exist [OHL72].

For the generalized analyzing powers rotation invariance together with parity conservation yield the following relation [KNU72]

$$\tau_{kqKQ}^{k'q'K'Q'}(p_3, \theta_3, \phi_3; p_4, \theta_4, \phi_4)$$

$$= (-)^{\sum_{j=1}^{4} k_j} \left[\tau_{kqKQ}^{k'q'K'Q'}(p_3, \theta_3, -\phi_3; p_4, \theta_4, -\phi_4) \right]^*, \quad (6.32)$$

which simplifies for the coordinate system introduced above, in which $\phi_3 = 0$ and $\phi_4 = \pi$

$$\tau_{kqKQ}^{k'q'K'Q'} = (-)^{\sum_{j=1}^{4} k_j} [\tau_{kqKQ}^{k'q'K'Q'}]^*. \tag{6.33}$$

This means that all polarization-transfer coefficients are either real or imaginary. From the polarization of the outgoing particle 3

$$t_{k'q'} = \frac{\tau_{0000}^{k'q'00}}{\tau_{0000}^{0000}}, \tag{6.34}$$

also follows

$$t_{k'q'} = (-)^{k'} t_{k'q'}^*. \tag{6.35}$$

For $k' = 1$ (vector polarization) only Im $(t_{11}) = -\frac{1}{2}\sqrt{3}p_{y'}$ is $\neq 0$, i.e. the polarization vector points perpendicular to the scattering plane.

In Refs. [KNU72, OHL81] also the case of a reaction with three particles in the exit channel is discussed. In this case in general no restrictions of the number of observables by parity conservation apply. Exceptions are:

- The three particle momenta and the beam form a plane. The system then behaves like a two-particle reaction.
- Only one particle is detected (this is a kinematically incomplete measurement, in which averaging over the momenta of the unobserved particles takes place). Here again the transfer coefficients will be either purely real or purely imaginary thus reducing their number by a factor 2.

Time-reversal invariance leads to relations between observables of the forward and the backward reaction. One such relation is the principle of "detailed balance", which requires equality (up to phase-space factors) of the cross sections of both. For polarization observables similar relations result, e.g. the equality of the vector analyzing power for the forward reaction with polarized projectile a and the exit channel polarization of the ejectile b in the backward reaction produced with an unpolarized beam (or target)—which normally has to be measured in a second scattering. Since in elastic scattering projectile a and ejectile b are identical a double scattering experiment will yield A_y^2 absolutely (but not the sign of A_y). A widely used example for an analyzer reaction for protons is ^{4}He$(p, p)^4$He.

Reference [OHL72] gives (for Cartesian observables) a detailed description of the formalism especially for the polarization-transfer and spin-correlation coefficients and (counting) rules for the restrictions imposed by parity conservation and time-reversal invariance. Systems with the spin structures

- Polarization transfer:
$$-\vec{\tfrac{1}{2}} + A \to \vec{\tfrac{1}{2}} + B$$
$$-\vec{1} + A \to \vec{\tfrac{1}{2}} + B$$

- $-\vec{\tfrac{1}{2}} + A \to \vec{1} + B$
- $-\vec{1} + A \to \vec{1} + B$

• Spin correlations:

- $-\vec{\tfrac{1}{2}} + \vec{\tfrac{1}{2}} \to b$ or B
- $-\vec{1} + \vec{\tfrac{1}{2}} \to b$ or B
- $-\vec{1} + \vec{1} \to b$ or B

are discussed there.

6.6 Examples

In the following only a few examples for the observables of important spin systems will be discussed.

6.6.1 Systems with Spin Structure $1/2 + 0 \longrightarrow 1/2 + 0$

Examples ^{4}He$(p,p)^4$He, ^{4}He$(n,n)^4$He, ^{1}H$(p,p)^1$H. The form of the transfer matrix for spin structure $1/2 + 0 \longrightarrow 1/2 + 0$ is

$$\mathbf{M}(\mathbf{k}_{\text{in}}, \mathbf{k}_{\text{fin}}) = A + B(\sigma\hat{\mathbf{y}}) \tag{6.36}$$

with A = non-spinflip amplitude, B = spinflip amplitude. The description will be in right-handed coordinate systems either in the c.m. or the lab. system. Table 6.1 shows the possible (and partly redundant) observables of these systems. For elastic scattering there is a substantial reduction of the number of independent observables by:

- Parity conservation (**P**): The expectation values of some observables vanish, e.g. the longitudinal analyzing power A_z;
- Time-reversal invariance (**T**): **T** connects observables of the forward and backward reaction.

With these symmetries the number of independent observables (experiments) is reduced from 15 possible to 3 independent "true" polarization experiments + the measurement of the differential and the total cross sections (see Table 6.2). The number of complex amplitudes is $N = 2$, thus the number of real amplitudes is $2N - 1 = 3$, equal to the minimum number of necessary independent polarization experiments. Of these there are at most 4 (including the unpolarized cross section) allowing for an additional relation between the observables. After identifying the transfer coefficients in the lab. system with the "Wolfenstein" parameters a well-known relation between different observables reads

$$p_{y'}^2 + R^2 + A^2 = 1. \tag{6.37}$$

6.6 Examples

Table 6.1 Table of the observables of a spin system $1/2 + 0 \longrightarrow 1/2 + 0$

Ejectile	Beam			
	1	p_x	p_y	p_z
$I(\theta)$	$\|A\|^2 + \|B\|^2$		$2\,\text{Re}(A^*B)$	
	I_o		$I_o A_y$	
$p_{x'}(\theta)$		$\|A\|^2 - \|B\|^2$		$2\text{Im}(A^*B)$
		$I_o K_x^{x'}$		$I_o K_z^{x'}$
$p_{y'}I(\theta)$	$2\text{Re}(A^*B)$		$\|A\|^2 + \|B\|^2$	
	$I_o p_y$		$I K_y^{y'}$	
$p_{z'}I(\theta)$		$-2\text{Im}(A^*B)$		$\|A\|^2 - \|B\|^2$
		$I_o K_z^{x'}$		$I_o K_z^{z'}$

Table 6.2 Reduction of the number of independent observables by parity conservation (**P**) and time-reversal invariance (**T**)

	Beam unpolarized	Beam polarized
Differential cross section	$(\frac{d\sigma}{d\Omega})_0$ (1)	A_i (3) $\overbrace{(1)}^{\mathbf{P}}$
Ejectile polarization	p_i (3) $\overbrace{(1)}^{\mathbf{P}}$	$K_i^{j'}$ (9) $\overbrace{(5)}^{\mathbf{P}}$ $\overbrace{(2)}^{\mathbf{T}}$

6.6.2 Systems with Spin Structure $1/2 + 1/2 \longrightarrow 1/2 + 1/2$

Examples are the NN system, reactions such as $^3\text{He}(p,p)^3\text{He}$, $^3\text{H}(n,n)^3\text{H}$, etc. A very detailed description of the formalism of elastic scattering of these systems has been given in Ref. [BYS78] where the connection between observables and M-matrix elements is made. Table 6.3 lists all possible polarization observables for this spin system. Thus in principle there are 255 possible polarization experiments + measurement of the unpolarized differential cross section (+ measurement of the total cross section). For elastic scattering parity conservation and time-reversal invariance will reduce this number to 25 for identical particles (such as in p–p scattering), and to 36 linear independent experiments for non-identical particles.

The form of the transfer matrix for systems with spin structure $1/2 + 1/2 \longrightarrow 1/2 + 1/2$: (elastic scattering) is

Table 6.3 Observables for the spin structure $1/2 + 1/2 \longrightarrow 1/2 + 1/2$ with the designation of the classes of experiments as B for "beam", T for "target", u for "unpolarized", p for "polarized" are X_{pqik} with p, q, i, and k, each with values s, n, or l (polarized) or o (unpolarized) with the indices k for the target, i for the beam, q for the recoil nucleus, and p for the ejectile. ℓ stands for "longitudinal", n for "normal" (along the scattering normal) and s (or m) for "sideways" (perpendicular to ℓ and n)

Observable	Bu, Tu	Bp, Tu	Bu, Tp	Bp, Tp
Differential cross section	$I_{oooo}(1)$	$A_{ooio}(3)$	$A_{oook}(3)$	$A_{ooik}(9)$
Ejectile polarization	$p_{pooo}(3)$	$D_{poio}(9)$	$K_{pook}(9)$	$M_{poik}(27)$
Recoil polarization	$p_{oqoo}(3)$	$K_{oqio}(9)$	$D_{oqok}(9)$	$N_{oqik}(27)$
Polarization correlation	$C_{pqoo}(9)$	$C_{pqio}(27)$	$C_{pqok}(27)$	$C_{pqik}(81)$

$$\mathbf{M}(\vec{k}_{\text{in}}, \vec{k}_{\text{fin}}) = 1/2\big[(a+b) + (a-b)(\sigma_1\mathbf{n})(\sigma_2\mathbf{n}) + (c+d)(\sigma_1\mathbf{m})(\sigma_2\mathbf{m})$$
$$+ (c-d)(\sigma_1\mathbf{l})(\sigma_2\mathbf{l}) + e(\sigma_1 + \sigma_2)\mathbf{n}$$
$$+ f(\sigma_1 - \sigma_2)\mathbf{n}\big].$$

Here σ_i are the (Cartesian) Pauli spin operators and $\mathbf{m}$, $\mathbf{n}$, $\mathbf{l}$ the basis vectors of a right-handed c.m. coordinate system with:

$$\mathbf{l} = \frac{\vec{k}_{\text{fin}} + \vec{k}_{\text{in}}}{|\vec{k}_{\text{fin}} + \vec{k}_{\text{in}}|} \qquad \mathbf{m} = \frac{\vec{k}_{\text{fin}} - \vec{k}_{\text{in}}}{|\vec{k}_{\text{fin}} - \vec{k}_{\text{in}}|} \qquad \mathbf{n} = \frac{\vec{k}_{\text{fin}} \times \vec{k}_{\text{in}}}{|\vec{k}_{\text{fin}} \times \vec{k}_{\text{in}}|}. \tag{6.38}$$

In pp scattering, after considering parity conservation, time-reversal invariance and the Pauli principle there are $N = 5$, in np scattering $N = 6$ invariant, independent complex amplitudes (of a total of 16 possible ones).

Thus in a complete experiment $2N - 1$ real quantities have to be determined by at least as many independent experiments:

- for pp: 9
- for np: 11.

6.6.3 Systems with Spin Structure $\vec{\tfrac{1}{2}} + \vec{1}$ and Three-Nucleon Studies

This system is in principle very important because the three-nucleon system $N + d$ is—after the NN system—the most important system for the test of fundamental interactions such as meson-exchange or effective-field theory NN input into Faddeev-like calculations, including tree-body and Coulomb forces. However, a very limited number of different polarization observables has been measured to date. These are nucleon and deuteron vector analyzing powers A_y and iT_{11}, as well as deuteron analyzing powers T_{2q} of elastic scattering, and analyzing powers of the breakup reactions $N + d \to 3N$. Observables of elastic scattering of the system: polarized spin-1/2 on unpolarized spin-1 particles including a phase-shift parametrization have been discussed in Ref. [SEY69].

6.6.4 Systems with Spin Structure $\vec{1} + \vec{1}$ and $\vec{\frac{1}{2}} + \vec{\frac{1}{2}}$ and the Four-Nucleon Systems

The four-nucleon system is the smallest nuclear system with a rich structure consisting of excited states and different channels with several clusterings $1 + 3$ as well as $2 + 2$. Its importance reaches from theoretical approaches using again the fundamental NN interactions via Faddeev-Yakubowsky equations and including three-nucleon and four-nucleon forces as well as the Coulomb interaction, to applications in fusion-energy research. Polarization effects in these reactions are relatively large, and a significant number of different observables has been collected, especially at low (fusion and astrophysically relevant) energies. In Sect. 19.3 the role of spin correlations of the $D + D$ reactions in fusion energy will be discussed.

6.6.5 Practical Criteria for the Choice of Observables

In practice more sets of experiments than minimally necessary are chosen for the following reasons:

- Consistency checks provided by relations between observables.
- Resolution of possible discrete ambiguities caused by the bilinear form of the equations relating the M matrix to the observables.
- Unavoidable experimental errors require that a fit procedure with more observables than fit parameters (matrix elements, phase shifts, etc.) is necessary.

Criteria for the selection of suitable polarization observables:

- Redundancy: observables should be linearly independent of each other (of course they depend on each other via different combinations of matrix elements).
- Technical realizability.
- Availability e.g. of a polarized target.
- Ease of orientation of the polarization in the beam and target into three orthogonal directions.
- Avoidance of the complicated three or four-spin observables.
- Avoidance of measuring a longitudinal polarization component in spin transfer (which needs spin rotation by magnetic field).
- "Sensitivity" of all observables to the amplitudes, small covariances between different observables (this may be important when determining reaction amplitudes in a fit procedure).

The measured (polarization) observables have to be compared to predictions of some (preferably the best available) theory for the description of a nuclear reaction. As an "interface" between theory and experiment a single observable could be used. When more (and different) observables have been measured it is better to extract common basic quantities such as the transfer-matrix elements, the S or T matrix elements or—for elastic scattering—a related parametrization such as phase shifts for comparison with the theory. This would also permit the prediction of unmeasured quantities from experimental data alone or in comparison with the theory. In the following the partial-wave analysis will be discussed.

6.7 Exercises

6.1 Show how an $(\vec{\ell} \cdot \vec{s})$ force causes a left-right asymmetry of the cross section if the incident spin-1/2 beam is (partially or fully) polarized.

6.2 Likewise show how an $(\vec{\ell} \cdot \vec{s})$ force causes an unpolarized spin-1/2 beam to become polarized.

6.3 Derive the figure of merit, defined as some polarization quantity (such as a beam or target polarization, an analyzing power etc.) squared times an intensity related quantity (such as a cross section, a beam intensity, or the density of polarized nuclei in a target etc.), e.g. $p_y^2 \cdot I$, by the requirement of minimal measurement time to reach a given experimental precision.

6.4 We define a right-handed coordinate system for reactions with polarized spin-1/2 projectiles such that

$$\hat{z} = \frac{\vec{k}_{\text{in}}}{|\vec{k}_{\text{in}}|}, \qquad \hat{y} = \frac{\vec{k}_{\text{in}} \times \vec{k}_{\text{out}}}{|\vec{k}_{\text{in}} \times \vec{k}_{\text{in}}|}, \quad \text{and} \quad \hat{x} = \hat{y} \times \hat{z}. \tag{6.39}$$

Show that under parity conservation only the component $A_y \neq 0$ can exist whereas A_x and A_z vanish.

6.5 Explain why in the $^3\text{H}(\vec{d}, n)^4\text{He}$ reaction at $E_{\text{lab}} = 107$ keV, which is completely dominated by an s-wave, $J^\pi = 3/2^+$ resonance in ^5Li, $A_y = 0$, and only tensor analyzing powers $A_{ik;i,k=x,y,z} \neq 0$ may exist.

6.6 Why is a similar behavior true for the $^3\text{He}(\vec{d}, p)^4\text{He}$ reaction at $E_{\text{lab}} = 430$ keV? Which differences between the two mirror reactions would you nevertheless expect because of the different lab. energies?

References

[BAK61] Basel Convention in [BAS61] (1961)
[BAS61] P. Huber, K.P. Meyer (eds.), *Proc. Int. Symp. on Polarization Phenomena of Nucleons*, Basel, 1960. Helv. Phys. Acta Suppl., vol. 6 (Birkhäuser, Basel, 1961)
[BYS78] J. Bystricki, F. Lehar, P. Winternitz, J. Phys. **39**, 1 (1978)
[DAR71] S.E. Darden, in [MAD71] (1971), p. 39

References

[EBE74] D. Fick (ed.), *Proc. Meeting on Polarization Nucl. Phys.*, Ebermannstadt, 1973. Lecture Notes in Phys., vol. 30 (Springer, Berlin, 1974)
[FIC67] D. Fick, Z. Phys. **199**, 309 (1967)
[GWU11] http://gwdac.phys.gwu.edu
[HGS10] H. Paetz gen. Schieck, Eur. Phys. J. A **44**, 321 (2010)
[HGS12] H. Paetz gen. Schieck, *Nuclear Physics with Polarized Particles*. Lecture Notes in Physics, vol. 842 (Springer, Heidelberg, 2012)
[HOF66] H.M. Hofmann, D. Fick, Z. Phys. **194**, 163 (1966)
[KNU72] L.D. Knutson, Nucl. Phys. A **198**, 439 (1972)
[KOE68] W.E. Köhler, D. Fick, Z. Phys. **215**, 408 (1968)
[MAD71] H.H. Barschall, W. Haeberli (eds.), *Proc. 3rd Int. Symp. on Polarization Phenomena in Nucl. Reactions*, Madison, 1970 (University of Wisconsin Press, Madison, 1971)
[MAK71] Madison Convention in [MAD71], xxv (1971)
[NNO13] Theor. High-Energy Phys. Group, Radboud University, Nijmegen, The Netherlands, http://nn-online.org
[OHL72] G.G. Ohlsen, Rep. Prog. Phys. **35**, 717 (1972)
[OHL73] G.G. Ohlsen, P.W. Keaton Jr., Nucl. Instrum. Methods **109**, 41 (1973), and p. 61
[OHL81] G.G. Ohlsen, R.E. Brown, F.D. Correll, R.A. Hardekopf, Nucl. Instrum. Methods **179**, 283 (1981)
[PET67] Cl. Petitjean, P. Huber, H. Paetz gen. Schieck, H.R. Striebel, Helv. Phys. Acta **40**, 401 (1967)
[PRZ06] B.v. Przewoski et al., Phys. Rev. C **74**, 064003 (2006)
[SAID13] SAID Partial-Wave Analysis Facility, Physics Dept., The George Washington University, CNS DAC Services, Center for Nuclear Studies, http://gwdac.phys.gwu.edu
[SEY69] R.G. Seyler, Nucl. Phys. A **124**, 253 (1969)
[SIM74] M. Simonius, in [EBE74]
[SPE83] F. Sperisen, W. Grüebler, V. König, Nucl. Instrum. Methods **204**, 491 (1983)
[SYD93] L. Sydow, S. Vohl, S. Lemaître, P. Niessen, K.R. Nyga, R. Reckenfelderbäumer, G. Rauprich, H. Paetz gen. Schieck, Nucl. Instrum. Methods A **327**, 441 (1993)
[SYD98] L. Sydow, S. Vohl, S. Lemaître, H. Patberg, R. Reckenfelderbäumer, H. Paetz gen. Schieck, W. Glöckle, D. Hüber, H. Witała, Few-Body Syst. **25**, 133 (1998)
[WEL63] T.A. Welton, in *Fast Neutron Physics II*, ed. by J.B. Marion, J.J. Fowler (Interscience, New York, 1963), p. 1317

Chapter 7
Partial Wave Expansion

Especially at low energies the partial-wave expansion of the observables is useful. One advantage is that—since the Legendre functions are eigenfunctions of the angular momentum—the influence of and dependence on different angular momenta in the reaction can be studied. When dealing with the nuclear part of the interaction—due to the short range of nuclear forces—the expansion can be truncated after a few terms; the centrifugal barrier prevents higher angular momenta from contributing. The problem with incident charged particles is that the Coulomb interaction, due to its long range, requires a very large number of partial waves.

7.1 Neutral Particles

The most general expansion for two-particle reactions between neutral particles was published by Welton [WEL63]. It describes the (spherical) tensor moments of the exit channel as function of the tensor moments of the entrance channel, as prepared. It includes the case of unpolarized cross sections (as discussed in Chap. 8) as a special case. By Heiss [HEI72] it was extended to elastic scattering of charged reaction partners and by Hofmann, Aulenkamp, Nyga in addition to the case of identical particles. Here the final result of Welton (without that symmetrization) will be given with the modification that the definition of the tensor moments follows that of Lakin [LAK55] and therefore complies with the *Madison Convention* [MAK71]. In order to avoid confusion with expressions of the R-matrix theory [LAN58], here the R and $\mathcal{R}$ of Welton have been renamed T and $\mathcal{T}$, the tensor moments are designated as introduced in the present text: $t^{kq,KQ}$ instead of $t^{q\gamma,Q\Gamma}$ for the exit channel, $t_{k'q',K'Q'}$ instead of $t_{q'\gamma',Q'\Gamma'}$ for the entrance channel.

$$t^{kq,KQ} = (2k_{\text{in}})^{-2}(\hat{t}\hat{I})^{1/2}$$

$$\times \sum \begin{Bmatrix} i & I & s_1 \\ k & K & t \\ i & I & s_2 \end{Bmatrix} \begin{Bmatrix} i' & I' & s'_1 \\ k' & K' & t' \\ i' & I' & s'_2 \end{Bmatrix} \begin{Bmatrix} l_1 & s_1 & J_1 \\ l & t & L \\ l_2 & s_2 & J_2 \end{Bmatrix} \begin{Bmatrix} l'_1 & s'_1 & J_1 \\ l' & t' & L \\ l'_2 & s'_2 & J_2 \end{Bmatrix}$$

$$\times \langle l_1 l_2 00 | l0 \rangle \langle l'_1 l'_2 00 | l'0 \rangle \langle lt 0 \Lambda | L \Lambda \rangle$$

$$\times \langle l't'0\Lambda' | L\Lambda' \rangle \langle kKqQ|t\Lambda \rangle \langle k'K'q'Q'|t'\Lambda' \rangle$$

$$\times T^{J_1^{\pi_1}} T^{J_2^{\pi_2}*} D^L_{\Lambda'\Lambda}(\phi,\theta,0)$$

$$\times (\hat{i}'\hat{I}')^{-1/2}$$

$$\times t_{k'q',K'Q'}, \tag{7.1}$$

where the meaning of the different symbols is as follows:

- Primed quantities: entrance channel, unprimed ones: exit channel.
- Alternatives in each channel are distinguished by 1 and 2.
- Particle spins: i, I, i', I'.
- Channel spins: s, s'.
- Orbital angular momenta: ℓ, ℓ', total angular momentum: J (the only conserved angular momentum).
- Rank and component of the tensor moments: k, q and K, Q.
- Sums are over all indices except $k, K, q, Q; I, i, I', \iota'$.
- $\hat{I}$ means $2I + 1$.
- Symbols in wavy brackets are the 9j symbols (see [BRI71] and Sect. 22.3).
- Symbols like $\langle l_1 l_2 00 | l0 \rangle$ are the Clebsch-Gordan coefficients (see [BRI71] and Sect. 22.3).
- The matrix elements $T^{J_1^{\pi_1}}$ and $T^{J_2^{\pi_2}}$, resp., are defined in a representation with the asymptotically good quantum numbers as

$$T^{J_1^{\pi_1}} = \langle \alpha' \ell' s' | \mathcal{T} | \alpha \ell s \rangle, \tag{7.2}$$

where $\mathcal{T} = \mathcal{S} - 1$ defined in spin space with $\mathcal{S}$ being the usual S-matrix.

Already from the abbreviated form

$$t \propto \sum_{1,2} B(1,2) T_1 T_2^* D^L_{\Lambda\Lambda'} \cdot t', \tag{7.3}$$

some general conclusions can be derived:

The $B(1,2)T_1 T_2^* D^L_{\Lambda\Lambda'}$ are components of the generalized analyzing powers $T^{kq,KQ}_{k'q',K'Q'}$. By interchanging indices $1 \leftrightarrow 2$ one finds that $B(2,1) = (-)^{k+K+k'+K'}$. $B(2,1)$ and

$$t \propto \sum_{1,2} \frac{1}{2}[T_1 T_2^* B(1,2) + T_1^* T_2 B(2,1)] D^L_{\Lambda'\Lambda} \cdot t'$$

$$= \sum_{1,2} 1/2 [T_1 T_2^* B(1,2) + (-)^{k+K+k'+K'} T_1^* T_2 B(1,2)] D^L_{\Lambda\Lambda'} \cdot t'. \tag{7.4}$$

7.1 Neutral Particles

For

$$k + K + k' + K' = \begin{Bmatrix} \text{even} \\ \text{odd} \end{Bmatrix} \quad \text{only} \quad \begin{Bmatrix} \text{Re} \\ \text{Im} \end{Bmatrix} (T_1 T_2^*) \tag{7.5}$$

will appear. If e.g. only the incident *beam* is polarized and no outgoing polarization is measured ($k' = 1, k = K = K' = 0$), the analyzing power is

$$A_y \propto i T_{11} \propto \text{Im}\left(T_1 T_2^*\right). \tag{7.6}$$

Thus polarization effects of odd rank (e.g. the vector analyzing power or the vector polarization) vanish if

- the matrix elements are purely real. This will be the case e.g. in Born approximation with a real potential;
- only a single matrix element contributes: $T_1 T_1^* - T_1^* T_1 = 0$. This is the case for an isolated resonance with only one value of the orbital angular momentum (if no tensor force couples angular momenta of equal parity) and with no direct background contribution. An example is the $3/2^+$ resonance of the $^3\text{H}(d, n)^4\text{He}$ reaction at $E_d = 107$ keV;
- all matrix elements have the same phase: with $T_1 = r_1 e^{i\phi}, T_2 = r_2 e^{i\phi}$: $T_1 T_2^* = r_1 r_2 = $ real;
- only one value of ℓ exists and is zero;
- only one intermediate state with $J_1 = J_2 = 0$ or $1/2$ exists. In the last two cases the angular distribution of the unpolarized cross section is isotropic: from $\ell'_1 = \ell'_2 = \ell = \ell + t(=0) = L = 0 \to \sigma_0$ is isotropic;
- there is no interaction distinguishing (for one ℓ) between the two possible different values of J. A vector $\vec{\ell} \cdot \vec{s}$ force is e.g. necessary for producing vector polarization or analyzing power, resp., otherwise the above condition $T_1 \neq T_2$ is not fulfilled.

Some additional conclusions can be drawn:

- Parity conservation reduces the number of possible tensor moments.
  Example: The outgoing tensor moment $t_{00} = 0$ with incident tensor moment t'_{10}, polarized in the z direction, and therefore the (longitudinal) analyzing power of a parity conserving reaction $A_z \propto T_{10} = 0$. This is due to the one property of the CG coefficient $\langle \ell'_1 \ell'_2 00 | \ell' 0 \rangle$, which is only $\neq 0$ if $\ell'_1 + \ell'_2 + \ell'$ is even, and analogously for $\langle \ell_1 \ell_2 00 | \ell 0 \rangle$. Parity conservation requires that $\ell'_1 + \ell'_2 + \ell_1 + \ell_2$ be even, resulting in $L = \ell$ and $\ell + \ell'$ = even. However, with $k' = 1, q' = 0, k = q = 0$ and therefore $t' = 1, L = \ell' + t'$ we have $\ell = \ell'_1 \pm 1$ and $\ell + \ell' = 2\ell' \pm 1 = $ odd in contradiction to the above.
- The complexity (defined as the maximum possible order L of the functions $D^L_{\Lambda'\Lambda}$ or Y^Λ_L or P^Λ_L) of angular distributions can be obtained as

$$-L_{\max} \leq J_1 + J_2$$
$$-L_{\max} \leq \ell_1 + \ell_2 + k + K$$
$$-L_{\max} \leq \ell'_1 + \ell'_2 + k' + K'$$

- It is evident that for the unpolarized cross section $t_{00,00}$ $\Lambda = \Lambda' = 0$, i.e. there is no ϕ dependence and the D_{00} are reduced to simple Legendre Polynomials $P_L(\cos\theta)$, a few of which are depicted in Fig. 4.11.

Especially for s waves ($\ell_1 = \ell_2 = L = 0$) the angular distribution becomes isotropic. This also holds for $J_1 = J_2 = 0$ or $= 1/2$. Inspection of the relevant 9j symbol shows that, as in the case of $J_1 = J_2 = 1/2$, $\ell = L = 0$ is the maximum possible value:

$$\begin{Bmatrix} \ell_1 & s_1 & J_1 \\ \ell & t & L \\ \ell_2 & s_2 & J_2 \end{Bmatrix} = \begin{Bmatrix} \ell_1 & s_1 & 0 \\ \ell & 0 & \ell \\ \ell_2 & s_2 & 0 \end{Bmatrix}. \tag{7.7}$$

7.2 Charged Particles

The case including the Coulomb interaction in elastic scattering has been treated by Heiss [HEI72] in such a way that instead of one expression for an observable there are now three: the pure nuclear-interaction term, the pure Rutherford term, and an interference term between both. This last one is—due to the long range of the Coulomb force—the one, which may cause problems when truncating higher partial waves where they should be included up to very high ℓ values (corresponding to a large screening (or cut-off) radius for the interactions) while for the pure nuclear term very few low-ℓ partial waves may suffice. The pure Coulomb term is written down in closed form—it is just the Rutherford scattering, at least when dealing with the monopole term of the Coulomb force, i.e. between point charges.

The most general equation relating outgoing tensor moments with incident ones for the scattering of charged particles thus has three parts:

$$\begin{aligned} t^{kq,KQ} = (2k_{\text{in}})^{-2} &\Big\{ 4\pi \delta_{\alpha\alpha'}\delta_{iI,i'I'}|C_\alpha(\theta)|^2 \sum B_1\big(kqKQ; k'q'K'Q'; L\Lambda\Lambda'\big) \\ &+ (4\pi)^{1/2} \delta_{\alpha\alpha'}\delta_{iI,i'I'} \sum B_2\big(\ell s_2 kqKQ; \ell' s_2' k'q'K'Q'; L\Lambda\Lambda'; I\big) \\ &\times \big[iC(\theta)T^* + (-)^{k+k'+K+K'}(iC(\theta)T^*)^*\big] \\ &+ \frac{1}{2} \sum B_4\big(\ell_1 s_1 \ell_2 s_2 kqKQ; \ell_1' s_1' \ell_2' s_2' k'q'K'Q'; L\Lambda\Lambda'; J_1 J_2\big) \\ &\times \big[(T_1 T_2^*) + (-)^{k+k'+K+K'}(T_1 T_2^*)^*\big] \Big\} D^L_{\Lambda\Lambda'}(\phi,\theta,0) t_{k'q',K'Q'}. \end{aligned} \tag{7.8}$$

The sums run over all arguments of the B coefficients; the B coefficients are defined as the Rutherford term

$$B_1 = \delta_{kk'}\delta_{KK'}\langle kKq'Q'|L\Lambda\rangle\langle kKqQ|L\Lambda'\rangle, \tag{7.9}$$

the interference term

$$B_2 = (-)^{s_2+s_2'-2I}\hat{I}\hat{k}'\hat{K}'(\hat{\ell}\hat{\ell}'\hat{s_2}\hat{s_2'})^{1/2}$$

$$\times \sum_{tt's}(\hat{t}\hat{t}')^{1/2}\begin{Bmatrix} i & I & s \\ k & K & t \\ i & I & s_2 \end{Bmatrix}\begin{Bmatrix} i & I & s \\ k' & K' & t' \\ i & I & s' \end{Bmatrix} W(stI\ell; s_2 L)W(st'I\ell'; s_2'L)$$

$$\times \langle t\ell \Lambda' 0|L\Lambda'\rangle\langle t'l't'\Lambda 0|L\Lambda\rangle\langle kKqQ|t\Lambda'\rangle\langle k'K'q'Q'|t'\Lambda\rangle, \tag{7.10}$$

and the pure nuclear term (the Welton formula)

$$B_4 = \left(\frac{\hat{i}\hat{I}}{\hat{k}\hat{K}}\right)^{1/2}\left(\frac{\hat{i'}\hat{I'}}{\hat{k'}\hat{K'}}\right)^{1/2} F(\ell_1\ell_2s_1s_2J_1J_2L\Lambda'; kKqQ)$$

$$\times F(\ell_1'\ell_2's_1's_2'J_1J_2L\Lambda; k'K'q'Q') \tag{7.11}$$

with

$$F(\ell_1\ell_2s_1s_2J_1J_2L\Lambda; kKqQ) = (\hat{\ell}_1\hat{\ell}_2\hat{s}_1\hat{s}_2\hat{J}_1\hat{J}_2\hat{k}\hat{K})^{1/2}(-)^{\ell_1+L}$$

$$\times \sum_{\ell t}\langle l_1l_200|\ell 0\rangle\langle lt0\Lambda|L\Lambda\rangle\langle kKqQ|t\Lambda\rangle$$

$$\times \begin{Bmatrix} i & I & s_1 \\ k & K & t \\ i & I & s_2 \end{Bmatrix}\begin{Bmatrix} \ell_1 & s_1 & J_1 \\ \ell & t & L \\ \ell_2 & s_2 & J_2 \end{Bmatrix}. \tag{7.12}$$

The quantities in these equations are as in Eq. (7.1). In addition,

$$C(\theta) = (4\pi)^{-1/2}\eta_S \csc^2\left(\frac{\theta}{2}\right)\exp\left\{-2i\eta_S \ln\left[\sin\left(\frac{\theta}{2}\right)\right]\right\} \tag{7.13}$$

is the Coulomb (Rutherford) amplitude describing the long-range part of the interaction, and the W coefficients are Racah coefficients equivalent to 6j symbols (see [BRI71] and Sect. 22.3). The formalism of Welton/Heiss has been transformed repeatedly into computer programs to compute the relations between incoming and outgoing spin-tensor moments as functions of the T-matrix elements as well as for fitting these tensor moments to observables. Details can be found in Ref. [HGS12].

7.3 Exercises

7.1. Verify that for a reaction with neutral unpolarized particles in the entrance channel the unpolarized differential cross section has the form given in Chap. 8 and discuss its angular complexity.
7.2. Do the same for the vector analyzing power of a reaction with Spin-1/2 projectiles on a spin-zero target.
7.3. What will be the modifications with identical particles in the entrance channel?

7.4. (a) Rework the conclusions between Eqs. (7.5) and (7.7), drawn from the general formula Eq. (7.1).
  (b) Apply this to case of one (dominant) s-wave matrix element plus one (weaker) p-wave element. Show how a polarization observable such as the vector analyzing power (which vanishes for pure s-waves) may be more sensitive to smaller contributions in the presence of a strong one than the differential cross section.

References

[BRI71] D.M. Brink, G.R. Satchler, *Angular Momentum* (Clarendon Press, Oxford, 1971)
[HEI72] P. Heiss, Z. Phys. **251**, 159 (1972)
[HGS12] H. Paetz gen. Schieck, *Nuclear Physics with Polarized Particles*. Lecture Notes in Physics, vol. 842 (Springer, Heidelberg, 2012)
[LAK55] W. Lakin, Phys. Rev. **98**, 139 (1958)
[LAN58] A.M. Lane, R.G. Thomas, Rev. Mod. Phys. **30**, 145 (1958)
[MAD71] H.H. Barschall, W. Haeberli (eds.), *Proc. 3rd Int. Symp. on Polarization Phenomena in Nucl. Reactions*, Madison, 1970 (University of Wisconsin Press, Madison, 1971)
[MAK71] Madison Convention in [MAD71], xxv (1971)
[WEL63] T.A. Welton, in *Fast Neutron Physics II*, ed. by J.B. Marion, J.J. Fowler (Interscience, New York, 1963), p. 1317

Chapter 8
Unpolarized Cross Sections

8.1 General Features

In the following some formal features of unpolarized cross sections will be discussed (for more reading see e.g. Refs. [NEW66, GOL64, ROD67, MOT65, JOA83, SEG83] and many other common textbooks cited in Chaps. 1 and 4). In the simplest case we deal with spinless particles and purely elastic scattering. Modifications arise in the following cases:

- If inelastic and other absorption channels are open the differential cross section contains a term $e^{2i\eta_\ell} - 1$ instead of $e^{2i\delta_\ell} - 1$ (the phase becomes complex, see Sect. 8.2). For heavy-ion reactions with fusion and other exit channels this expression is just a transmission coefficient $T_{\ell SJ}$, that may be calculated in the framework of the optical model or simplified by sharp-cutoff models, see Chap. 13 and Sect. 11.6.2.
- If the spins of the projectile (I_1) and target (I_2) nuclei are $\neq 0$, the particles are unpolarized, and no polarization is measured, then $2\ell + 1$ has to be replaced by

$$\frac{(2J+1)}{(2I_1+1)(2I_2+1)} \tag{8.1}$$

  and δ_ℓ by $\delta_{\ell SJ}$, likewise also η_ℓ. S is the channel spin, coupled from I_1 and I_2 and has the values $|I_2 - I_1| \leq S \leq I_2 + I_1$. The cross section terms have to be summed over ℓ, S, and J where J is the total angular momentum coupled from S and ℓ. The formula (8.1) is derived by counting the relative number of spin channels contributing to the number of total angular-momentum states.
- In the case of identical particles exchange symmetry (see Sect. 3.4) requires that the wave functions must be properly (anti-)symmetrized. With $I_1 = I_2 = I$ and the *channel spin* S we have $0 \leq S \leq 2I$ and the integrated cross section is

$$\sigma_J = \frac{4\pi}{k^2} \frac{(2J+1)}{(2I+1)^2} \sum_{\ell,J} [1 + (-1)^{\ell+S}] T_{\ell SJ}. \tag{8.2}$$

- The general case of spin-polarized particles is treated in detail in Chaps. 5 and 6.

In this chapter we return to the simplest case. In the framework of partial-wave expansions the differential cross section reads

$$\frac{d\sigma}{d\Omega} = |f|^2 = ff^* = \frac{1}{k^2}\left|\sum_{\ell=0}^{\infty}(2\ell+1)\left(e^{2i\delta_\ell}-1\right)P_\ell(\cos\theta)\right|^2$$

$$= \frac{1}{k^2}\sum_{\ell=0}^{\infty}(2\ell+1)\sum_{\ell'=0}^{\infty}(2\ell+1)(2\ell'+1)e^{i[\delta_\ell(k)-\delta_{\ell'}(k)]}\sin\delta_\ell\sin\delta_{\ell'}P_\ell P_{\ell'}$$

$$= \frac{1}{k^2}\sum_{L=|\ell-\ell'|}^{\ell+\ell'}\sum_{\ell'=0}^{\infty}\sum_{\ell=0}^{\infty}(2\ell+1)(2\ell'+1)\langle\ell\ell'00|L0\rangle^2 e^{i[\delta_\ell(k)-\delta_{\ell'}(k)]}$$

$$\times \sin\delta_\ell\sin\delta_{\ell'}P_L(\cos\theta). \tag{8.3}$$

Here the relation

$$P_\ell P_{\ell'} = \sum_{L=|\ell-\ell'|}^{\ell+\ell'}\langle\ell\ell'00|L0\rangle^2 P_L(\cos\theta) \tag{8.4}$$

was used.

From the differential cross section the *integrated cross section* is obtained by integration over the entire solid angle. Assuming ϕ independence (e.g. when no spin polarization is involved) this results in

$$\sigma_{\text{int}} = 2\pi\int\frac{d\sigma}{d\Omega}\sin\theta d\theta$$

$$= \frac{4\pi}{k^2}\sum_{\ell=0}^{\infty}(2\ell+1)\sin^2\delta_\ell = \sum_{\ell}\sigma_\ell \tag{8.5}$$

after using the relation $\int_{-1}^{+1}P_\ell(\cos\theta)P_{\ell'}(\cos\theta)d(\cos\theta) = \frac{2}{2\ell+1}\delta_{\ell\ell'}$. The integrated cross section is thus an incoherent sum over partial cross sections belonging to ℓ, and they show a weighting factor of $2\ell+1$. The maximal integrated partial cross section such as in the maxima of resonances, for which $\delta_\ell(k) = (n+1/2)\pi$; $n = 0, \pm 1, \pm 2, \ldots$ increases linearly with ℓ

$$\sigma_{\ell,\max}(k) = \frac{4\pi}{k^2}(2\ell+1). \tag{8.6}$$

On the other hand the contribution of higher ℓ-values (at least for short-range potentials) is limited by the effects of the centrifugal barrier. Semi-classically, for a "range" d of the potential only partial waves can contribute, for which

$$\ell \leq kd. \tag{8.7}$$

8.2 Inelasticity and Absorption

Fig. 8.1 Increase of the integrated cross section for neutrons with ℓ for different reaction mechanisms. ℓ—in a semi-classical picture—may be assigned to increasing interaction distances

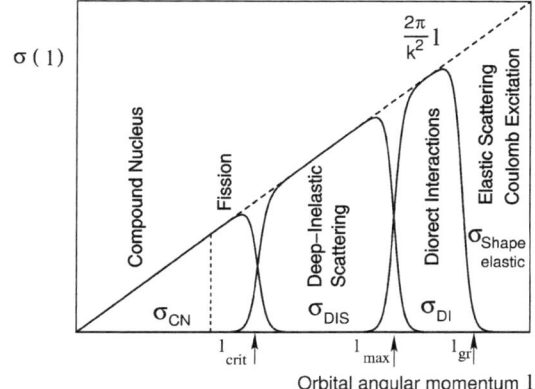

In a quantum-mechanical picture this follows from the behavior of the Bessel functions, see Fig. 10.7. The first maximum of

$$j_\ell(kr) \text{ is located at } r_0 \approx \ell/k. \tag{8.8}$$

For small r

$$j_\ell(kr) \propto r^\ell \quad \text{i.e. for } d \leq \ell/k, \, j_\ell(kr) \text{ and } \delta_\ell \tag{8.9}$$

are small. This upper limit for $\ell_{\max}$, often: ℓ_{gr} (gr for "grazing"), is only reached in reactions occurring at the potential surface i.e. in *direct* processes. In *compound nuclear* processes, which proceed rather in the nuclear center the upper limit for ℓ is lower (ℓ_{crit}). For *dissipative* processes like "incomplete fusion" or "deep-inelastic scattering" $\ell_{\max}$ is between ℓ_{crit} and ℓ_{gr}. This behavior as function of ℓ is depicted in Fig. 8.1. Such considerations apply especially for heavy-ion reactions with strong absorption as long as the wave aspects are insignificant, see e.g. [NOE76].

For reactions that pass through a definite intermediate state (e.g. an isolated resonance in the interaction of particles without spins) a special conclusion can be drawn. The CG-Koeffizient $\langle \ell\ell'00|L0\rangle$ (see Sect. 22.3) has the property of being $\neq 0$ only when $\ell + \ell' + L$ is even. This means here: Because $\ell = \ell'$ is L even and because L can only assume even values from $|\ell - \ell'|\ldots\ell + \ell'$ in the Legendre expansion with $L_{\max} = 2\ell$, the angular distribution contains only even Legendre polynomials, which are symmetric around $\pi/2$. $\ell = J$ of the intermediate state is given by half of the "complexity value" L of the angular distribution. This helps in the determination of the total spin J of isolated resonances in nuclear and particle physics (example: the Δ resonance, see Sect. 11.5.2 and Chap. 5).

8.2 Inelasticity and Absorption

Inelastic processes are all those processes that are non-elastic, i.e. those, which take particle flux out of the direct elastic channel. Among these are all *reactions* in the narrower sense (thus without direct elastic scattering, shape-elastic scattering), but also compound-elastic scattering, which goes via intermediate states, and also induced nuclear fission and—at higher energies—the production of particles above the corresponding thresholds. For these processes a few general statements may be made without having to consider details of the reaction mechanism.

The flux of the incoming particles (entrance channel α) is normally distributed among several exit channels (β). We shall use the usual notation for reactions $\eta_{\ell,\beta} \equiv \eta_\alpha^\ell$, for purely elastic scattering $\eta_\ell \equiv \eta_{\ell,\alpha}$, where η is the scattering function defined above, and S is the equivalent S-matrix. The S-matrix has a few properties that can be useful here. E.g. the following facts apply

- Unitarity: $\sum_\beta |S_{\beta,\alpha}^\ell|^2 = 1$
- Reciprocity (under time-reversal invariance): $S_{\alpha,\beta}^\ell = S_{\beta,\alpha}^\ell$.

Applied to η this implies:

$$\sum_\beta |\eta_{\ell,\beta}|^2 = 1, \tag{8.10}$$

where β runs over all open channels including the elastic (entrance) channel α. With this the *absorption cross section* σ_{abs} can be defined as sum over all non-elastic cross sections:

$$\sigma_{\text{abs}} = \sum_{\beta \neq \alpha} \sigma_\beta = \frac{4\pi}{k^2} \sum_\ell (2\ell+1) \sum_{\beta \neq \alpha} |\eta_{\ell,\beta}|^2$$

$$= \frac{4\pi}{k^2} \sum_{\ell=0}^{\infty} (2\ell+1)\bigl[1 - |\eta_{\ell,\alpha}|^2\bigr]. \tag{8.11}$$

Simplified this reads

$$\sigma_{\text{abs}} = \frac{\pi}{k^2} \sum_\ell (2\ell+1)\bigl[1 - |\eta_\ell|^2\bigr]. \tag{8.12}$$

On the other hand the elastic scattering cross section is (after extending the form derived above for one channel to the general case of a scattering function for several channels):

$$\sigma_{\text{el}} = \frac{4\pi}{k^2} \sum_\ell (2\ell+1)|\delta_{\alpha,\beta} - \eta_{\ell,\beta}|^2, \tag{8.13}$$

which again can be simplified to:

$$\sigma_{\text{el}} = \frac{\pi}{k^2} \sum_{\ell=0}^{\infty} (2\ell+1)|1 - \eta_\ell|^2. \tag{8.14}$$

8.2 Inelasticity and Absorption

Fig. 8.2 Relation between elastic scattering and absorption (for uncharged particles and valid for each ℓ separately). Plotted is
$\alpha = \frac{\sigma_{el,\ell}}{(2\ell+1)\pi/k^2} = |1 - \eta_\ell|^2$
against
$\beta = \frac{\sigma_{r,\ell}}{(2\ell+1)\pi/k^2} = 1 - |\eta_\ell|^2$.
The possible range is given by
$\alpha = 2[1 - \sqrt{1-\beta}\cos(2\mathrm{Re}\delta_\ell)] - \beta$,
the limiting curve by the values $\mathrm{Re}\,\delta_\ell = 0, \pi$

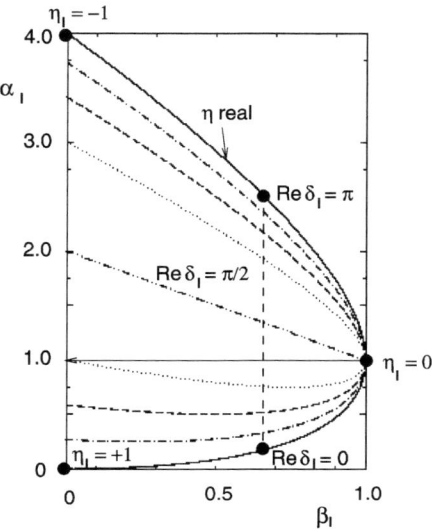

For the sum, the *total* (integrated) cross section, one obtains:

$$\sigma_{\mathrm{tot}} = \frac{2\pi}{k^2}\sum_{\ell=0}^{\infty}(2\ell+1)\big[1 - |\eta_\ell|\cos(2\mathrm{Re}\delta_\ell)\big]. \tag{8.15}$$

The relations between the elastic and the absorption cross section may be plotted graphically, see Fig. 8.2 and show e.g. that there may be elastic scattering without absorption (thus for $|\eta_\ell| = 1$) (below the corresponding threshold energies), but that absorption is always accompanied by elastic scattering. It is plausible that elastic scattering amplitude and thus the cross section reach their maximum values, which are given by the unitarity of the S-matrix of $\frac{4\pi}{k^2}(2\ell+1)$ only for vanishing absorption. The maximum value of the absorption cross section $\frac{\pi}{k^2}(2\ell+1)$ is obtained for $\eta_\ell = 0$ such that also for the total cross section the upper limit is $\frac{4\pi}{k^2}(2\ell+1)$.

Formally absorption is described by assuming the potential causing scattering to be complex:

$$V = U + iW. \tag{8.16}$$

This Ansatz is the basis of the *optical model* of elastic scattering and of the *optical potential* (for more detail see Chap. 10.3). The imaginary part W describes the attenuation of the incident particle flux by absorption into channels other than the elastic channel. This absorption can be understood in the following way: after the Schrödinger equation is set up with a complex potential, the complex-conjugate equation is formed, then both equations are multiplied with Ψ^* and Ψ, resp. Then, subtracting both equations from one another a continuity equation is produced for a stationary state, i.e. with $d\rho/dt = 0$

$$\nabla^2 \Psi + 2\mu/\hbar^2 [E - (U + iW)]\Psi = 0 \quad | \cdot \Psi^*,$$
$$\nabla^2 \Psi^* + 2\mu/\hbar^2 [E - (U + iW)]\Psi^* = 0 \quad | \cdot \Psi, \quad (8.17)$$

$$\underbrace{\Psi^* \nabla^2 \Psi - \Psi \nabla^2 \Psi^*}_{\nabla \vec{j}} = \underbrace{\frac{2i\mu W}{\hbar^2} \Psi \Psi^*}_{-k v \rho}. \quad (8.18)$$

The r.h.s. term characterizes the particle sink with $W = \frac{1}{2} v \hbar k = \frac{1}{2} \frac{\hbar v}{\lambda}$ and $W < 0$. The wave number is also complex and can be written

$$k = \frac{2\mu}{\hbar^2}[E - (U + iW)]^{1/2} = \sqrt{\frac{2\mu}{\hbar^2}(E - U)} \sqrt{1 - i\frac{W}{(E - U)}} = k_0 + \frac{i}{\lambda}. \quad (8.19)$$

For a plane wave this means e.g.:

$$e^{ikz} = e^{ik_0 z} \cdot e^{-z/\lambda} = e^{ik_0 z} \cdot e^{-z \frac{2W}{\hbar v}} \quad (8.20)$$

i.e. an amplitude attenuation with a *mean free path* $\lambda = \hbar v / 2W$.

Formally one may also identify the complex wave number with a complex scattering phase shift

$$\text{Im}(\delta) = \text{Im}[(k_{\text{pot}} - k_{\text{free}})d], \quad (8.21)$$

i.e. the imaginary part of the phase shift also describes the absorption. The scattering function has the following form:

$$\eta_\ell = e^{2i(\text{Re}(\delta_\ell) + i\text{Im}(\delta_\ell))} = e^{-2\text{Im}(\delta_\ell)} e^{2i\text{Re}(\delta_\ell)} = |\eta_\ell| e^{2i\text{Re}(\delta_\ell)} = \rho_\ell e^{2i\text{Re}(\delta_\ell)}. \quad (8.22)$$

The absolute value of the scattering amplitude is reduced by the attenuation term $|\eta_\ell|$. In a phase-shift analysis therefore e.g. above the particle-production threshold in the NN interaction the phases always have to be complex, which e.g. doubles the number of fitting parameters.

8.3 Low-Energy Behavior of the Scattering

The low-energy behavior of the scattering (i.e. for $k \to 0$) is characterized by the dominance by s-wave scattering. The following discussion thus will be limited to scattering with $\ell = 0$ and of neutral projectiles.

8.3 Low-Energy Behavior of the Scattering

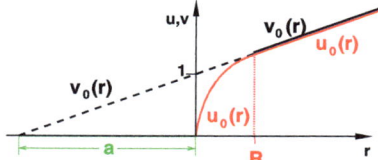

Fig. 8.3 The geometrical interpretation of the scattering length as the intersection distance on the kr axis of the asymptotic wave function. It is obvious that at or near a resonance when the resonance matching condition requires a horizontal tangent at $r = 0$ the scattering length assumes large values. Thus the large negative np scattering length a_{np} is indicative of the slightly unbound singlet deuteron state

8.3.1 Scattering Length a

For $k \to 0$ the asymptotic wave function in the extra-nuclear space converges against a linear function of k:

$$\sin(kr) \to kr. \tag{8.23}$$

By matching this function at the nuclear surface $r4 = R$ continuously to the wave function in the nuclear interior and normalizing it such that it has the value 1 for $r = 0$ one obtains a straight line, which intersects the kr axis at $r = a$. Figure 8.3 shows this geometrical interpretation of the scattering length. On the one hand

$$v_0 = \frac{\sin(kr + \delta_0)}{\sin \delta_0} = 1 + kr \cot \delta_0, \tag{8.24}$$

on the other—with appropriate normalization

$$v_0 = \text{const}(r - a) = -1/a(r - a) = 1 - (r/a). \tag{8.25}$$

The comparison of both forms of the same quantity yields $-(r/a) = \lim_{k \to 0}(kr \cot \delta_0)$ and thus the definition of the *scattering length*:

$$a = -\lim_{k \to 0} \frac{\tan \delta_0(k)}{k}. \tag{8.26}$$

With it the s-wave scattering amplitude can be expressed:

$$\begin{aligned} f_0 &= \frac{1}{k} e^{i\delta_0} \sin \delta_0 = \frac{1}{k} \frac{\sin \delta_0}{e^{-i\delta_0}} \\ &= \frac{1}{k} \frac{\sin \delta_0}{\cos \delta_0 - i \sin \delta_0} = \frac{1}{k(\cot \delta_0 - i)} \\ &= \frac{1}{(\frac{k}{\tan \delta_0}) - ika}, \end{aligned} \tag{8.27}$$

from which follows:

$$\lim_{k \to 0} f_0 = -a \quad \text{and} \quad \lim_{k \to 0} \sigma = 4\pi a^2. \tag{8.28}$$

The absolute value of the scattering length can be evaluated by the low-energy limit of the total cross section, however not its sign. For charged particles the sign may be obtained from the sign of the interference with the Coulomb amplitude or from a complete scattering-phaseshift analysis. A refinement of the scattering length concept, provided by an expansion in an energy region above 0, is the effective range that connects the s-wave phaseshift with a and r_{eff}

$$k \cot \delta_0 = -\frac{1}{a} + \frac{1}{2} r_{\text{eff}} k^2 \tag{8.29}$$

and allows additional comparisons with NN predictions.

The geometrical interpretation of the scattering length $r = a$ by the intersection of the wave function on the r axis shows that it is a quantity especially sensitive to the behavior of the wave function at the nuclear surface i.e. where the logarithmic derivatives of the wave functions of the interior and external spaces are matched. If the wave function has a horizontal tangent there—this is the case for a state, which at energy zero just becomes bound—a increases $a \to \infty$. Very small differences in the potential may effectuate very large changes of a; this is a *magnifying-glass or lever effect*. The measured charge-independence (isospin) breaking on the order-of-magnitude of $\Delta V/V \approx 1$ % is responsible for the differences of the NN $^1S_0$ scattering lengths

-
  $a_{nn} = (-18.7 \pm 0.6)$ fm (for discrepancies see below)

  a_{pp} (Coulomb corrected) $= (-17.1 \pm 0.2)$ fm

  $a_{np} = (-23.715 \pm 0.015)$ fm

  A significant difference between a_{nn} and a_{pp} would mean a (small) charge-symmetry breaking, see Sect. 3.3.3.

- There is an unresolved discrepancy in a_{nn} between $2N$ and $3N$ methods (an argument for three-nucleon forces):
  - From D$(n, nn)p$ with (-16.73 ± 0.47) fm and
  - from D$(\pi^-, \gamma)2n$ with (-18.59 ± 0.40) fm.

- Another open discrepancy between results with similar methods (for a discussion see e.g. [SLA07]:
  - $a_{nn} = (-16.27 \pm 0.4)$ fm [VWI06, HUH00] vs.
  - (-18.7 ± 0.6) fm [GON06].

- The scattering length for the $^3S_1$ NN system (only for np with $T = 0$) is $a_{np} = (+5.423 \pm 0.005)$ fm, indicating a strong attraction (i.e. one bound state, the deuteron).

Fig. 8.4 Hard-sphere scattering phase shifts (in deg) for $\ell = 0, 1, 2, 3,$ and 4

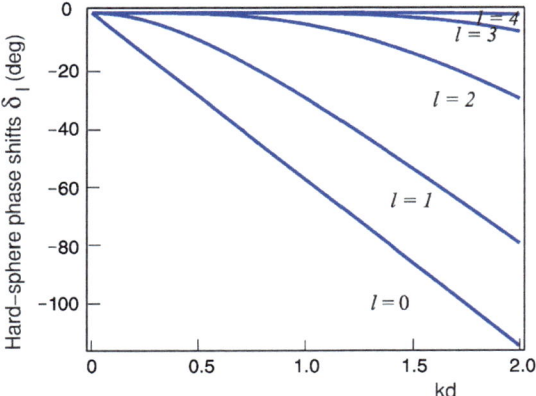

8.3.2 Analytically Solvable Models for the Low-Energy Behavior

Example: Scattering of Neutrons from a Hard Sphere Hard-sphere scattering is scattering from a completely reflecting nucleus. There is no absorption and no penetration of the wave function into the infinitely high potential region at $r \leq d$. Thus, the wave function can only have a node at $r = d$. In the exterior region the wave function must fulfill the condition:

$$\tan \delta_\ell = -\frac{j_\ell(kd)}{n_\ell(kd)}, \tag{8.30}$$

which leads to:

$$\sin^2 \delta_\ell = \frac{\tan^2 \delta_\ell}{1 + \tan^2 \delta_\ell} = \frac{1}{j_\ell^2(kd) + n_\ell^2(kd)} j_\ell^2(kd) = P_\ell(kd) \cdot j_\ell^2(kd). \tag{8.31}$$

From the asymptotic form of $j_\ell(kr)$ and $n_\ell(kr)$ for $k \to \infty$ the behavior of the scattering phases for large kd is derived:

$$\delta_\ell \xrightarrow[kd \gg 1]{} -(kd - \ell\pi/2) \tag{8.32}$$

and from this the total scattering cross section

$$\sigma_\ell = 4\pi d^2 (2\ell + 1) \frac{1}{(kd)^2} P_\ell(kd) j_\ell^2(kd) = \frac{4\pi}{k^2} (2\ell + 1) P_\ell(kd) j_\ell^2(kd). \tag{8.33}$$

Figure 8.4 shows the hard-sphere scattering phase shifts and demonstrates that a repulsive potential is characterized by (increasingly) negative scattering phases.

For pure s waves ($\ell = 0$)

$$\tan \delta_0 = -\frac{\sin(kd)}{\cos(kd)} = -\tan(kd), \tag{8.34}$$

$$\delta_0 = -kd, \tag{8.35}$$

and

$$a = d. \tag{8.36}$$

For this special case the scattering length is equal to the potential radius and

$$\sigma_0 = 4\pi d^2 = 4 \cdot \sigma_{\text{class}}. \tag{8.37}$$

Example: Scattering of Neutrons from a Rectangular Potential Well The potential is given by:

$$U(r) = \begin{cases} -U_0 (U_0 > 0) & r < d, \\ 0 & r > d. \end{cases} \tag{8.38}$$

The radial equation

$$\left[\frac{d^2}{dr^2} + \kappa^2 - \ell(\ell+1)/r^2\right] u_\ell(r) = 0 \tag{8.39}$$

with

$$\kappa = \sqrt{k^2 + U_0} \tag{8.40}$$

(wave number in the potential well) has the regular solution in the range of the potential $r \le d$

$$u_\ell(r) = r C_\ell j_\ell(\kappa r) \tag{8.41}$$

or

$$R_\ell(r) = C_\ell j_\ell(\kappa r), \tag{8.42}$$

in the external region $r \ge d$

$$u_\ell(r) = C_\ell j_\ell(kr + \delta_\ell) = r A_\ell \left[j_\ell(kr) - \tan \delta_\ell n_\ell(kr)\right] = r R_\ell(kr). \tag{8.43}$$

The matching of R_ℓ and R'_ℓ at the edge of the potential $r = d$ is done such that the logarithmic derivatives

$$L_\ell = \left(\frac{dR_\ell}{dr} \frac{1}{R_\ell}\right)_{r=d} \tag{8.44}$$

of the internal and the external solutions are equated:

$$\kappa \cdot \frac{j'_\ell(\kappa d)}{j_\ell(\kappa d)} = \frac{j'_\ell(kd) - \tan \delta_\ell n'_\ell(kd)}{j_\ell(kd) - \tan \delta_\ell n_\ell(kd)}. \tag{8.45}$$

When restricting to s waves, with $j_0(x) = \sin x/x$ and $n_0(x) = -\cos x/x$

$$\tan \delta_0 = \frac{k \tan(\kappa d) - \kappa \tan(kd)}{\kappa + k \tan(kd) \tan(\kappa d)} \tag{8.46}$$

8.4 Exercises

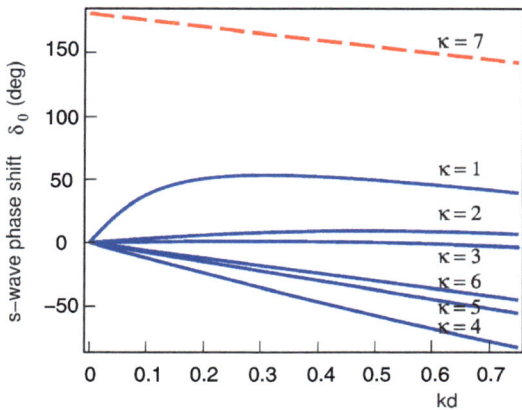

Fig. 8.5 *s*-wave scattering phase shift in a square-well potential for different potential depths U_0, characterized by κ, and potential width d. The *lower continuous curves* belong to potential depths that are not sufficiently deep to maintain a bound state whereas the *upper dashed curve* is such that there is just one bound state (like in the case of the deuteron)

and

$$\delta_0 = -kd + \arctan\left[\frac{kd}{\kappa d}\tan(\kappa d)\right]. \tag{8.47}$$

For small k ($kd \ll 1$)

$$\delta_0 = kd\left(\frac{\tan \kappa d}{\kappa d} - 1\right) + n\pi. \tag{8.48}$$

The behavior of the s phase depends strongly on the depth of the potential well, namely on whether the depth is sufficient to support one or more bound states or not. This is illustrated in Fig. 8.5.

There is an ambiguity of the scattering phases modulo π, which in a phase-shift analysis (at one energy) cannot be resolved. It is customary to assign $\delta_0 \to 0$ for $k \to 0$ for a potential without a bound state, and $\delta_0 \to n\pi$ for potentials with n bound states (see also *Levinson's theorem* [LEV49, JOA83]). During the traversal of an isolated purely elastic resonance δ progresses by $\Delta\delta = \pi$, with $\Delta\delta = \pi/2$ at the resonance energy. The total scattering phase, which contains also the potential-scattering phase, increases appropriately.

From the s-wave phase shift we can derive the scattering length a (see Sect. 8.3.1)

$$a = \left[1 - \frac{\tan(\sqrt{U_0} \cdot d)}{\sqrt{U_0} \cdot d}\right] \cdot d \tag{8.49}$$

and the corresponding cross section

$$\sigma = 4\pi a^2. \tag{8.50}$$

8.4 Exercises

8.1 The square-well potential with depth V_0 and width (range) d is a simple model for the central neutron-proton interaction. It shows a number of features that do

Fig. 8.6 Graphical solutions of Eq. (8.51)

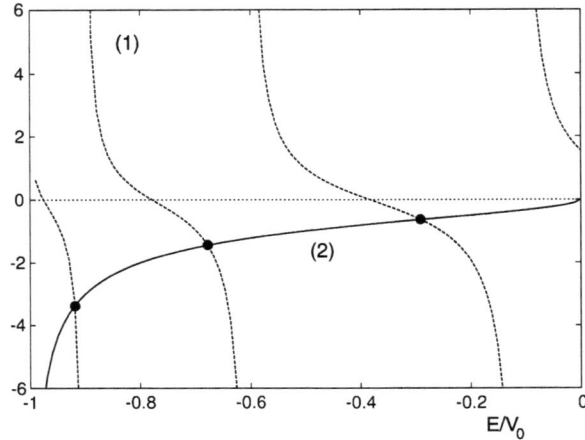

not depend strongly on the exact shape of the potential employed, especially when no spins or higher partial waves are considered. It can be applied to scattering states (total energy $E > 0$) as well as to bound states of a two-nucleon system such as the np system.

(a) Verify Eq. (8.46) as general phase-shift solution of the radial equation (8.39). Apply this to low energies ($\ell = 0$) and derive scattering length and cross section.

(b) In order to work with the true potential V we redefine the (reduced) potential $U = \frac{2\mu V}{\hbar^2}, \kappa = \frac{\sqrt{2\mu(V_0 - E)}}{\hbar}$, and $k = \frac{\sqrt{2\mu E}}{\hbar}$. A square-well potential is defined as $V = -V_0$ in the interior region $r \leq d$ and $E - V = V_0 - |E| > 0$ and $V = 0$ in the exterior region $r \geq d$ with $E - V = -|E| > 0$. Matching the logarithmic derivatives of the wave functions at $r = d$ leads to a transcendental equation that may be solved graphically or numerically, see Fig. 8.6:

$$\cot\left[\frac{d}{\hbar}\sqrt{2\mu(V_0 - |E|)}\right] = -\sqrt{\frac{|E|}{V_0 - |E|}}. \tag{8.51}$$

What is the number i of bound states for the given potential depth $-V_0$? (Note: i is equal to he number of crossings of the function (2) with the negative branches of function (1)).

(c) What is the condition for threshold values of V_0 and d for the transition from no to one bound state (*zero-energy resonance*)?

(d) Write a program (e.g. in FORTRAN, C, MAPLE, or MATHEMATICA) to solve Eq. (8.51) numerically for $V_0 = -50$ MeV, $d = 1.4$ fm. Try the same with a Woods-Saxon potential of equal depth and width, and diffuseness of 0.2 fm and a Yukawa potential $V(r) = g^2 \frac{\exp(-r/r_0)}{r}$, where g is the pion-nucleon coupling constant ≈ 0.3 $\hbar c$ and the radius parameter $r_0 = 1.4$ fm, the approximate range of the nuclear force.

8.2 Apply this to the real case of the $n-p$ system (more specifically: the $^3S_1$ system), which has a bound state, the deuteron. With the known binding energy of $E_{BE} = -2.2245$ MeV and a somewhat arbitrarily assumed value of $d = 2.1 \times 10^{-15}$ m calculate the potential depth $-V_0$ and the number of bound states.

8.3 Calculate the wave function of the deuteron $u_0(r)$ under these simplifying assumptions (the real deuteron or more generally the np interaction has a tensor-force component providing for a D-state admixture of a few percent to the S state). The probability of finding the neutron at a distance between r and $r + dr$ from the proton, is $|u_0(r)|^2 dr$. The absolute normalization of the wave function is obtained by the condition $\int |u_o(r)|^2 dr = 1$ together with the matching condition at the boundary $r = d$. Plot $|u_0(r)|^2$. Discuss how the long tail of this function with $r_{\text{rms}} \approx 5.4$ fm which is much larger than the radius from the systematics of $R = 1.2 \cdot A^{1/3}$ (see Sect. 2.4) suggests, can be related to the strength of the binding potential. Make a connection to the radii of halo nuclei, cf. Sect. 2.4.2.

References

[GOL64] M.L. Goldberger, K.M. Watson, *Collision Theory* (Wiley, New York, 1964)
[GON06] D.E. Gonzales Trotter et al., Phys. Rev. C **73**, 034001 (2006)
[HUH00] V. Huhn et al., Phys. Rev. C **63**, 014003 (2000)
[JOA83] C. Joachain, *Quantum Collision Theory*, 3rd edn. (North-Holland, Amsterdam, 1983)
[LEV49] N. Levinson, Kgl. Danske Videnskab. Selskab. Mat.-fys. Medd. **254**(9) (1949)
[MOT65] N.F. Mott, H.S.W. Massey, *The Theory of Atomic Collisions* (Clarendon Press, Oxford, 1965)
[NEW66] R.G. Newton, *Scattering Theory of Waves and Particles* (McGraw-Hill, New York, 1966)
[NOE76] W. Nörenberg, H.A. Weidenmüller, *Introd. to the Theory of Heavy Ion Collisions*. Lecture Notes in Physics, vol. 51 (Springer, Heidelberg, 1976)
[ROD67] L.S. Rodberg, R.M. Thaler, *Introd. to the Quantum Theory of Scattering* (Academic Press, New York, 1967)
[SEG83] E. Segrè, *Nuclei and Particles* (Benjamin, Reading, 1983)
[SLA07] I. Šlaus, Nucl. Phys. A **790**, 199c (2007)
[VWI06] W. von Witsch, X. Ruan, H. Witała, Phys. Rev. C **74**, 014001 (2006)

Chapter 9
The Nucleon-Nucleon Interaction

The nucleon-nucleon-interaction is the prototype for the action of the nuclear forces. With their complete "understanding" one can hope to also understand the structure and interactions of complex nuclei, at least in the sense of *effective* interactions. The description of the NN interaction in the framework of meson-exchange models has become so good that the latest *precision potentials* can be fitted to the wealth of data with reduced χ^2-values near 1. The status of the NN interaction up to 1976 is summarized in Ref. [BRO76].

In addition, this allows to test these potentials in the next-heavier three-nucleon systems. This is possible because these systems can be described exactly in terms of two-(and more-) body interactions such as the NN forces. The formalism was already given by Faddeev [FAD61] and extended by Alt, Grassberger, and Sandhas into the AGS equations, see Ref. [ALT67], see also [GLO96]. Present-day *Faddeev* calculations are considered numerically exact such that even additional new phenomena such as *three-body forces*, which are not forbidden can be searched for and investigated. A new development is the application of *chiral perturbation theory or effective-field theories EFT* in order to base the description on the more fundamental basis of quantum chromodynamics QCD. It is interesting that the results are very similar to those of the meson-exchange approach. Here only the NN-system results will be discussed. The role of isospin or isospin breaking in connection with the NN scattering lengths has been discussed above, see Sect. 3.3.3.

9.1 The Observables of the NN Systems

For the determination of the NN scattering phases, which, above the pion-production threshold, are complex, a multitude of (especially polarization-) observables have to be measured (for more detail see Chap. 5). In principle, there are 256, which, by conservation laws (P and T invariance) and exchange symmetry (for the nn and pp systems), are reduced to 36 or 25, respectively, linearly independent measurable observables. Because the scattering is completely described by 5 (or 6, resp.)

complex scattering amplitudes the measurement of only 9 (pp, nn) or 11 (np) quantities, selected appropriately from these, would suffice. The ones, which are easiest to measure are, besides the differential cross section, analyzing powers, the polarization transfer, and spin-correlation coefficients.

For NN these experiments in limited energy regions may be considered *complete*. In practice, more observables than those considered minimally necessary for a solution of the equations for a determination of the $M(T)$ matrix, must be determined. The usual parametrization of the total set of data is that in the form of scattering-phase shifts for partial waves for each j (coupled from the particle spins and orbital angular momenta). Also the direct parametrization by scattering amplitudes, i.e. the elements of the M matrix, has been used.

9.1.1 NN Observables

Here only a selection of typical observables will be shown. One basic observable is the total cross section $\sigma(E)$. Figure 9.1 shows the smooth behavior of pure potential scattering without resonances in this energy range (resonances such as the Δ, see Fig. 11.8, an excited nucleon state, have been found at higher energies, others, more exotic resonances such as *dibaryon resonances*, possible 6-quark states, have been searched for in excitation functions but could not be identified so far, see also Fig. 4.6). The neutron-proton scattering at the very low energies is very different from that in the pp case where the Coulomb scattering is dominant; it shows the interference region at the lowest energies, the constant s-wave cross section at low-to-intermediate energies and the onset of higher partial waves, then the opening of inelastic channels. The np cross section towards very low energies (i.e. for cold or ultra-cold neutrons) increases beyond the constant value in the thermal and epithermal regions because of the neutron's wave properties. This is shown in Fig. 9.2. Both are almost constant but at the very highest energies so far achieved (at the LHC) the elastic as well inelastic pp integrated cross sections rise again. The behavior of angular distributions also gives a first impression of the reaction mechanism of the NN interactions as shown in Figs. 9.3 and 9.4. The pp cross section angular distributions are dominated by two features, one being the symmetry about 90° due to the identity of the entrance-channel particles, the other the interference between the two reaction mechanisms of nuclear and Coulomb scattering. The latter becomes stronger with lower energies and the interference minimum is very dominant around $\approx$400 keV and 90°. The interference is destructive, which signifies that—with the repulsive Coulomb force—the nuclear force must be attractive in this energy region. The nuclear part (central region) is nearly isotropic as is expected for s-wave scattering. The np angular distributions are shown for comparison in Fig. 9.4. Despite many years of experimental work the precision of the np data is substantially lower than that of pp data. Figure 9.5 shows the analyzing powers of pp and np scattering at 25 MeV. The large differences between the two systems—for the purely nuclear (hadronic) interaction very similar observables are expected

9.1 The Observables of the NN Systems

Fig. 9.1 Excitation functions of the total cross sections of pp and np scattering. The figures show the similarity of the pp and np cross sections (outside the Coulomb-dominated lower energies) due to isospin conservation. The low-energy cross section of pp scattering is dominated by Coulomb whereas in np scattering we have the pure NN interaction. The low- to very-low-energy regions of the np scattering display several quite different features and are shown in the following figure

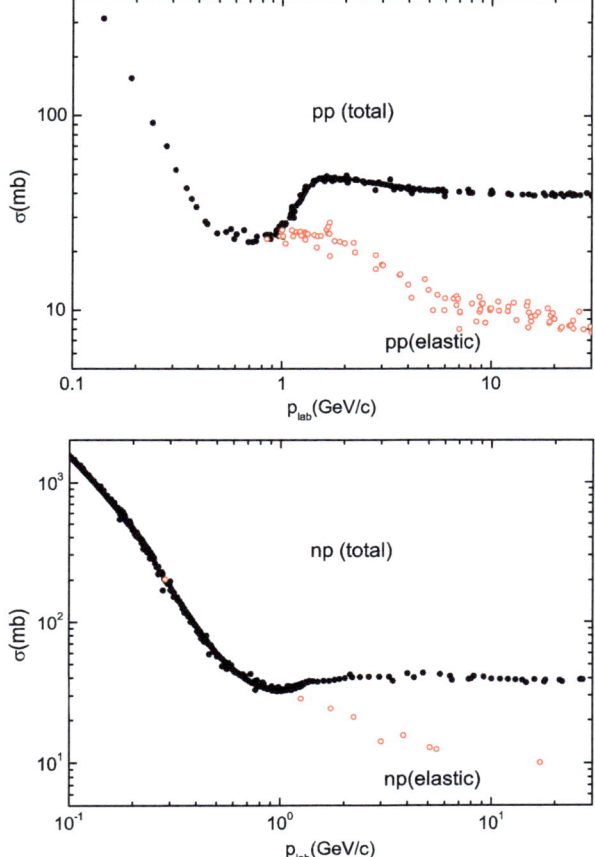

Fig. 9.2 The low-energy behavior of np scattering

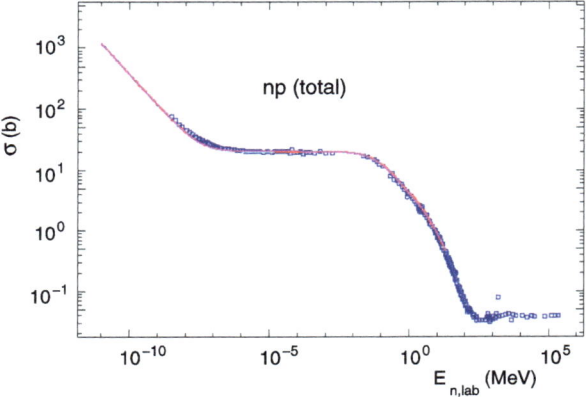

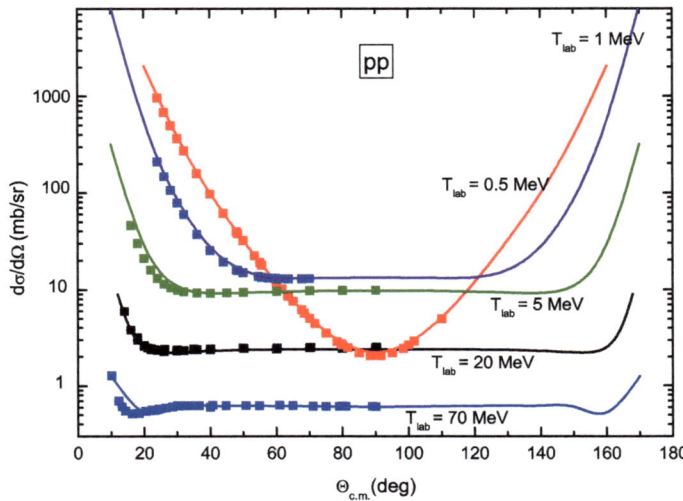

Fig. 9.3 The *pp* cross section angular distributions at a few selected energies. The data have also been selected to give a clearer picture, and the *curves* show recent fits to the data base via a phase-shift analysis and using the Nijmegen93 NN potential. Note the very deep interference minimum of the 0.5 MeV data

due to isospin symmetry—come about by the interference with the Coulomb interaction in the *pp* case. The analyzing powers are very small because at these energies the interaction is *s*-wave dominated and polarizations are due to interference of *S*- with higher waves. High precision of the measurements is required. Especially the polarized neutrons must be produced in primary reactions such as the $D(d,n)$ reaction (see below) by spin transfer. Thus, count rates are low, and multiple scattering and detector efficiencies must be taken into account. Whereas complete sets of NN observables exist at medium energies (several 100 MeV) very few or no data exist at tandem VdG energies, especially for *np* and for two-spin observables. As an example Fig. 9.6 shows the transfer observables $K_y^{y'}$ and $K_x^{x'}$ at 25 MeV.

9.1.2 NN Scattering Phases

These and all other observables at all energies where data exist are continuously subjected to common phase-shift analyses. The scattering phases obtained by χ^2 minimization parametrize the data in an angle-independent form as functions of energy. Starting at the lowest particle-production threshold (the pion threshold) the phases must be complex because of the imaginary absorption term, see Figs. 9.7 and 9.8.

9.1 The Observables of the NN Systems   149

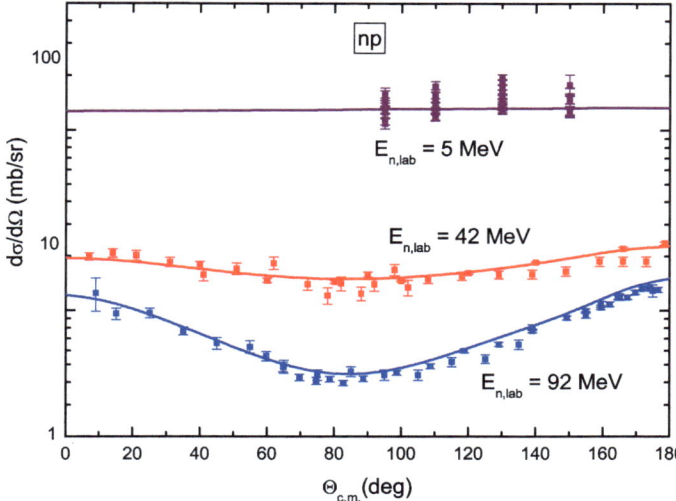

Fig. 9.4 The *np* cross section angular distributione at a few selected energies. The data are also selected to give a clearer picture, and the *curves* show recent fits to the data base via a phase-shift analysis and using the Nijmegen93 NN potential. Note the near-isotropy typical for *s*-wave scattering and slow beginning of anisotropy at higher energies with the onset of *p*-wave scattering. The approximate symmetry around 90° can be interpreted as due to the NN interaction being an exchange force with approximate isospin conservation

9.1.3 NN Interaction as Exchange Force

All fundamental interactions are explained by the exchange of virtual particles. In the case of the NN interaction the exchange particles are mesons (or pairs of mesons). The masses of the different exchange particles determines the range of the different contributions of the nuclear force via the uncertainty relation. The lightest massive virtual particles, the pions, are responsible for the nuclear forces "far out", the heavier ones for the forces "further in". Various modern (precision) exchange potentials (typically "CD BONN", "NIJMEGEN", "ARGONNE 18" etc.) describe die data with $\chi^2_{red} \approx 1$. Figure 9.9 shows as a Feynman diagram the exchange process and the radial behavior of the central part of the nuclear force with an attractive outer, and further in, of a repulsive part ("hard core") that is usually attributed to the quark-quark or quark-gluon interactions. The conservation laws or symmetries (parity, time reversal, exchange symmetry, isospin etc.) allow only certain forms of the nuclear interaction (central, spin-spin, spin-orbit, tensor force etc.). Experimentally these different force components contribute differently in different observables. An example is the quadrupole moment of the only bound NN system (that of the deuteron) that requires a *D*-wave admixture to the *s*-wave ground state, which can only be explained by the action of a *tensor force*. The *spin-spin* force explains the energy difference between the bound *np* triplet (the deuteron) and the

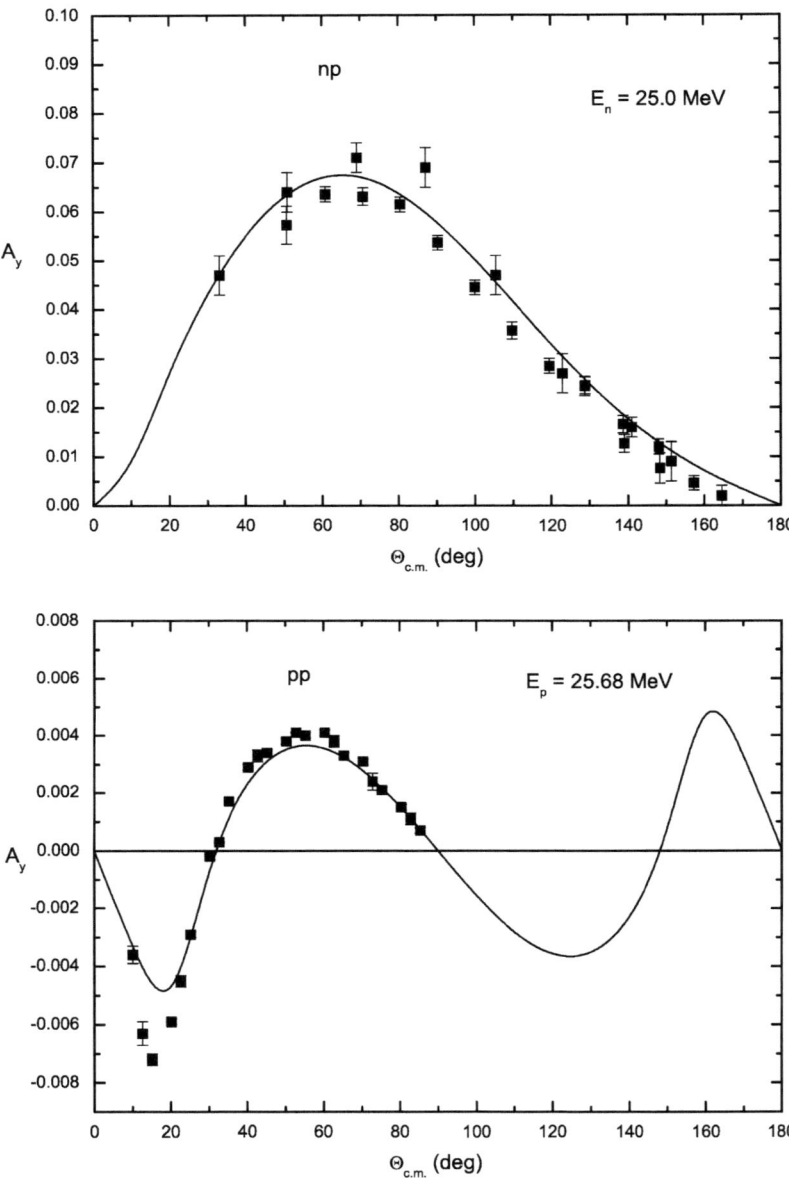

Fig. 9.5 Angular distributions of the analyzing power of the *pp* and *np* scattering at 25 MeV. Note the smallness of the effects and the precision of the measurements necessary. The *lines* are results of an NN phase-shift analysis. All data from the NN database CNS/SAID [CNS13]

9.1 The Observables of the NN Systems 151

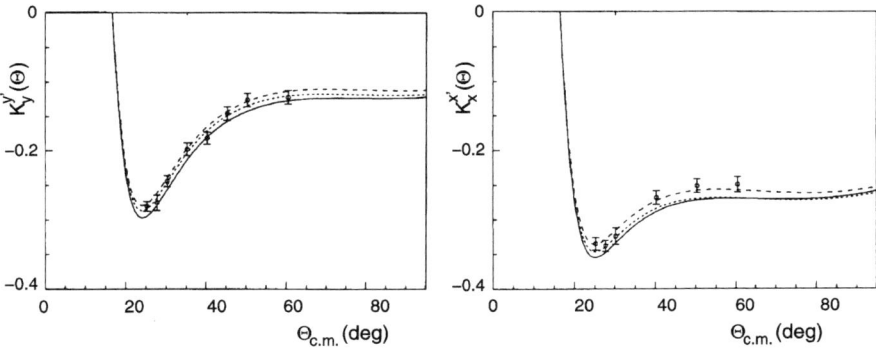

Fig. 9.6 Angular distributions of the polarization transfer coefficients $K_y^{y'}$ and $K_x^{x'}$ of the pp scattering at 25 MeV. The *lines* are results of calculations with the Bonn (*solid*), Paris (*short-dashed*), and Nijmegen (*long-dashed*) NN potentials. All data from [KRE94]

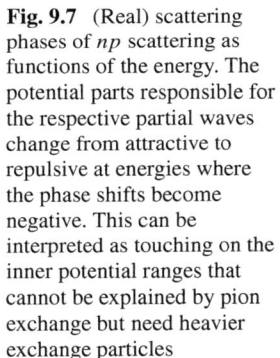

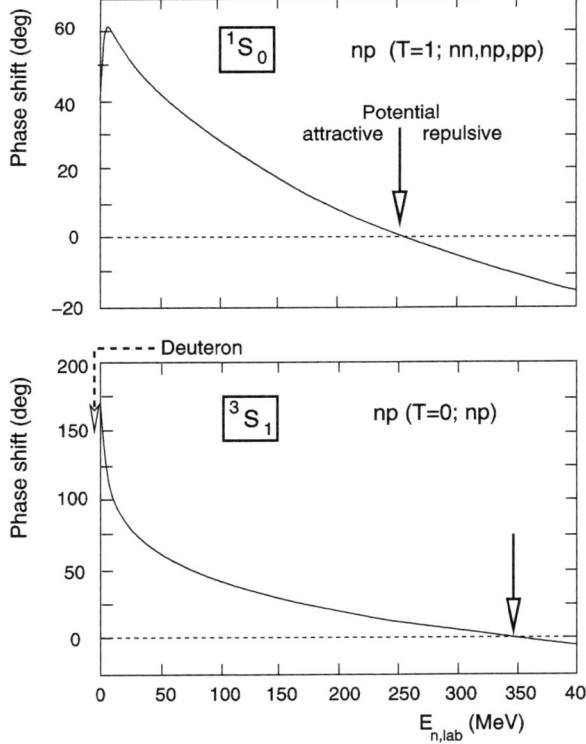

Fig. 9.7 (Real) scattering phases of np scattering as functions of the energy. The potential parts responsible for the respective partial waves change from attractive to repulsive at energies where the phase shifts become negative. This can be interpreted as touching on the inner potential ranges that cannot be explained by pion exchange but need heavier exchange particles

unbound singlet np scattering system (the singlet-deuteron d^*). The *spin-orbit* potential acts predominantly in polarization observables such as the analyzing power. Table 9.2 shows these contributions.

Fig. 9.8 Inelasticity: Imaginary part (with $|\eta_0| = \exp(-2\mathrm{Im}(\delta_0)))$ of the $^1S_0$ scattering phase of s-wave np scattering. At the pion threshold the phases become complex

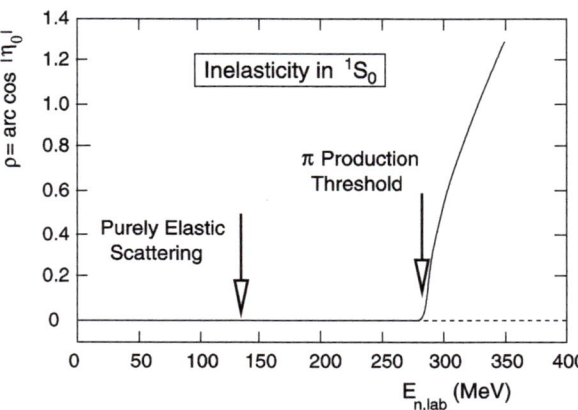

Fig. 9.9 Feynman diagram of the principle of the nucleon-nucleon force, as following from boson exchange, and shape of the NN potential showing the different exchange bosons and the ranges, in which they dominate (qualitatively)

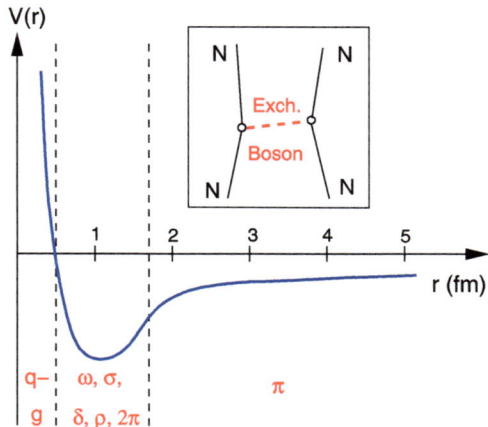

Table 9.1 Role of the exchange bosons in the Bonn NN potential. The meaning of the abbreviations is: no sign: attraction, −: repulsion, C: Central, T: Tensor, LS: Spin-orbit, SS: Spin-Spin

Boson	Main potential contribution	Range
π	C, T, LS	Long
η	C, T, LS	Medium
ω	C (-), SS, T, LS	Medium
ρ	C, T(-), LS	Short to medium
$2\pi \equiv \sigma$	C, LS	Medium
Correlated 2π		

9.2 Few-Nucleon Systems

With the progress in high-speed computing one of the basic goals of nuclear theory gets into reach, the "exact" calculation of nuclear properties and reactions with the fundamental NN (or even q-g interactions) as input. The success of NCSM (no-core

9.2 Few-Nucleon Systems

Table 9.2 NN-force contributions. Nucleon-spin dependent quantities are marked

Form compatible with conservation laws/invariance principles: $V = \sum_i V_i \cdot \mathcal{O}_i$

Contributions	$\mathcal{O}_i$	
Central	1	
Spin-orbit	$L \cdot S$	$S = \frac{1}{2}(\sigma_1 + \sigma_2)$
Tensor	$\frac{3(\sigma_1 \cdot r_1)(\sigma_2 \cdot r_2)}{r^2} - \sigma_1 \cdot \sigma_2$	
Spin-spin	$\sigma_1 \cdot \sigma_2$	
Quadratic Spin-orbit	$(L \cdot S)^2$	

shell model) and other methods in calculating ground and excited state properties of heavier light nuclei is evident. Important steps in this direction are numerically-exact calculations of the three- and four-nucleon systems. Recent developments are the application of *Effective Field Theories, Chiral-Perturbation Theories* based on low-energy approximations to QCD.

9.2.1 The Two-Nucleon System

The most fundamental few-nucleon system is the NN system, which already has a rich structure such as

- It contains three scattering systems (nn, np, and np in the spin singlet state) and one bound-state system, the deuteron (np in the spin-triplet state).
- It is a testing ground for isospin symmetry.
- It provides the interplay between the different components of the hadronic nuclear and the Coulomb forces.
- The observables of the NN systems and the scattering phase shifts derived from them form the basic input into calculations of systems with $A > 2$. The quality of the least-squares fits of different theoretical (but still phenomenological from the QCD viewpoint) models for NN potentials based on meson exchange is such that $\chi_{\text{red}} \approx 1$. Consequently, these NN parameters have been used in the most advanced Faddeev [FAD61] and Faddeev-Yakubovsky [YAK67] calculations for three- and four-nucleon observables. The methods applied are considered to be numerically exact [GLO83, GLO96]. Experimentally, features of the NN interaction can be studied also in special situations of reactions with more than two outgoing particles, especially three-body breakup reactions, see Sect. 9.2.2. An example is the *Final-State Interaction FSI* of the np as well as the nn systems at $E_{\text{rel}} \approx 0$ allowing the determination or confirmation of the pertinent scattering lengths a_{np} and a_{nn} (see Fig. 9.12). The nn scattering length is practically impossible to obtain by direct nn scattering (attempts to use nuclear explosions failed so far). In systems where the FSI leads through a resonant state its properties may be studied: An example is the reaction

$$d + \alpha \to {}^5\text{Li}^* + n \to \alpha + p + n \tag{9.1}$$

with the two charged particles measured in coincidence, i.e. in a kinematically complete experiment. This method was generalized by the THM (the *Trojan-Horse Method* [BAU04]).

9.2.2 The Three-Nucleon System

The three-nucleon system is—after the $2N$ system—the next more-complicated system governed by nucleon-nucleon forces. With *numerically exact* calculations in the framework of the *Faddeev* formalism it was possible to describe successfully the $3N$ system by a sum of NN interactions. The observables that must be described are not only those of the scattering system ($d + N$ elastic scattering, deuteron breakup, and capture reactions), but also the binding energies of the two three-nucleon isotopes ^{3}H and ^{3}He.

For many observables the theoretical description is rather good. However, for a few observables (at energies below 30 MeV) there are still discrepancies with the predictions of the theory, becoming smaller at higher energies. They are:

- The predicted binding energy of the triton (^{3}H) is about 1 MeV too low.
- Different single observables of the scattering system are not well described (despite a relatively good description of many others):
  - The analyzing power A_y of the elastic scatterings ^{2}H$(p, p)^2$H and ^{2}H$(n, n)^2$H ("the analyzing-power puzzle");
  - The vector analyzing power iT_{11} of the elastic scattering ^{1}H$(d, d)^1$H;
  - The cross sections of the breakup reaction ^{2}H$(p, pp)n$ and ^{2}H$(n, nn)^1$H in certain configurations such as the "space-star", the "SCRE (symmetric constant relative energy)", and possibly the quasi-free (QFS) configurations.

A recent reference addressing especially these discrepancies is Ref. [SAG10].

9.2.3 Elastic Scattering in the Three-Nucleon System

Only one result for elastic scattering of the $N + d$ system with this discrepancy will be mentioned here. Figure 9.10 shows an example at one energy of the discrepancy of A_y and a plot of its systematic behavior with energy.

9.2.4 Kinematics of Three-Nucleon Breakup Reactions

Whereas the elastic scattering in these systems is like in all other two-body scattering reactions there is the additional possibility of particle breakup. Here the case of the three-nucleon system will be discussed. The schemes are

$$n + d \to n + p + n \tag{9.2}$$

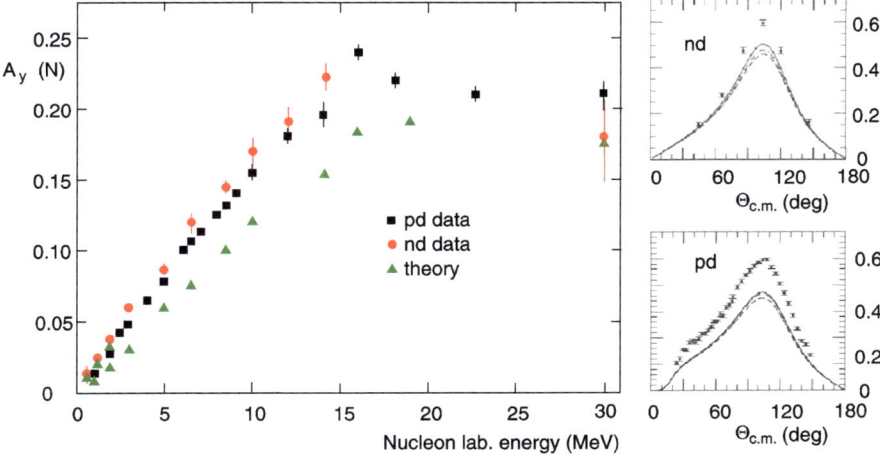

Fig. 9.10 Illustration of the A_y puzzles of the elastic scattering of nucleons from ^{2}H by comparing the experimental analyzing powers at 3 MeV in the maximum with theoretical predictions (the nd data are slightly higher than the pd data). The examples (*right*) show the discrepancies for neutron as well as proton scattering off deuterons (in the maximum). The *left figure* shows the systematic behavior of the A_y puzzle for nd as well as pd data with energy up to 30 MeV. The data points and theoretical values were taken from the maxima of the angular distributions near 120°. All calculations with quite different approaches: Faddeev calculations with different NN potentials, with and without three-body forces, with and without the Coulomb force, and, finally, using EFT, scatter around values that are far too low, i.e. so far no theoretical approach was capable of reproducing the angular distributions

or

$$p + d \to p + p + n \tag{9.3}$$

or more generally

$$1 + 2 \to 3 + 4 + 5. \tag{9.4}$$

We assume the detectors to be point detectors, which often approximates the behavior of small solid-state detectors, or the single-pixel elements of large (e.g. $\approx 4\pi$) detectors. For the (two-body) elastic scattering the measurement of the energy under polar and azimuthal angles θ, ϕ of one ejectile determines the entire kinematics uniquely, also for the corresponding recoil particle, with the two outgoing particles lying in a plane: the three parameters just satisfy the momentum and energy conservation laws.

For a breakup reaction with three outgoing particles the measurement with one such detector is "kinematically incomplete" (or "inclusive"). Four parameters, energy and the three momentum components of one ejectile are insufficient to fix the outgoing kinematics completely, i.e. the measured events of one ejectile are averaged over the variables of the other two. However, with two detectors in coincidence (a "kinematically complete" or "exclusive" measurement), e.g. with two proton detectors in Eq. (9.4) six momentum components are known whereas five would suffice to satisfy energy and momentum conservation. Thus, the problem is onefold overdetermined. Therefore, all events of a given reaction channel must lie on a line, the kinematical locus embedded in a continuum of background events. These could come, e.g., from electronic noise (or in reactions with more particles) from a four-body breakup. This overdetermination can be used to separate true events from any background. It also has the consequence that, in general, there exist relations between kinematical parameters such as detector angles and ejectile energies. The form of the kinematical curve can be derived from such a relation between the kinetic energies T_i of two outgoing particles:

$$k(T_i, T_j) = 0; \quad i \neq j = 3, 4, 5. \tag{9.5}$$

The relation is obtained from the invariance of the square of the four-momentum of the unobserved particle 5:

$$P^\mu = \left(\frac{E}{c}, \vec{p}\right) = (\gamma_1 m_1 c + m_2 c, \gamma_1 m_1 \vec{v}_1)$$
$$P_5^2 = P_5^\mu P_{5\mu} = m_5^2 c^2 = [P - (P_3 + P_4)]^2. \tag{9.6}$$

Explicitly this leads to

$$0 = 1/2\left(m_3^2 c^2 + m_4^2 c^2 - m_5^2 c^2 + M^2 c^2\right)$$
$$- ET_3/c^2 + \sqrt{(E/c)^2 - M^2 c^2}\sqrt{(T_3/c^2) - m_3^2 c^2} \cos\theta_3$$
$$- ET_4/c^2 + \sqrt{(E/c)^2 - M^2 c^2}\sqrt{(T_4/c^2) - m_4^2 c^2} \cos\theta_4$$
$$+ T_3 T_4/c^2 - \sqrt{(T_3/c)^2 - m_3^2 c^2}\sqrt{(T_4/c^2) - m_4^2 c^2} \cos\theta_{34} \tag{9.7}$$

with

$$M^2 c^2 = m_1^2 c^2 + m_2^2 c^2 + 2m_2 T_1,$$
$$T_i = \gamma_i m_i c^2 = T_i + M_i c^2, \quad \text{and}$$
$$\cos\theta_{34} = \cos\theta_3 \cos\theta_4 + \sin\theta_3 \sin\theta_4 \cos(\phi_3 - \phi_4). \tag{9.8}$$

At a typical tandem VdG energy of 20 MeV $T_1/m_1 c^2 \approx 0.02$ thus the differences between relativistic and non-relativistic descriptions are in the one-percent range. Non-relativistically the analogous relation between kinetic energies in the lab. system is again obtained from energy and momentum conservation and contains the Q-value of the reaction considered:

$$0 = T_1(m_1 - m_5) - m_5 Q + (m_5 + m_3)T_3 + (m_5 + m_4)T_4$$
$$- 2\sqrt{m_1 m_3 T_1 T_3} \cos\theta_3 - 2\sqrt{m_1 m_4 T_1 T_4} \cos\theta_4$$
$$+ 2\sqrt{m_3 m_4 T_3 T_4} \cos\theta_{34}. \tag{9.9}$$

9.2 Few-Nucleon Systems

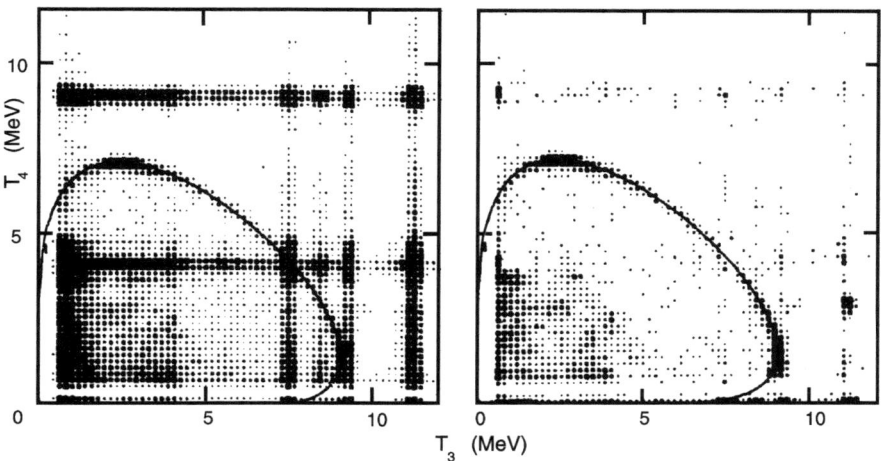

Fig. 9.11 Kinematical curves of the two coincident protons of the reaction ^{2}H$(p, pp)n$ in the T_4 vs. T_3 plane: raw data (*left*) and data after background subtraction using cuts on the time-of-flight difference spectrum. The spots with enhanced intensities are caused by the quasi-two-particle p_3-n and p_4-n final-state interactions

In the momentum, i.e. the $(\sqrt{T_3}, \sqrt{T_4})$ space this equation describes an ellipse. A typical set of kinematical curves is shown in Fig. 9.11. In general, in the c.m. system, three outgoing particles will be ejected in all possible directions $\{\theta_i, \phi_i\}_{i=1,3}$, i.e. in all of phase space (in the lab. system and depending on the Q values the emission may be restricted to a forward cone). In order to register the interactions between all particles completely a 4π arrangement of many single detectors or position-sensitive devices like in intermediate and high-energy physics are required. Among other reasons mostly the limited resources of smaller laboratories led to investigations of special kinematical configurations. "Classically" these were:

- The final-state interaction (FSI), e.g. between two of three outgoing particles. Examples are the *nn*, *pp*, or *np* FSI in the $d + p$ breakup reaction, especially at relative energies between the two particles near zero, which is selected by choosing appropriate angles. This is important for the determination of the respective scattering lengths a_{nn}, a_{np}, and a_{pp}. Especially when an intermediate, longer-lived state is formed in a two-particle subsystem of the three-particle output channel, sequential decay may occur which allows the study of that subsystem. This method is closely related to the *Trojan-Horse Method* which was especially devised to study astrophysical two-particle reactions in the very-low-energy regime where the problems with the strong Coulomb interaction can be circumvented. Applications to reactions in inverse kinematics with radioactive (or rare) ion beams have been discussed. For a survey see Ref. [BAU04].
- The quasi-free scattering (QFS), e.g. the *pp* or *nn* QFS in the $d + p$ breakup that is characterized by the third particle being a spectator, i.e. at rest in the c.m. system.

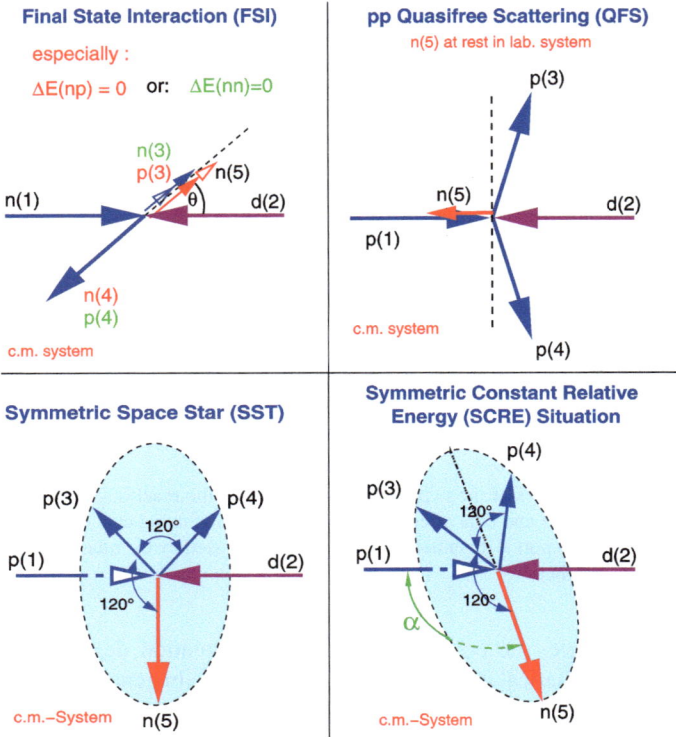

Fig. 9.12 Four "classical" situations of the $N+d$ deuteron breakup reactions

- The collinearity (COL) situation where the third particle is at rest in the lab. system with the idea that for the two other particles colliding "head-on" the effects of two-body forces would be weak and those of three-body forces would appear more clearly (which was never really proved).
- All forms of true three-particle reactions, but again restricted to a finite number of situations such as the space-star (SST) where the momenta of all three particles lie in a plane whose orientation could be arbitrary.

Figure 9.12 shows schematically four typical breakup situations for the case of the $3N$ breakup $p(1) + d(2) \to p(3) + p(4) + n(5)$ or $n(1) + d(2) \to n(3) + n(4) + p(5)$. The ideal intensity distributions of particles from special (quasi-two-body) interactions (such as the FSI of two subsystems) reaction mechanisms appear as points on the kinematical curve. Usually the intensity distribution along the kinematical curve (which is spread around the theoretical locus by finite experimental resolution) is projected onto this curve and plotted as function of the path length $S(T_3, T_4)$ after removing background.

9.2.5 Results for the Three-Nucleon Breakup Reaction

The results (cross sections $\frac{d^3\sigma}{dSd\Omega_3 d\Omega_4}$ or polarization observables as functions of $S(T_3, T_4)$) are compared to theoretical predictions using advanced Faddeev methods with NN meson-exchange potentials or effective-field approximations as input. For both the results are very similar and two cases of excellent agreement as well as continuing disagreement are shown here in Fig. 9.13, see also Ref. [HGS01]. Figure 9.14 shows two examples of breakup analyzing powers which are quite small (s-wave dominance). Figure 9.15 shows the systematic behavior of the unsolved discrepancies of the data of the symmetric space-star (SST) situations of the ^{2}H(p, pp)n as well as the ^{2}H(n, nn)^{1}H reactions, as compared to "numerically exact" Faddeev calculations with the CD Bonn NN potential with ([DEL08], triangles) and without (line) the Coulomb interaction. The results in the framework of EFT input are not very different, and the addition of three-body forces do not remedy the situation. In summary, the calculations with and without the Coulomb interaction and/or three-body forces disagree with the data of both types of reactions. For a recent evaluation of the situation see e.g. Ref. [KAL12]. All attempts so far to remedy the discrepancies discussed above by adding additional phenomenological three-body forces or using the more fundamental EFT approach have failed at the low energies. At higher (intermediate) energies above 100 MeV the introduction of three-body forces has improved the small existing discrepancies. One has to conclude that the three-body force, at least of the types tested so far, becomes more important only at higher energies whereas the remaining low-energy problems are still unsolved. It is not only an experimental fact, but also suggested by EFT that these three-nucleon forces—at least at low reaction energies—are quite small (<1 %). The indications that at intermediate energies effects of three-body forces cannot be neglected suggest that these forces may be of short range. For a careful evaluation of the role of three-nucleon interactions in few-nucleon systems see Ref. [KAL12].

9.2.6 Recent Progress in Few-Nucleon Reactions

Treatment of the Coulomb Force For many years the full treatment of few-body reactions was hampered by the fact that the Coulomb force could not be exactly included when charged particles were involved such as in the $d + p$ system. The description was therefore always approximate in the sense that protons were treated as uncharged. The problem, however, was that the experiments with protons had higher quality (as to statistical as well as systematic errors) than the corresponding neutron experiments. Isospin breaking would in principle disallow this procedure and the comparison between isospin mirror channels is interesting by itself. Although for many observables the Coulomb-free treatment yielded good agreement

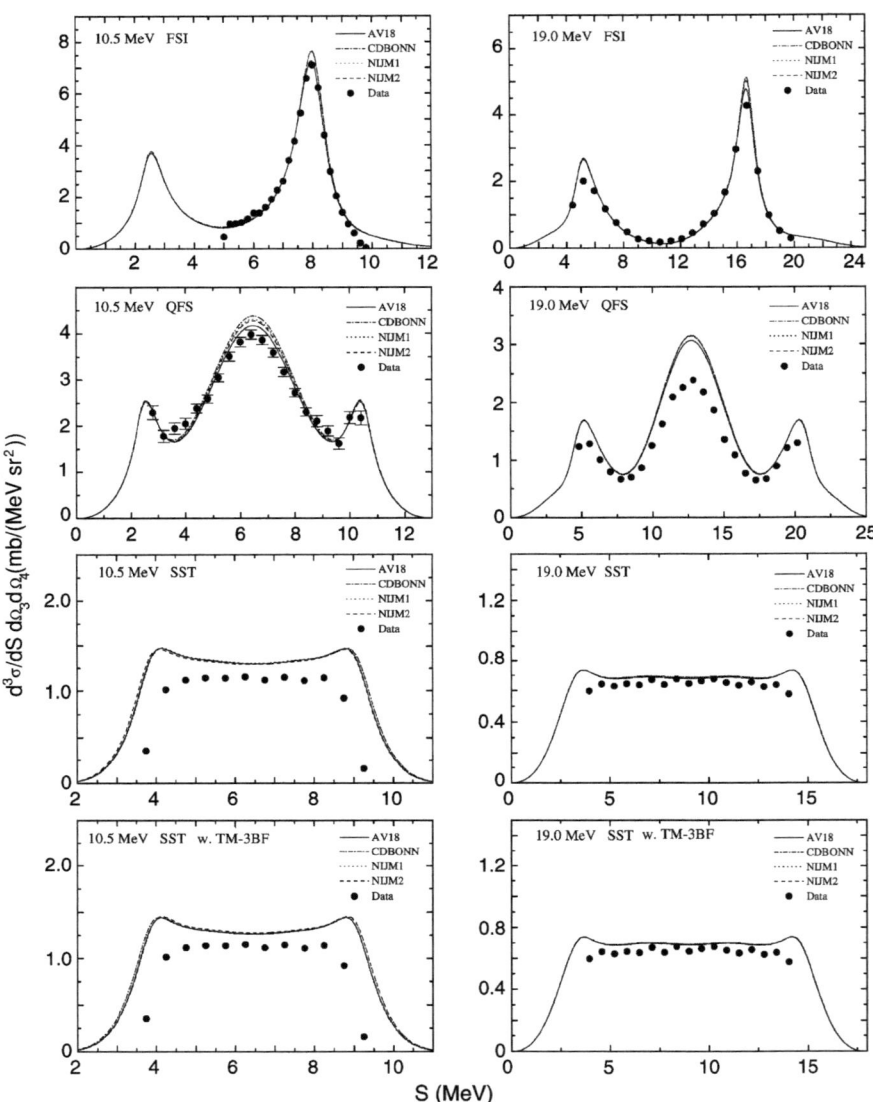

Fig. 9.13 Examples for different "special" configurations of the breakup reaction ^{2}H$(p, pp)n$. The data [HGS01] are compared to predictions in the framework of the Faddeev theory using different NN interactions fitted to NN data, as input. The inclusion of the Tucson-Melbourne three-body force does not change the results significantly

with the proton-induced data, conclusions about the Faddeev calculations, the influence of three-body forces, or their interference with the Coulomb force were always somewhat uncertain. Therefore it is a great achievement to be able to include the Coulomb force in the Faddeev calculations.

9.2 Few-Nucleon Systems

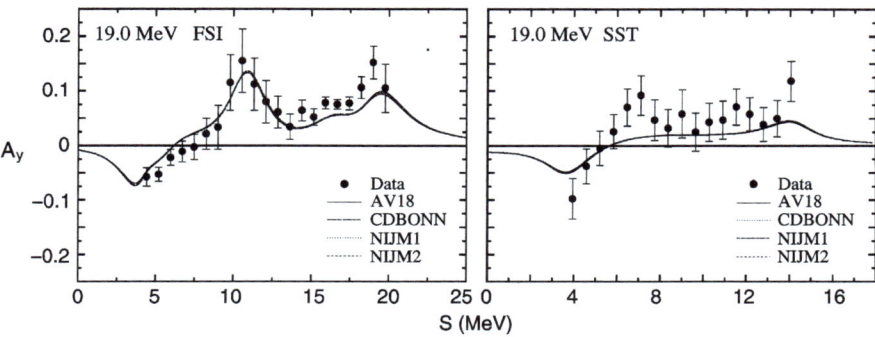

Fig. 9.14 Examples of analyzing powers for two different "special" configurations of the breakup reaction ^{2}H$(p, pp)n$ [HGS01]. The data are compared to predictions in the framework of the Faddeev theory using different NN interactions fitted to NN data, as input

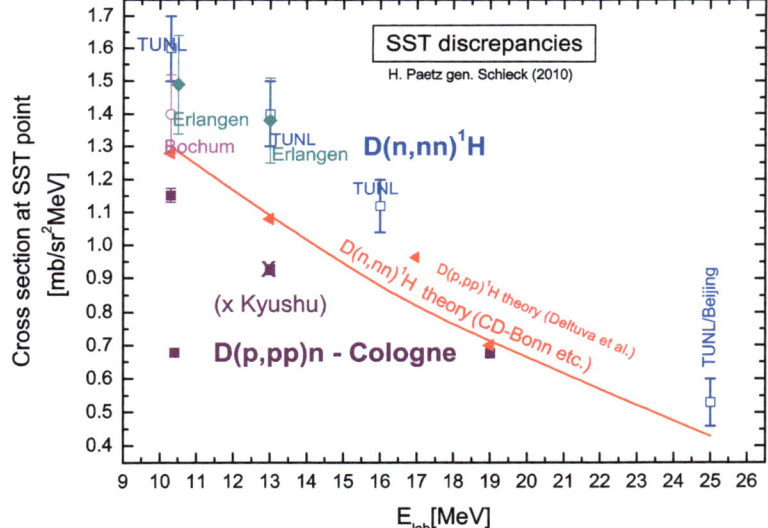

Fig. 9.15 Low-energy behavior of the kinematically complete ^{2}H$(p, pp)n$ and ^{2}H$(n, nn)^1$H reactions in identical SST situations. The data (*points* with *error bars* from different laboratories TUNL(Duke U.) [SET96, CRO01, MAC04], Bochum [STE89], Erlangen [STR89, GEB93], TUNL/Beijing [ZHO01], Cologne [HGS01] and references therein, and Kyushu U. [SAG10]) are compared to predictions in the framework of the Faddeev theory using the CD Bonn NN interactions fitted to NN data, e.g. from the Bochum-Cracow group (H. Witała et al. [GLO96]) and A. Deltuva [DEL09], as input

Effective-Field/Chiral Perturbation Theory The fact that *three-body forces* cannot be excluded by physical laws leads to the postulate of such forces, which e.g. are known in molecules (*Axilrod-Teller force*), also for nuclei, see e.g. Ref. [HAM13]. Such forces can be integrated in Faddeev calculations (an example: the Tucson-

Melbourne force) and thus their effects may be studied explicitly. On the other hand, the full and exact integration of the Coulomb force—because of its long range and subsequent convergence problems—into the microscopic calculations has been difficult and succeeded satisfactorily only recently, see e.g. Ref. [DEL08].

The theories aim at deriving the nuclear forces from the quark-gluon interaction described by QCD. The typical energies of nuclear physics are in the "nonperturbative" region, in which exact solutions in the framework of QCD appear difficult or impossible. The "chiral perturbation theory", however, has already provided approximate solutions at different levels characterized by their relation to the leading order of approximation, e.g. present predictions for the $3N$ systems are on N^3LO, i.e. "next-to-next-to-next-to-leading-order". (For a survey of the present status of the field see Ref. [KAL12] and references therein). They have been successful in describing equally well the observables of the $3N$ systems but—remarkably—also with the exceptions discussed above. In view of the claim of the calculations being exact, these low-energy discrepancies. which seem to resist all approaches to solutions so far, have to be taken seriously and point to some unknown parameter in the theories. Generally, a direct influence of the quarks or gluons and their properties on phenomena of nuclear physics seems almost absent—with exceptions such as in isospin breaking by the u-d quark mass difference.

9.2.7 Other Few-Nucleon Systems

For the four-nucleon systems the appropriate formalism (the *Faddeev-Yakubovsky* equations) [YAK67] was published long ago, but even without the Coulomb force the exact calculations meet large calculational difficulties. EFT predicts that the possible four-nucleon forces will be smaller than the three-nucleon forces, probably by an order of magnitude, and may, for the time being, neglected. For the description of the nuclear interactions between heavier nuclei one has to rely on "effective" interactions derived from the more fundamental NN interaction. The predictions of ground-state and low-excited state properties of light nuclei of the p and sd shells using different models with these interactions have been surprisingly successful. The properties of bound and low-excited states of light nuclei seem to require the three-nucleon force for their description. The same seems to be true for heavier nuclei, e.g. the neutron-rich Ca isotopes, such that the evolution of the shell structure towards the dripline requires a three-body force component [HAG12].

9.3 Exercises

9.1. The square-well potential model of Chap. 8, describing the deuteron, can be equally applied to the s-wave ($\ell = 0$) np scattering ($E > 0$).

   (a) Explain why partial waves with $\ell > 0$ are not important below ≈20 MeV.

9.3 Exercises

(b) Starting with Eqs. (8.46) and (8.47) with $n = 0$ (no bound states) calculate for very small E or $k \to 0$ δ_0 and σ_0 of the np system, using similar potential parameters as for the deuteron, e.g. $V_0 = 36$ MeV, $d = 2.1 \cdot 10^{-15}$ m. The resulting value of $\sigma_0(np)$ will disagree strongly with the experimental value of $\sigma_{0,\text{exp}} = 20.4$ b, see the typical ≈constant cross section in the s-wave region in Fig. 9.2. Discuss how the potential describes features of the deuteron quite well, but fails for np scattering.

(c) The discrepancy between $\sigma_{0,\text{exp}}$ and that as calculated above suggests that the potential derived for the spin-triplet system $^3S_1$ is insufficient. Evidence shows that n and p scatter strongly also in the spin-singlet state $^1S_0$ which does not contribute to the np bound state. If we consider the cross sections and their relation to scattering lengths and assume that the cross sections for both states add up incoherently, then

$$\sigma_{\text{tot}}(np) = 4\pi \left(\frac{3}{4} a_t^2 + \frac{1}{4} a_s^2 \right). \tag{9.10}$$

Calculate e.g. the singlet scattering length a_s after using $\sigma_{0,\text{exp}}$ and the triplet scattering length a_t as input and compare with the established values, see e.g. Sect. 2.3.

9.2. The sign of the scattering lengths cannot be established from a total-cross section measurement alone, but needs additional information such as from interference terms (see the following exercise) or isospin arguments. Interpret the large negative value of $a_s(pn)$ as to a possible singlet state of the deuteron (the d^*).

9.3. Neutrons can be captured by hydrogen (i.e. protons) to form ^{2}H according to

$$n + p \to {}^2\text{H} + \gamma. \tag{9.11}$$

What is the threshold neutron energy for this reaction? Explain the terms: "binding energy" of the deuteron and "Q-value". How could you determine the neutron rest mass by this or the inverse reaction?

9.4. What is the minimum lab. energy of a proton for the breakup of a deuteron (at rest) by the

$$p + d \to p + p + n \tag{9.12}$$

reaction? What, if a deuteron is the projectile bombarding the proton at rest?

9.5. Prove the following properties of the NN tensor force

(a) Its operator

$$S_{12} = \frac{3(\vec{\sigma}_1 \vec{r})(\vec{\sigma}_2 \vec{r})}{r^2} - \vec{\sigma}_1 \vec{\sigma}_2 \tag{9.13}$$

vanishes when integrated over the full 3D space. It was defined such that any central-force component vanishes.

(b) It vanishes in the spin-singlet state ($S = 0$) of the np system, thus only contributing to the np spin-triplet states ($S = 1$, the deuteron and the triplet np scattering observables).

9.6. Calculate the eigenvalues of

- the spin-spin operator $(\vec{s}_1 \vec{s}_2)$ and
- the spin-orbit operator $(\vec{\ell} \cdot \vec{s})$.

9.7. For neutron energies, low enough that their de-Broglie wavelength is $> d$, the distance of the two protons in the H_2 molecule, they scatter from both nuclei coherently. The hydrogen comes in two possible spin states: *ortho-* and *parahydrogen* are two states of the molecule H_2 with different states of the two proton spins: $S = 0$ (↑↓)(para) and $S = 1$ ↑↑ (ortho). Their mixture in H_2 depends on temperature, being mostly para at low temperature (99.9 % at 20 K) and a statistical mixture of 75 % ortho to 25 % para at room temperature. Coherent scattering from parahydrogen means $\sigma_{\text{para}} = 4\pi (a_1 + a_2)^2$ (instead of $\sigma_{\text{para}} = 4\pi (a_1^2 + a_2^2)$), leading to an interference term in σ. This term allows the determination of the sign of the scattering lengths, which in turn are related to the binding strength of the np interaction, different for singlet or triplet p-n scattering via the spin-spin force. The neutron scattering from each proton can occur in the p-n singlet or triplet states with expectation values $\langle \vec{\sigma}_n \cdot \vec{\sigma}_p(i) \rangle = -3$ (singlet) or $= 1$ (triplet) with corresponding scattering lengths a_s and a_t.

For σ_{ortho} the contributions of the singlet and triplet p-n states to the mixed states of H_2, leading to total spins of $3/2$ and $1/2$, have to be considered. Experimentally only σ_{para} and σ of the scattering from the 75 % ortho/25 % para mixture of H_2 at room temperature or from the free proton can be measured.

(a) Derive the values of the p-n ortho cross section and of the singlet and triplet p-n scattering lengths. Use the following definitions: The combined scattering length of scattering of the neutron from one proton can be expressed by $a_{1,2} = a_s(1 - \vec{\sigma}_n \cdot \vec{\sigma}_p) + a_t(3 + \vec{\sigma}_n \cdot \vec{\sigma}_p)$ (the same for both protons). The total cross section is thus $\sigma = 4\pi(a_1 + a_2)^2$. With $\vec{\sigma}_1 + \vec{\sigma}_2 = \vec{\sigma}_{H_2} = 0$ we obtain

$$\sigma_{\text{para}} = \pi (a_s + 3a_t)^3. \tag{9.14}$$

Show that a similar Ansatz for ortho-H_2 leads to

$$\sigma_{\text{ortho}} = \sigma_{\text{para}} + 2\pi (a_t - a_s)^2. \tag{9.15}$$

(Note: $\langle \vec{\sigma}_n \cdot \vec{\sigma}_{H_2} \rangle = 0$ due to their spins being uncorrelated; use the expectation values $\langle \vec{\sigma}_n \cdot \vec{\sigma}_{H_2} \rangle$ for total spin states $S = 1/2$ and $3/2$ (spin triplet H_2+ neutron spin) to calculate $\langle \vec{\sigma}_n \cdot \vec{\sigma}_{H_2} \rangle^2$. $\vec{\sigma}_i$ designates the spin 1/2 Pauli operators).

Experimental results are: $\sigma_{\text{ortho}} = 155$ b and $\sigma_{\text{para}} = 5$ b at ≈ 0.8 meV. Calculate the singlet and triplet scattering lengths a_s and a_t, compare them

with values from p-n scattering experiments and the deuteron bindig energy, and discuss their signs in view of spin-dependent parts of the nuclear force.
(b) Show that the p-n triplet and singlet are bound/unbound and discuss this spin-dependent part for the NN force.
(c) Show that the measured cross section values are in agreement with a neutron spin of 1/2, but not of 3/2.

9.8. The three spin-triplet states $^3D_1$, $^3D_2$, and $^3D_3$ would be degenerate with the spin-spin and central forces alone. The phase-shift analysis of n-p scattering shows, however, that they are different at all energies: Which additional force is required to explain this (see exercises above)?

References

[ALT67] E.O. Alt, P. Grassberger, W. Sandhas, Nucl. Phys. B **2**, 167 (1967)
[BAU04] G. Baur, S. Typel, Prog. Theor. Phys. Suppl. **154**, 333 (2004)
[BRO76] G.E. Brown, A.D. Jackson, *The Nucleon-Nucleon Interaction* (North-Holland, Amsterdam, 1976)
[CNS13] http://nn-online.org/NN/ (2013)
[CRO01] A. Crowell, Ph.D. thesis, Duke U (2001)
[DEL08] A. Deltuva, A.C. Fonseca, P.U. Sauer, Annu. Rev. Nucl. Part. Sci. **58**, 27 (2008)
[DEL09] A. Deltuva, Phys. Rev. C **80**, 064002 (2009)
[FAD61] L.D. Faddeev, Sov. Phys. JETP **12**, 1014 (1961)
[GEB93] K. Gebhardt et al., Nucl. Phys. A **561**, 232 (1993)
[GLO83] W. Glöckle, *The Quantum-Mechanical Few-Body Problem* (Springer, Berlin, 1983)
[GLO96] W. Glöckle, H. Witała, D. Hüber, H. Kamada, J. Golak, The three-nucleon continuum: achievements, challenges, and applications. Phys. Rep. **274**, 107 (1996)
[HAG12] G. Hagen, M. Hjorth-Jensen, G.R. Jansen, R. Machleidt, T. Papenbrock, Phys. Rev. Lett. **109**, 032502 (2012)
[HAM13] H.-W. Hammer, A. Nogga, A. Schwenk, Colloquium: three-body forces: from cold atoms to nuclei. Rev. Mod. Phys. **85**, 197 (2013)
[HGS01] H. Paetz gen. Schieck, H. Witała, J. Golak, J. Kuroś, R. Skibinski, Few-Body Syst. **30**, 81 (2001)
[KAL12] N. Kalantar-Nayestanaki, E. Epelbaum, J.G. Messchendorp, N. Nogga, Rep. Prog. Phys. **75**, 016301 (2012)
[KRE94] W. Kretschmer, J. Albert, M. Clajus, P.M. Egun, A. Glombik, W. Grüebler, P. Hautle, P. Nebert, A. Rauscher, P.A. Schmelzbach, I. Šlaus, Phys. Lett. B **328**, 5 (1994)
[MAC04] R. Macrì Ph.D. thesis, Duke U (2004)
[SAG10] K. Sagara, Few-Body Syst. **48**, 59 (2010)
[SET96] H. Setze et al., Phys. Lett. B **388**, 229 (1996)
[STE89] M. Stephan et al., Phys. Rev. C **39**, 2133 (1989)
[STR89] J. Strate et al., Nucl. Phys. A **501**, 51 (1989)
[YAK67] O.A. Yakubovsky, Sov. J. Nucl. Phys. **5**, 937 (1967)
[ZHO01] Z. Zhou et al., Nucl. Phys. A **468**, 545c (2001)

Chapter 10
Models of Reactions—Direct Reactions

Models of real reactions have, by definition, limited ranges of applicability. There is a hierarchy of excitations that have to be described by different reaction models. Direct reactions are in principle those, which excite only a small number of degrees of freedom thus also involving only few nucleons in the reaction. The limiting case are single-particle excitations in reactions, which therefore proceed fast and peripheral. Typical structures in excitation functions are therefore wide ($\approx$ MeV). Compound-nuclear reactions are at the other end of the spectrum of the possible excitations and involve many, in the limit all nucleons in the interaction. They require the attainment of a state of thermal equilibrium of all degrees of freedom. Necessarily this is connected with long equilibration times and—via the uncertainty relation—narrow structures in the excitation functions. Between these extremes of the hierarchy of excitations (for illustration see Fig. 1.4) we have a number of different processes that, depending on their contexts are called *semi-direct, pre-equilibrium, pre-compound, doorway, hallway etc.* processes.

10.1 Generalities

Direct reactions have a main feature, which is their *time scale*. They occur in times on the order-of-magnitude of the passage time of the projectiles through the target nucleus.

10.2 Elastic Scattering

The most important reaction model for the description of the direct ("shape") elastic scattering is the *optical model*. It is based on the idea that at somewhat higher energies absorption may be assumed and thus the potential (and subsequently also the scattering phase shifts) must be complex (see also Sect. 8.2).

10.3 Optical Model

As standard literature on the optical model only a few references will be given here. The references [HOD63, HOD67, MAR70] cover much of the field. The relevant radial Schrödinger equation for protons (spin $s = 1/2$) reads

$$\left[\frac{d^2}{dr^2} + k^2 - \frac{\ell(\ell+1)}{r^2} + Vf(r) + iWg(r) - V_C(r)\right.$$

$$\left. + (V_{\text{s.o.}} + iW_{\text{s.o.}})h(r) \cdot \left\{\begin{array}{c}\ell \\ -(\ell+1)\end{array}\right\}\right]u_{\ell j}^{(\pm)}(kr) = 0. \quad (10.1)$$

Here: V, W are real and imaginary parts of the central potential, V_C the Coulomb potential, and $V_{\text{s.o.}}$, $W_{\text{s.o.}}$ real and imaginary parts of the spin-orbit potential, especially important for die description of polarization observables. For neutrons the Coulomb term vanishes. The solution of this equation is part of the total scattering function

$$\Psi = \frac{1}{kr}\sum_{\ell j\lambda} i^\ell \left[4\pi(2\ell+1)\right]^{1/2} \langle \ell 0 s\mu | jm\rangle \langle \ell \lambda s\nu | jm\rangle u_{\ell j}^{(\pm)}(kr) Y_\ell^\lambda(\theta,\phi) \chi_s^\mu e^{i\sigma_\ell}$$

(10.2)

with

$$u_{\ell j} \to_{r\to\infty} \frac{1}{2i}\left[e^{-i(kr-\eta_S \ln 2kr - \ell\pi/2)} - e^{2i\delta_{\ell j}} e^{i(kr-\eta_S \ln 2kr - \ell\pi/2 + 2\sigma_\ell)}\right]. \quad (10.3)$$

$\delta_{\ell j}$ are the complex nuclear scattering phases, $\sigma_\ell = \arg\Gamma(1+\ell+i\eta_S)$ the Coulomb scattering phases, $\eta_\ell^j = e^{2i\delta_{\ell j}}$ the "reflection coefficients", $\eta_S = Z_1 Z_2 e^2/\hbar v$ the Coulomb (or Sommerfeld) parameter (see Eq. (2.2)), and $k = \sqrt{2\mu E_{\text{kin}}^{\text{c.m.}}/\hbar^2}$ the entrance-channel wave number.

The potential form factor $f(r)$ is defined in analogy to the shape of the usual nuclear density or potential distributions that are used in the classical shell model (*Woods-Saxon form*). The absorption occurs predominantly at the nuclear surface. Thus, as form for $g(r)$ at low energies one chooses the derivative of the Woods-Saxon-form factor, and for the spin-orbit term h the *Thomas form* $g(r)/r$. At higher energies more absorption in the nuclear volume is plausible, which is taken into account by a gradual transition from the surface absorption to volume absorption.

The best sets of parameters have been obtained by fits with χ^2 minimization to a large number of data sets of cross sections as well as analyzing powers. The latter are important for fixing the $(\vec{L} \cdot \vec{S})$ potential and removing typical ambiguities in the potential parameters. A few of these parameter sets have become standards for the optical model. For nucleon scattering the parametrization most used is that of Greenlees and Becchetti [BEC69], especially because they provide a global set of parameters (i.e. valid over a large region of the periodic table). However, in special cases e.g. near doubly-magic nuclei this set is not as good as a single fit. It is interesting that the depth of the real potential corresponds closely to that of the shell model

Fig. 10.1 Form factors of the optical model. *Upper*: Woods-Saxon form of the real part f. *Center*: Derivative Woods-Saxon form $g = f'$ of the imaginary part. *Lower*: Sliding-transition form of the surface-to-volume imaginary part as function of energy

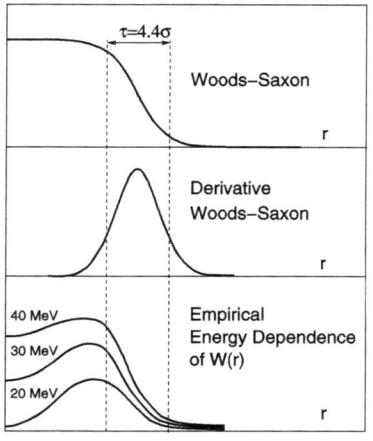

potential, similarly for the LS term. For light projectiles consisting of A nucleons (deuterons, α particles etc.) one has potential depths that are A-fold the nucleon potential depths. For heavy-ion scattering there are quite different approaches, partly with very shallow potentials. Figure 10.1 shows the form factors and the behavior of imaginary potentials with energy. It can only be mentioned in passing that steps to found the optical potentials on more microscopic grounds have been undertaken by creating *folding potentials*. In these the potential of one nucleon of the projectile with the target nucleus is folded with the nucleon density of the projectile nucleus and vice versa, or the same for both nuclei (*double folding potentials*).

10.4 Direct (Rearrangement) Reactions

The multitude of these reactions may be classified:

- Reactions without change of the mass number

    - Elastic potential scattering (see above; description by the optical model).
    - Direct inelastic scattering [$(p, p'\gamma)$, (α, α'), ...]. It leads preferentially to collective nuclear excitations (such as rotation, vibration etc.).
    - Quasi-elastic (charge-exchange) processes ((p, n), (n, p), $(^3\text{He}, t)$, $(^{14}\text{N}, ^{14}\text{C})$, ...). These lead e.g. to isobaric-analog states of the target nucleus.

- Reactions with change of the mass number

    - Pickup reactions [One-nucleon transfer: (p, d), $(d, ^3\text{He})$, (d, t), ..., few- or multi-nucleon transfer: (p, α), $(d, ^6\text{Li})$, ...].
    - Stripping reactions [One-nucleon transfer: (d, p), (d, n), $(^3\text{He}, d)$, ...) few- or multi-nucleon transfer: $(^6\text{Li}, d)$, (α, p), $(^3\text{He}, p)$, ...].
    - Knockout reactions [(p, α), (p, p'), ...].

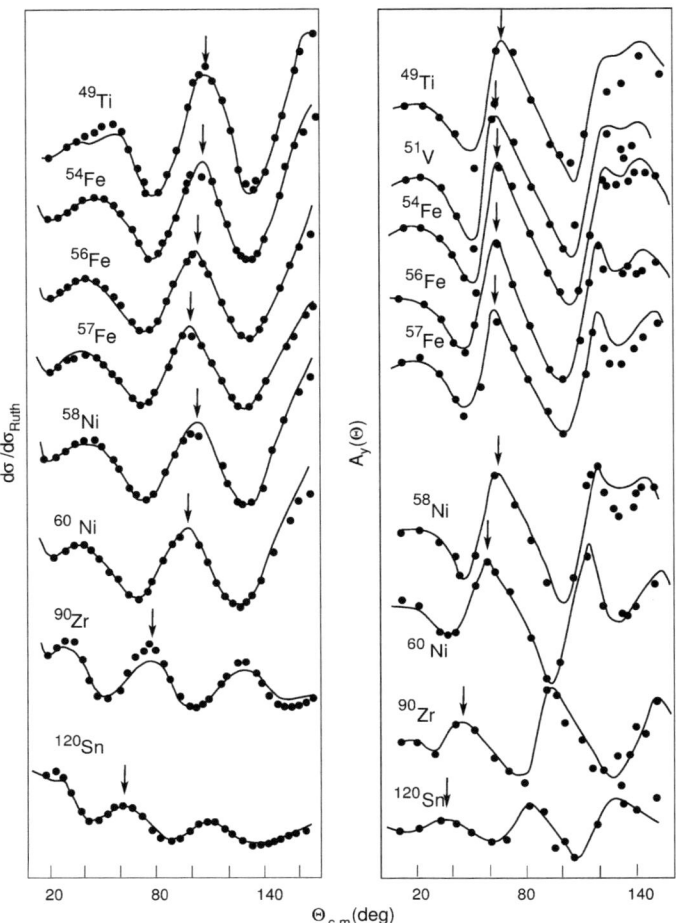

Fig. 10.2 Global fit of the optical model to elastic scattering data of 14.5 MeV polarized protons for a large nuclear mass range. The cross sections are normalized to the Rutherford cross sections (i.e. to 1 at 0°), the analyzing powers are 0 at 0°. The *arrows* indicate the systematic variation of characteristic diffraction maxima with the target mass. Adapted from Ref. [BEC69]

- Direct breakup processes like knockout with few-particle exit channels [$(p, pp), (\alpha, 2\alpha), \ldots$].
- Induced fission is a special case of a rearrangement reaction resulting in larger debris.
- Processes of higher order (multi-step processes via excited intermediate states, coupled channels).

Here only the simplest case of the stripping reaction will be discussed. The many details of direct interaction processes are subjects of a large number of books, see e.g. Refs. [AUS70, SAT83, GLE63, GLE83]. Standard computer codes such as

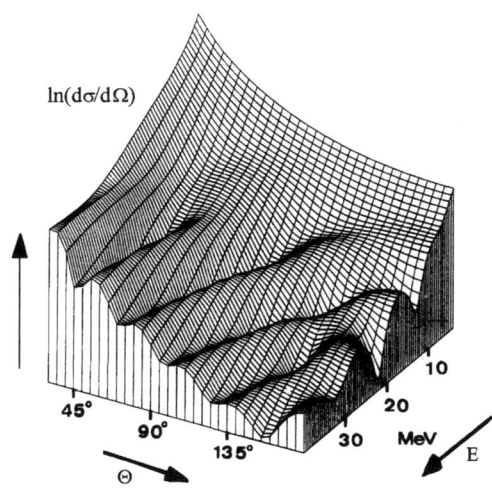

Fig. 10.3 Angular and energy dependence of the cross section of elastic proton scattering from ^{90}Zr calculated with standard Greenlees-Becchetti parameters of the optical model. The interference structure of the angular distributions may be interpreted as "resonant" (single-particle) structures of the excitation function with widths typical for fast (i.e. *direct*) processes. They are also analogous to diffraction structures in classical optics

DWUCKn (distorted wave code) and CHUCKn (coupled channels distorted wave code) [KUNZ] are available. For induced nuclear fission see Sect. 11.4.1. Figure 10.2 shows the systematics of the optical-model description with mass number. Figure 10.3 depicts the diffraction-like structures of the OM with energy and angle.

10.5 Stripping Reactions

Already a semi-classical Ansatz provides a qualitative picture of the angular distributions of stripping reactions. It explains the expected behavior with the assumption of a rapid process that is localized at the nuclear rim and is non-equilibrated. The wave-number vector of the incoming deuterons is $\vec{k}_d$, those of the transferred nucleon and of the outgoing nucleons are $\vec{k}_n$ and $\vec{k}_p$, respectively. They form a momentum diagram, from which the connection between a preferred scattering angle θ and the transferred momentum and also the angular momentum can be deduced:

$$p_n R = \hbar k_n R = \hbar \ell_n. \tag{10.4}$$

For small θ

$$\theta_0 \approx \frac{k_n}{k_d} = \frac{\ell_n}{k_d R}. \tag{10.5}$$

Because of the quantization of ℓ there are discrete values of θ increasing with ℓ. This qualitative picture is not changed when calculating the angular distributions quantum-mechanically. As an example for the reaction ^{52}Cr$(d, p)^{53}$Cr the angles of

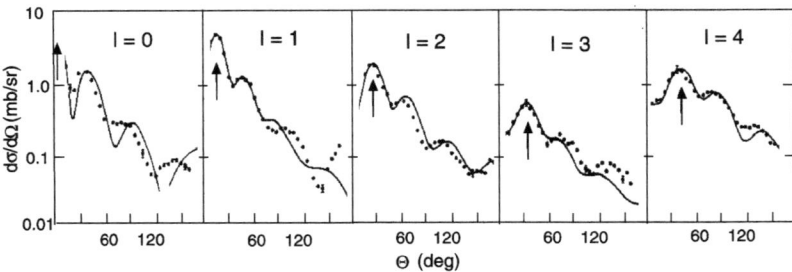

Fig. 10.4 Characteristic and systematic features of the stripping maximum as function of the transferred orbital angular momentum ℓ. The *arrows* indicate the increase of the reaction angle of the stripping peak with ℓ. Adapted from Ref. [EHR67]

the stripping maximum in calculated in different ways are

$$
\begin{array}{cccc}
 & \theta^{\text{DWBA}} & \theta^{\text{PWBA}} & \theta_{\text{s.c.}} \\
\ell = 0 & 0^0 & 0^0 & 0^0 \\
\ell = 1 & 18^0 & 13^0 & 13^0 \\
\ell = 2 & 34^0 & 19^0 & 26^0 \\
\ell = 3 & 49^0 & 30^0 & 39^0 \\
\ell = 4 & 64^0 & 40^0 & 52^0
\end{array}
\tag{10.6}
$$

The measured angular distributions of the cross sections show—in addition to diffraction structures—marked stripping maxima, which often allow the determination of the angular momentum of the transferred nucleon. An example is shown in Fig. 10.4 for the stripping reactions $\text{Zn}(d,p)$ from different Zn isotopes to different final states at $E_d = 10$ MeV, selected according to the transferred angular momentum ℓ.

Historically this feature was (and still is) important for the assignment of the final-nuclear states of a reaction to orbitals in the one-particle shell model. The energy relations in stripping reactions are such that the transferred nucleon near magic shells is preferentially inserted into low-lying shell-model states.

Because of the spin-orbit splitting the complete assignment requires that also the total angular momentum j of the transferred nucleons is known. A good method is the measurement of the analyzing power of the stripping reaction, i.e. the use of polarized projectiles. In many cases the distinction between the two possibilities for j can be made just from the sign of the analyzing power alone. An example is shown in Fig. 10.5 where the transitions to the $p_{1/2}$ and $p_{3/2}$ are clearly distinguished.

For an intuitive description of this fact there exists again a simple semi-classical model [NEW53]. It is assumed that the interaction happens at the nuclear surface and that we have a relatively strong absorption in the nuclear matter. Thus the front side of the nucleus directed towards the projectile contributes more strongly to the reaction than the backside of the target nucleus. In the front part the orbital angular momentum vector points upward perpendicularly to the reaction plane whereas in the back half the orbital angular momentum points down. If the incident deuteron is

10.5 Stripping Reactions

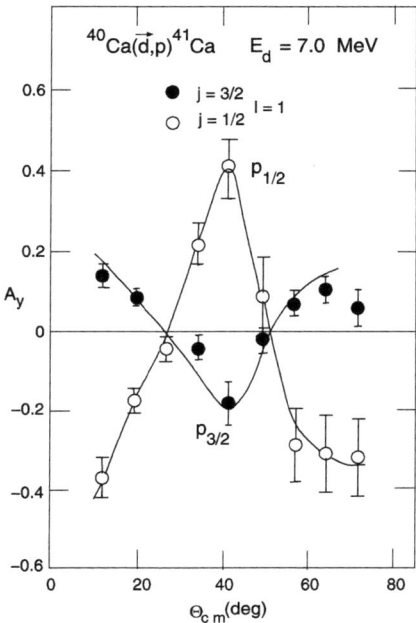

Fig. 10.5 Sensitivity (sign!) of the analyzing power of the stripping reaction to the total angular momentum j of the transferred nucleon. Adapted from Ref. [YUL68]

polarized up or down perpendicularly to the scattering plane—under the assumption of the existence of a spin-orbit force—the transferred nucleon in the scattering to the left ends preferentially in a state with $j = \ell + 1/2$ for the up case, in the down case with $j = \ell - 1/2$. The measured analyzing power

$$A_y = \frac{1}{P_d} \frac{N^{\text{up}} - N^{\text{down}}}{N^{\text{up}} + N^{\text{down}}}, \qquad (10.7)$$

will show opposite signs for the two cases. This behavior has been confirmed for many examples not only for stripping reactions. If one wants to know the degree, to which the transition considered is a single-particle transition the *spectroscopic factor* has to be determined. For that—at least approximately—quantitative theories are required (DWBA, CC-DWBA).

Because the single-particle strength is often strongly fractionated by the residual interaction, i.e. spread out over many states in a range of energies spectroscopic investigations on very many final nuclear states are necessary to make sure that all states belong to such a multiplet. Their strengths follow approximately a gaussian shape with a width proportional to the strength of the residual interaction—a typical strength-function and intermediate-structure behavior. Often these states are close together and high detector resolution to get a complete picture is necessary, i.e. one must be sure to observe all states in order to correctly compare with theories (*completeness*). Especially useful tools for this purpose are magnetic spectrographs with high resolution at tandem Van-de-Graaff accelerators, also with polarized particle beams (for more details see Sect. 17.3.3).

10.6 T Matrix and Born Series

10.6.1 Integral Equations

Starting from the Schrödinger equation

$$H\Psi = E\Psi, \tag{10.8}$$

which may be written

$$(E - H)\Psi = 0 \tag{10.9}$$

with the Ansatz

$$H = H_0 + U_\alpha \tag{10.10}$$

(where α designates the reaction channel formed by a and A) and

$$H_0 = H_a + H_A - \frac{\hbar^2}{2\mu}\nabla_\alpha^2, \tag{10.11}$$

two equations (an inhomogeneous and a homogeneous equation) result:

$$(E - H_0)|\Psi\rangle = U_\alpha|\Psi\rangle, \tag{10.12}$$

$$(E - H_0)|\Phi\rangle = 0. \tag{10.13}$$

The formal solution of the inhomogeneous equation is

$$\left|\Psi^{(+)}\right\rangle = \lim_{\epsilon \to 0} \frac{1}{E - H_0 + i\epsilon} U_\alpha \left|\Psi^{(+)}\right\rangle. \tag{10.14}$$

This solution already contains the boundary condition of an outgoing wave and the avoidance of poles in the complex E or k plane. It is an integral equation. The general solution of the inhomogeneous equation is the sum of this special solution and the general solution of the homogeneous equation (this is a plane incident wave):

$$\left|\Psi^{(+)}\right\rangle = |\Phi\rangle + \underbrace{\lim_{\epsilon \to 0} \frac{1}{E - H_0 + i\epsilon}}_{G_0} U_\alpha \left|\Psi^{(+)}\right\rangle. \tag{10.15}$$

This solution contains the solution function only implicitly. The solution is obtained iteratively with the plane wave as starting function. The resulting *Born series* may then be terminated at any iteration step.

$$\left|\Psi^{(+)}\right\rangle = \underbrace{|\Phi\rangle + G_0 U |\Phi\rangle}_{\text{1st Born approximation}} + G_0 U G_0 U |\Phi\rangle + \cdots. \tag{10.16}$$

10.7 Born Approximation

The form of G_0 is obtained from the solution of the inhomogeneous equation with a δ potential, the (*Green's function*)

$$G_0(\vec{r},\vec{r}') = -\frac{1}{4\pi} \frac{e^{ik|\vec{r}-\vec{r}'|}}{|\vec{r}-\vec{r}'|}. \tag{10.17}$$

With this we have the *explicit* form of the *Lippmann-Schwinger equation*

$$\Psi^{(+)} = e^{i\vec{k}\vec{r}} - \frac{1}{4\pi} \int \frac{e^{ik|\vec{r}-\vec{r}'|}}{|\vec{r}-\vec{r}'|} U\Psi(\vec{r}')d\vec{r}'. \tag{10.18}$$

Asymptotically with

$$\frac{1}{|\vec{r}-\vec{r}'|} \xrightarrow{r\to\infty} \frac{1}{r} \quad \text{and} \quad |\vec{r}-\vec{r}'| \xrightarrow{r\to\infty} r - \frac{\vec{r}}{r}\vec{r}' \tag{10.19}$$

the *asymptotic* Lippmann-Schwinger equation results

$$\Psi^{(+)}(\vec{k},\vec{r}) = e^{i\vec{k}_i\vec{r}} - \frac{e^{ikr}}{r}\left[\frac{\mu}{2\pi\hbar^2}\int e^{-i\vec{k}\vec{r}'}V(\vec{r}')\Psi^{(+)}(\vec{k},\vec{r}')d\vec{r}'\right]. \tag{10.20}$$

The comparison with the usual Ansatz leads to the scattering amplitude (with $V = (\hbar^2/2\mu)U$)

$$f(\theta,\phi) = -\frac{\mu}{2\pi\hbar^2}\int e^{-i\vec{k}\vec{r}'}V(\vec{r}')\Psi^{(+)}(\vec{k},\vec{r}')d\vec{r}' = -\frac{\mu}{2\pi\hbar^2}\langle\Phi|V|\Psi^{(+)}\rangle \tag{10.21}$$

and—after normalization—to the incident particle flux

$$f(\theta,\phi) = -\frac{\mu(2\pi)^2}{\hbar^2}\underbrace{\langle\Phi|V|\Psi^{(+)}\rangle}_{\text{T-Matrix }T_{fi}}. \tag{10.22}$$

The relation between T- and S-matrix is

$$S_{fi} = \delta_{fi} - i(2\pi)\cdot c \cdot \delta(\vec{k}_i,\vec{k}_f)\delta(E_i,E_f)\cdot T_{fi}. \tag{10.23}$$

Only for elastic scattering is $\delta_{fi} \neq 0$.

10.7 Born Approximation

10.7.1 First Born Approximation = PWBA = Plane Wave Born Approximation

As indicated in Eq. (10.16) the insertion of the incident plane wave Φ also as a start solution function in the Born series and truncating it after the first term

$$|\Psi^{(+)}\rangle = |\Phi\rangle + GV|\Phi\rangle \tag{10.24}$$

makes the integral equation directly solvable. In analogy to the procedure discussed above we obtain the scattering amplitude

$$f(\theta,\phi) = -\frac{(2\pi)^2\mu}{\hbar^2}\langle\Phi|V|\Phi\rangle \qquad (10.25)$$

and from this all observables. The result is the same as obtained by inserting plane waves into *Fermi's Golden Rule*.

10.7.2 Distorted Wave Born Approximation = DWBA

In the framework of this First Born approximation an improvement can be reached by using waves, which are "distorted" in the nuclear and Coulomb field. These waves contain diffraction as well as absorption contributions and are described by the optical model of elastic scattering for the relevant entrance and exit channels. An example: For the description of the transfer reaction ^{40}Ca$(d, p)^{41}$Ca the wave functions are needed that were obtained from best-fits of optical potentials to scattering data of ^{40}Ca$(d, d)^{40}$Ca and ^{41}Ca$(p, p)^{41}$Ca at the proper channel energies. Since ^{41}Ca targets are not available the scattering ^{40}Ar$(p, p)^{40}$Ar was measured, which—because of the weak A dependence of the optical-model parameters—is justified, see Fig. 10.6.

10.8 Details of the Born Approximations

Here one uses the first Born approximation, i.e. the first term of the Born series. Starting from Fermi's *Golden Rule* of perturbation theory, which predicts for the differential cross section

$$\frac{d\sigma}{d\Omega} = \frac{(2I_b+1)(2I_B+1)}{2\pi^2\hbar^4}\mu_i\mu_f\frac{k_f}{k_i}|T_{if}|^2, \qquad (10.26)$$

one has to make assumptions about the transition matrix element.

In the *Plane Wave Born Approximation* PWBA (also: Butler theory) for the incoming and outgoing waves plane waves are used. Since the radial wave functions are Bessel functions one finds a simple diffraction pattern for the cross section

$$\frac{d\sigma}{d\Omega} \propto \left[j_\ell(kR)\right]^2. \qquad (10.27)$$

For illustration Fig. 10.7 shows the few lowest-order spherical Bessel functions squared. The angle dependence of the stripping maximum is contained only in the momentum relation $k^2 = k_{\text{in}}^2 + k_{\text{out}}^2 - 2k_{\text{in}}k_{\text{out}}\cos\theta$. Only in a few simple cases the

10.8 Details of the Born Approximations

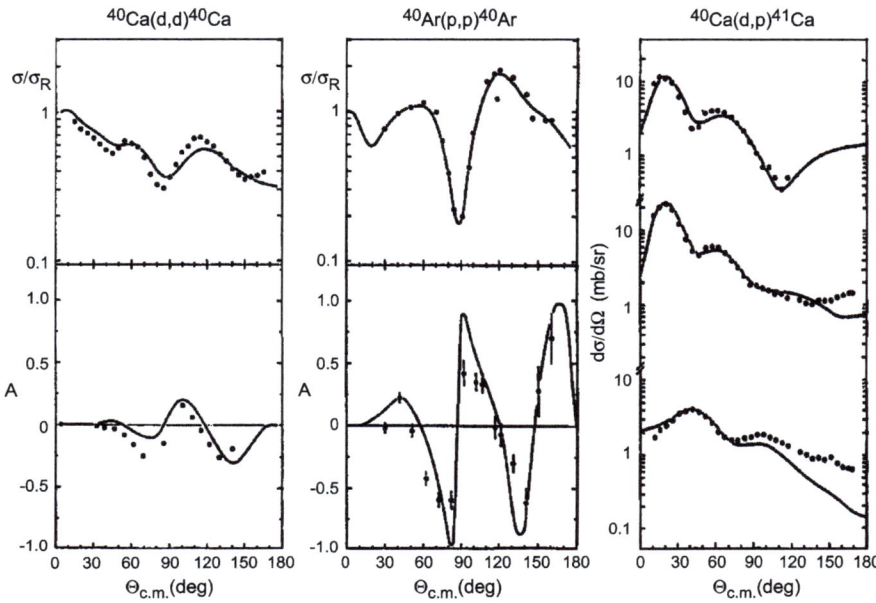

Fig. 10.6 Description of observables of the transfer reaction ^{40}Ca$(d, p)^{41}$Ca to several final states in ^{41}Ca by DWBA: $f_{7/2}$ ground state (*bottom right*), excited states $p_{3/2}$ at $E_x = 1.95$ MeV (*center*) and $p_{1/2}$ at $E_x = 3.95$ MeV (*top right*). The "distorted waves" were determined by fitting cross sections and polarization data of the elastic scatterings of the (d, p) entrance and exit channels by the optical model at the appropriate energies $E_d = 7.0$ MeV, $E_p = 12$ MeV (14.5 MeV for the analyzing power). From Ref. [YUL68]

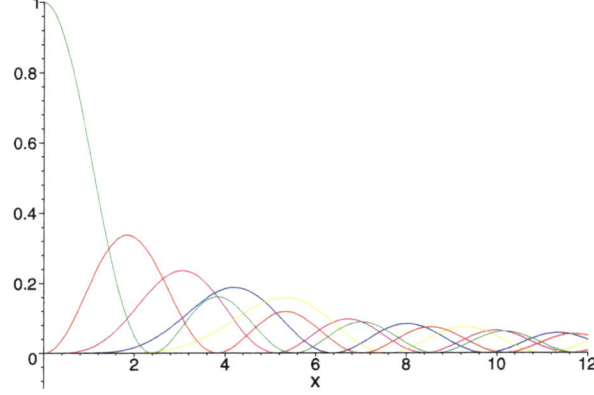

Fig. 10.7 The behavior of the squares of the lowest-order spherical Bessel functions $j_\ell(kr)$ as functions of $x = kr$

angular distribution near the maximum are satisfactorily described by PWBA. It also makes no statements about polarization observables and contains no information about nuclear structure.

Better results at least for forward angles are obtained with the *Distorted Wave Born Approximation* DWBA. It was formulated with a number of additional and far-reaching assumptions:

- In the entrance and exit channels *distorted* waves are used, i.e. the wave functions are the solutions obtained from fits of the optical model (OM) to the elastic-scattering data in each pertaining channel at the proper channel energy. E.g. for the description of the reaction $A(d, p)B$ the OM wave functions from the fit to the data of the scattering $A(d, d)A$ as well as of $B(p, p)B$ are needed. Thus—still in the first Born approximation—the diffraction and absorption of the incoming and outgoing waves in the nuclear (and eventually the Coulomb) field, as well as the effect of the LS potential (see also Fig. 10.6) are taken into account.
- The nuclear initial and final states are shell-model states.
- The finite range of the nuclear forces is taken care of by a *finite-range* or even *zero-range* approximation.
- The T matrix is expanded into partial waves belonging to fixed angular-momentum transfer.
- The transfer matrix is factorized into a nuclear-structure dependent and into a kinematical part.

Thus the cross section reads

$$\frac{d\sigma}{d\Omega} = \frac{\mu_a \mu_b}{\pi \hbar^4} \left(\frac{m_B}{m_A}\right)^4$$
$$\times \frac{2J_B + 1}{(2J_A + 1)(2s_a + 1)} \frac{1}{k_a k_b} \sum_{\ell s j} \left[|A_{\ell s j}|^2 \sum_m |\beta_{sj}^{\ell m}|^2\right]. \quad (10.28)$$

The experimental cross section is a product of a fit parameter, the *spectroscopic factor* $S_{\ell j}$, and a theoretical cross section calculated in the framework of the DWBA with the assumption of single-particle states:

$$\left(\frac{d\sigma}{d\Omega}\right)_{\exp}^{\ell j} = S_{\ell j} \left(\frac{d\sigma}{d\Omega}\right)_{\text{DWBA}}^{\ell j}. \quad (10.29)$$

In a stripping process the spectroscopic factor is the square of the amplitude of a fragment of a single-particle state of the final nucleus. Because of this fractionation (which in reality is caused by the residual interaction of the many other nucleons) into many states with equal quantum numbers the *strengths* of all these states have to be summed up. If a complete collection from all these states is possible one obtains the total strength, which can also be calculated because the number of nucleons N in a subshell is known. Therefore sum rules can be applied, e.g. for single-particle stripping $\sum S_{\ell j} = (2J + 1)$. Mathematically the spectroscopic factor is the overlap integral between the anti-symmetrized k-particle final-nuclear state $\Psi_A(i)$, into which the nucleon is inserted, and the single-particle configuration of the anti-symmetrized $(k - 1)$-particles target, the nuclear ground state (core), and the single-particle wave function of the transferred k-th particle $\Psi(j)$. It thus gives the probability, with which a certain state is present in this configuration. When averaging over the strength distribution of all states that are fractions of one single-particle state (e.g. while assuming a Breit-Wigner distribution function) the position of the average energy provides the energy of the single-particle state, whereas the width of the distribution is a measure of its lifetime, the *(spreading width)* $\Gamma^\downarrow$. It measures the decay of the single-particle state into the real nuclear states, split and spread out by the residual interaction, and thus its strength.

10.9 Exercises

10.1. Experiments in the ^{40}Ca$(d, p_i)^{41}$Ca* reaction for two of many transitions to final states in ^{41}Ca show marked stripping peaks at weakly but $\approx$ linearly varying reaction angle as function of the incident deuteron energy.

- $\theta_{c.m.}$ shifts from $\approx 20°$ at $E_d = 7$ MeV to $13°$ at $E_d = 12$ MeV for the transition to the state with $E_x = 4.19$ MeV.
- $\theta_{c.m.}$ shifts from $\approx 40°$ at $E_d = 7$ MeV to $32°$ at $E_d = 12$ MeV for the transition to the state with $E_x = 6.14$ MeV.

Find the orbital angular momentum ℓ transferred by the neutrons for the two transitions. Are the variations with the incident energy consistent with the assignments?

10.2. Similarly transitions in the ^{68}Zn$(d, p)^{69}$Zn reaction at $E_d = 10$ MeV show stripping peaks at angles θ.

E_x (MeV)	$\theta_{c.m.}$
0.0	$\approx 16°$
0.44	$\approx 42°$
0.55	$\approx 35°$
2.4	$\approx 0°$
2.65	$\approx 23°$

Assign the transferred orbital angular momenta ℓ to the different final states.

10.3. Apply the Newns (absorption) model to two different transitions with transferred $\ell = 1$ to assign $j = \ell \pm 1/2$ in ^{40}Ca$(d, p)^{41}$Ca to the final states with $E_x = 1.95$ MeV that has a positive analyzing power A_y in the stripping peak, and $E_x = 3.95$ MeV with $A_y < 0$. Check this also for two transitions in ^{208}Pb$(d, p)^{209}$Pb at $E_{lab.} = 12.3$ MeV, both to final states assigned to $\ell = 2$, but with different total angular momentum $j = \ell \pm 1/2$, at $E_x = 2.54$ MeV, with $A_y \approx +0.4$ and at $E_x = 1.57$ MeV with $A_y \approx -0.3$. (In the shell model the energy splitting of states with equal ℓ intodoublets with different j is a consequence of the $(\vec{\ell} \cdot \vec{s})$ potential).

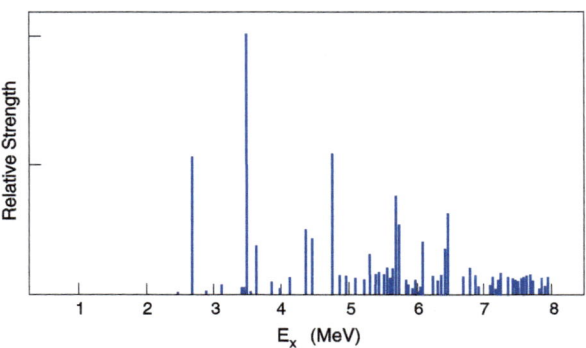

Fig. 10.8 Fractionation ("spreading") of a $d_{5/2}$ state in ^{61}Ni by the residual interaction, forming a strength function with a characteristic spreading width

10.4. An experimental evaluation of the ^{208}Pb$(d, p)^{209}$Pb transition to the $J^\pi = 9/2^+$ ground state of ^{209}Pb gave a spectroscopic factor of $S = 1.01 \pm 0.074$, whereas the spectroscopic factor for a transition in ^{207}Pb$(d, p)^{208}$Pb to the first excited state with $J^\pi = 3^-$ in ^{208}Pb at $E_x = 2.62$ MeV gave $S = 0.06$. However, for the neighboring particle-hole doublet states in ^{208}Pb with $J^\pi = 4^-, 5^-$ values of $S = 0.82$ and $S = 0.9$, respectively, were determined. Discuss the character of the different states.

10.5. Using a magnetic spectrograph many $d_{5/2}$ states ($\ell = 2$), clustering around a certain excitation energy E_0, in the reaction ^{60}Ni$(d, p)^{61}$Ni have been identified, see Fig. 10.8. Assuming that all states belong to one single-particle state (and that all states have been found and are thus fulfilling a sum rule) with the above quantum numbers, fractionated and spread out over a wider energy region, propose a method to determine the location of the s.p. state and the spreading width $\Gamma^\downarrow$ of the corresponding strength function (this is an "intermediate structure" with fine structure, see Chap. 12). Check the result with the s.p. shell model. Which interaction fractionates and spreads the s.p. state? What does the width $\Gamma^\downarrow$ tell us?

References

[AUS70] N. Austern, *Direct Nuclear Reaction Theory* (Wiley, New York, 1970)
[BEC69] F.D. Becchetti Jr., G.W. Greenlees, Phys. Rev. **182**, 1190 (1969)
[EHR67] D. von Ehrenstein, J.P. Schiffer, Phys. Rev. **164**, 1374 (1967)
[GLE83] N.K. Glendenning, *Direct Nuclear Reactions* (Academic Press, New York, 1983)
[GLE63] N.K. Glendenning, Nuclear stripping reactions. Annu. Rev. Nucl. Sci. **13**, 191 (1963)
[HOD63] P.E. Hodgson, *The Optical Model of Elastic Scattering* (Clarendon Press, Oxford, 1963)
[HOD67] P.E. Hodgson, The optical model of the nucleon-nucleus interaction. Annu. Rev. Nucl. Sci. **17**, 1 (1967)
[KUNZ]   P.D. Kunz, available from University of Colorado
[MAR70] P. Marmier, E. Sheldon, *Physics of Nuclei and Particles*, vol. II, 1087 ff. (Academic Press, New York, 1970)
[NEW53] H.C. Newns, Proc. Phys. Soc. Lond. A **66**, 477 (1953)
[SAT83] G.R. Satchler, *Direct Nuclear Reactions* (Clarendon Press, Oxford, 1983)
[YUL68] T.J. Yule, W. Haeberli, Nucl. Phys. A **117**, 1 (1968)

Chapter 11
Models of Reactions—Compound-Nucleus (CN) Reactions

11.1 Generalities

Resonances are a very general phenomenon in nature and therefore in all of physics. In classical physics they appear when a system capable of oscillations is excited with one or more of its eigenfrequencies, which—depending on the degree of damping—may lead to large oscillation amplitudes of the system, see e.g. Ref. [FEY77]. Nuclei are no exception. When tuning the system (changing the exciting frequency) these amplitudes pass through a resonance curve of *Lorentz form*. In particle physics most of the many known "particles" actually appear as resonances, i.e. as quantum states in the continuum, which decay with characteristic widths (or equivalently: lifetimes). For nuclear excitations the theoretical description goes back to G. Breit and E. Wigner [BRE36] and in its simplest form describes isolated resonances ("single-level formula"). More complicated cases are: interference with background from direct (fast) processes and overlapping states of equal quantum numbers (spins and parities) which require "multi-level formulas" because here the quantum-mechanical interference leads to level repulsion and asymmetric resonance shapes. The theory is equally applicable to compound as well as intermediate-structure resonances.

Resonances can be discussed in the energy picture where as function of energy excursions of Lorentz form (Breit-Wigner form) with a width Γ appear, but also in the complementary time picture where they appear as quantum states in the continuum, i.e. as states, which decay with finite lifetime τ. Between them there is the relation

$$\Gamma = \hbar/\tau. \tag{11.1}$$

In nuclear physics resonances appear in the continuum (i.e., in scattering situations, at positive total energy) when the projectile energy in the c.m. system plus the Q value of the reaction just equals the excitation energy of a nuclear state.

The excitation functions of observables such as the cross section show characteristic excursions from the smooth background when varying the incident energy. The background may be due to a *direct-reaction* contribution from Coulomb or

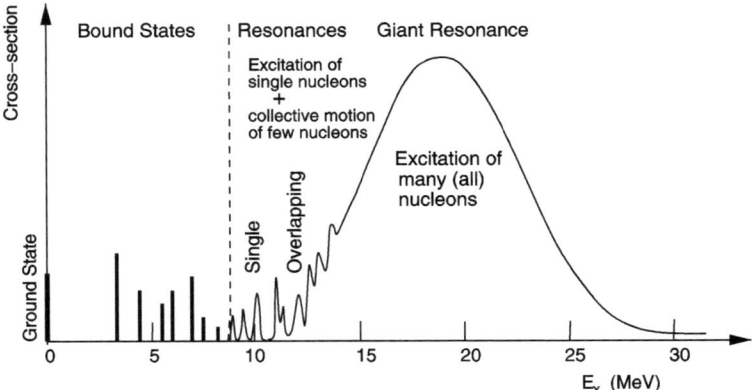

Fig. 11.1 The excitation of single resonances, overlapping resonances (with and without Ericson fluctuations) and giant resonances as functions of the energy in the continuum region above the bound-state energy

shape-elastic scattering or—in a region of high level density—may be the energy-averaged cross section of unresolved overlapping compound resonances; in this case resonant excursions would be due to doorway mechanisms. Likewise the scattering phases and scattering amplitudes change in characteristic ways over comparatively small energy intervals. Nuclei may be excited into collective modes such as rotations and/or vibrations of a part of the nucleons. At still higher energies new phenomena with high cross sections in charged-particle, neutron, γ, and π induced reactions appear involving up to all nucleons of a nucleus, the *Giant Resonances*, see Sect. 12.3. Figure 11.1 shows schematically the phenomena in different energy regions.

11.2 Theoretical Shape of the Cross Sections

A model assumption for resonances is—in contrast to direct processes—that the system goes via an intermediate state from entrance into the exit channel. For this case perturbation theory gives the following form of the transition matrix element

$$\langle \Psi_{\text{out}} | H_{\text{int}} | \Psi_{\text{in}} \rangle = \frac{\text{const}}{E - \tilde{E}_R}. \quad (11.2)$$

E_R is the energy of the nuclear eigenstate. However, since it is a state in the continuum it is not a stationary but one, which decays in time. Such states are best described by giving it a complex eigen-energy:

$$\tilde{E}_R = E_R + i\Gamma/2. \quad (11.3)$$

11.2 Theoretical Shape of the Cross Sections

The interpretation of the imaginary part is: the time development of a state has the form $e^{i\tilde{E}t/\hbar}$, on the other hand the state decays with a lifetime τ, whence

$$1/\tau = \text{Im}(\tilde{E}) = \Gamma/2. \tag{11.4}$$

The resonance amplitude thus has the form:

$$g(E) = \frac{F(E)}{E - E_R + i\Gamma/2}. \tag{11.5}$$

The meaning of $F(E)$ has to be determined. In the sense of Bohr's independence hypothesis formation and decay of a resonance are independent (i.e. decoupled). Therefore, one writes the amplitude as the product of the probability amplitude for its formation and its probability of decaying into the considered exit channel. In general, for one formation channel (the entrance channel c) there will be several exit channels c'.

The width Γ of the Breit-Wigner function is inversely proportional to the formation probability P and is the integral over the cross section in the energy range of the resonance:

$$P = \int \frac{\sigma_{aA} v_{aA}}{V} \cdot \frac{V p_{aA}^2 dp_{aA}}{2\pi^2 \hbar^3} = \int \frac{\sigma_{aA} k_{in}^2 dE}{2\pi^2 \hbar} \approx \frac{k_{in}^2 F(E_R)}{2\pi^2 \hbar} \int \frac{dE}{(E - E_R)^2 + \Gamma^2/4}$$

$$= \frac{k_{in}^2 F(E_R)}{\pi \hbar \Gamma}. \tag{11.6}$$

In equilibrium this is equal to the probability that the resonance re-decays into the entrance channel (purely elastic case). A measure for this is the partial width Γ_{aA} formed similarly as Γ, thus:

$$\Gamma_{aA}/\hbar = \frac{k_{in}^2 F(E_R)}{\pi \hbar \Gamma}, \tag{11.7}$$

$$F(E_R) = \frac{\pi}{k_{aA}^2} \Gamma_{aA} \Gamma. \tag{11.8}$$

By definition Γ is the sum of all partial widths over the open channels. Thus, the branching ratio for the decay into one definite channel $bB \equiv c'$ is equal to $\Gamma_{c'}/\Gamma$ and the *Breit-Wigner cross section* for the formation of the resonance via channel c and the decay via channel c' is

$$\sigma(E) = \frac{\pi}{k_{in}^2} \cdot \frac{\Gamma_c \Gamma_{c'}}{(E - E_R)^2 + \Gamma^2/4}. \tag{11.9}$$

This derivation is simplified and must be carried out—when there is interference with a direct background contribution and for the description of a differential cross section via a partial-wave expansion near a resonance—with complex *scattering* amplitudes. For elastic s-wave scattering it results in a resonant scattering amplitude of the form:

$$A_{\text{res}} = \frac{i\Gamma_{aA}}{(E - E_R) + i\Gamma/2}. \tag{11.10}$$

When a direct background is present, then, besides the pure resonance term and the pure direct (smooth) term, a typical interference term appears, which may be constructive or destructive. For σ we have then:

$$\sigma_{\text{tot}} = |A_{\text{res}} + A_{\text{pot}}|^2 = \sigma_{\text{res}} + \sigma_{\text{pot}} + 2\text{Re}\left(A_{\text{res}} A_{\text{pot}}^*\right) \tag{11.11}$$

where A_{pot} is the amplitude of the weakly energy-variable potential scattering.

11.3 Derivation of the Partial-Width Amplitude for Nuclei (s Waves Only)

The connection between the resonant scattering wave function and the wave function of the eigenstate of the nucleus is made by the *R-matrix theory*. Their basic features (for more details see [LAN58]) are approximately:

- The two wave functions and their first derivatives are matched continuously at nuclear radius (edge of the potential or similarly).
- The condition for a resonance is equivalent with the wave-function amplitude in the nuclear interior taking on a maximum value. This happens exactly if the matching at the nuclear radius occurs with a wave function with gradient zero (horizontal tangent).

Figure 11.2 illustrates the conditions for resonance. The two conditions may be summarized such that both logarithmic derivatives L (L_0 for pure s waves) at the nuclear radius are exactly zero. With the form of the wave function in the external region

$$u_0(r) = e^{-ikr} - \eta_0 e^{ikr}, \quad r > a \tag{11.12}$$

and the wave numbers in the external region k in the nuclear interior κ we obtain

$$L_0(E) = \left(\frac{a}{u_0}\frac{du_0}{dr}\right)_{r=a}, \tag{11.13}$$

which leads to the scattering function η_0 as function of L_0:

$$\eta_0 = \frac{L_0 + ika}{L_0 - ika} e^{-2ika}. \tag{11.14}$$

Inserting this scattering function into the known expressions for elastic scattering and absorption (and with $L_0 = \text{Re}(L_0) + i\text{Im}(L_0)$) the result is

11.4 Role of Level Densities

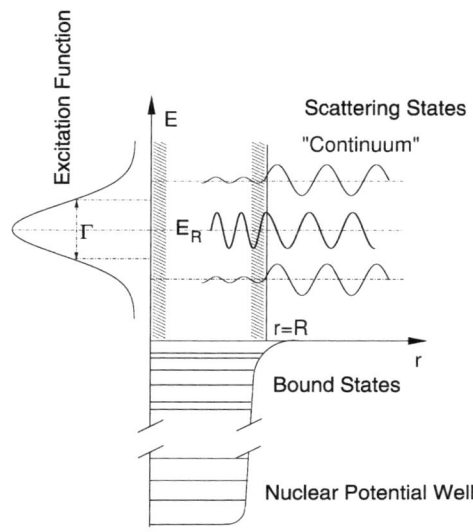

Fig. 11.2 Boundary conditions at the nuclear (potential) surface for the appearance of a resonance in the excitation function

$$\sigma_{el} = \frac{\pi}{k_{in}^2}|1 - \eta_0|^2 = \frac{\pi}{k_{in}^2}\left|\left[e^{2ika} - 1\right] - \frac{2ika}{\text{Re}(L_0) + i(\text{Im}(L_0) - ka))}\right|^2 \quad (11.15)$$

and

$$\sigma_{abs} = \frac{\pi}{k_{in}^2}\left(1 - |\eta_0|^2\right) = \frac{\pi}{k_{in}^2}\left[\frac{-4k_{in}\alpha\text{Im}(L_0)}{(\text{Re}(L_0))^2 + (\text{Im}(L_0 - ka)^2)}\right]. \quad (11.16)$$

By expanding $\text{Re}(L_0)$ in a Taylor series and terminating it after the first term, by comparison—besides obtaining the resonance-scattering amplitude (with $A_{pot} \propto e^{2ika} - 1$)—one obtains the results:

$$\sigma_{el} = \frac{\pi}{k_{in}^2}\left|\left(e^{2ika} - 1\right) + \frac{i\Gamma_{aA}}{(E - E_R) + i\Gamma/2}\right|^2 \quad (11.17)$$

$$\sigma_{abs} = \frac{\pi}{k_{in}^2}\frac{\Gamma_{aA}(\Gamma - \Gamma_{aA})}{(E - E_R)^2 + \Gamma^2/4}. \quad (11.18)$$

One sees that in agreement with our definition of *absorption* this encompasses all exit channels except the elastic channel.

11.4 Role of Level Densities

In nuclei the appearance and description of resonant states depends on the level density in the energy region considered. The number of possibilities to build a nuclear state from single particles and single holes (described in the shell model) increases

Fig. 11.3 Level-density parameter a in a Fermi-gas model as determined empirically from different sources such as counting of isolated neutron resonances. a follows approximately a $A/7.95$ or $A/8$ relation. The necessity of corrections such as shell corrections is evident

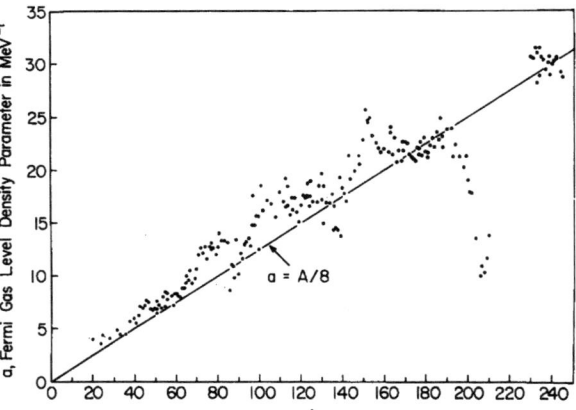

strongly with energy leading to an approximately exponential increase of the level density $\rho(E_x) = 1/D$ with excitation energy E_x (D is the (average) distance between neighboring levels). Also, because the number of configurations that can be formed from single-particle orbitals increases with the number of nucleons outside closed shells the level densities increase with A but also show the effects of shell closures and also deformations. The standard model for the statistical description of nuclear level densities (whose origins go back to Bethe [BET36, BET37, BET38]) is the *Backshifted Fermi-gas model*, e.g. [LAN54, HUI69, VON69, DIL73]. It describes the level density by an exponential term with excitation energy (often expressed as *nuclear temperature U*) and a spin-dependent term:

$$\rho(A, E, I, \pi) = \frac{2I + 1}{24} a^{1/2} \left(\frac{\hbar^2}{2\theta}\right)^{3/2} \frac{\exp 2[a(E - E_I)]^{1/2}}{(E - E_I)^2}, \tag{11.19}$$

where a is the *level density parameter* and $a \approx 7.95/A$, $E_I = (\hbar^2/2\theta)I(I+1)$ with E_I the energy of a rigidly rotating spherical nucleus with spin I. Figure 11.3 shows the level-density parameter a as function of the mass number A. This description in the framework of a statistical model was first published in Ref. [GIL65], and refinements can be found in the Reference Input Parameter Library [RIPL].

The appearance of observed quantities such as excitation functions of unpolarized cross sections but also of polarization observables depends not only on the basic reaction mechanism, but on the quantity Γ/D with $D = 1/\rho$ as well as the experimental resolution ΔE, which acts as an averaging interval. In order to see intermediate or coarse structures in the presence of fine structure also artificial averaging (with an averaging interval I) may be performed. Isolated resonances have Lorentz (Breit-Wigner) form whereas energy averaging leads to gaussian or other (trapezoidal) shapes. Both are then folded and must be disentangled to obtain the true resonance parameters. The direct observation of the true isolated-resonance shape requires $\Gamma/D < 1$. $\Gamma/D > 1$ leads to Ericson fluctuations that are only observable if ΔE is not much larger than Γ. The systematics of Γ is discussed in Sect. 11.5.4.

11.4 Role of Level Densities

11.4.1 Induced Nuclear Fission

Besides naturally occurring spontaneous fission of nuclei beyond a certain mass number, which was discovered in 1940, the process of induced fission was discovered by Otto Hahn et al. in 1939 [HAH39, HAHN39]. This form of fission may be induced by bombardment with slow or fast neutrons, γ's, but also with charged particles. A special variant are fusion reactions of two heavy projectiles/target nuclei leading to heavier (eventually superheavy) compound systems in which the decay by *alpha* emission and by fission compete, see also Sect. 15.3. Another variant is the fissioning of highly excited (rotating) nuclei due to centrifugal forces, see Fig. 13.5. Earlier surveys of fission are Refs. [WIL64, BRA72, PAU73, VAN73]. A recent text with many references is Ref. [KRA12].

Both processes, α emission and fission, also because of large Z values, are governed by the Coulomb barrier. Spontaneous fission occurs by tunneling through this barrier and has correspondingly large half-lives $T_{1/2}^f$ for nuclei near Uranium (examples: $7.04 \cdot 10^8$ a for ^{235}U; $4.468 \cdot 10^9$ a for ^{238}U). At larger mass numbers the Coulomb barrier is increasingly modified by shell-model effects (*shell corrections*, see e.g. [STR66]) leading to double- or triple-humped barriers, the phenomenon of *fission doorways* and shortened lifetimes.

If additional energy is provided induced fission occurs, especially by neutrons, which do not have to surmount a Coulomb barrier. Two basic possibilities exist: The negative neutron separation energy suffices to lift the energy of the system $n + A$ above the energy of the Coulomb barrier; then slow (thermal) neutrons will induce fission, such as for ^{235}U; or higher energy of the neutrons is required such as for ^{238}U. Generally there is a clear odd-even effect caused by the pairing of nucleons with more tightly bound configurations with nucleons paired with antiparallel spins compared to unpaired configurations. Of course, in detail the behavior is modified by the shell model. Figure 11.4 shows the energetic situation for two model cases of barrier shapes.

The true shape of the barrier is an individual property of each nuclide and depends on collective degrees of freedom such as deformation and shell-model degrees of freedom. The shapes shown in Fig. 11.4 have to be understood as a cut across a deformation landscape spanned above the standard two deformation parameters β and γ in the direction of the most probable path of the fissioning nucleus. Normal-deformed nuclei have typical axes ratios of 1.4:1, superdeformed nuclei in the second minimum about 2:1. New evidence for triple-humped fission barriers has been obtained with high-resolution (γ_n) (sub-barrier photofission) experiments on uranium isotopes. The third minimum of deformed nuclei corresponds to *hyperdeformed* shapes with axis ratios of $\approx$3:1 [CSI13] and is probed by its influence on intermediate-structure resonances decaying by transmission through the two barriers.

The fission process is interesting for its basic nuclear physics as well as for its applications which cannot be discussed in detail here, but for which ample references exist, especially on the technical and societal aspects of nuclear energy production

Fig. 11.4 Two schematic assumptions about the shape of the fission barrier. The *upper picture* is the simple liquid-drop model whereas the *lower picture* contains more (however schematic) information on realistic shapes of the double-humped barrier where a second minimum exists that is capable to support isomeric nuclear states. These states decay faster than from the first minimum and may form an intermediate structure, on which fine-structure states from the first-minimum fission are superimposed (*fission-doorway mechanism*, see Sect. 12.4)

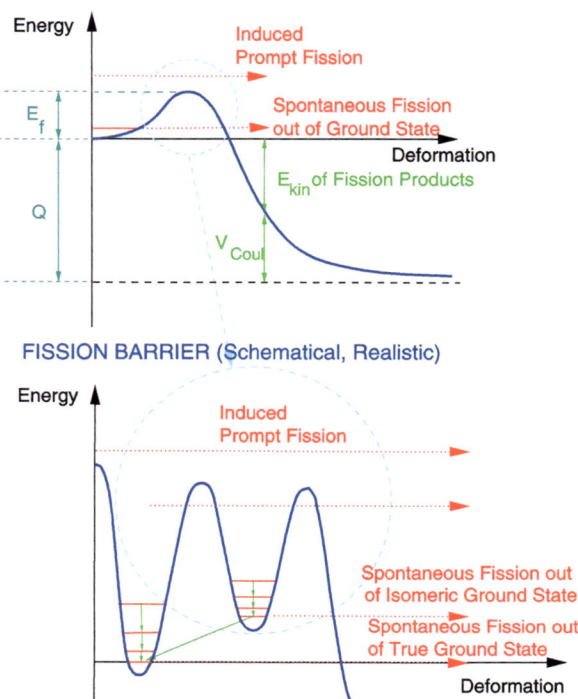

and nuclear weapons. Only a few keywords for some special features must suffice here:

- The (a)symmetric mass distributions of the fission products.
- The fission process is accompanied by neutron multiplication enabling a chain reaction used in reactors and weapons.
- The fission products—after the first step of the fission process—have high neutron excess and are therefore highly radioactive with partly long-lived isotopes.
- Transuranium nuclei produced by neutron capture and decays such as ^{239}Pu are also long-lived, radioactive, partly poisonous, and create enormous problems as nuclear waste which must be stored safely or transmuted (see also Sect. 19.9).

11.5 Single Resonances

11.5.1 General Features

The existence of (isolated) single resonances is visible by the following features:

11.5 Single Resonances

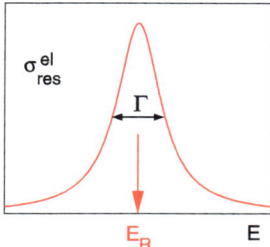
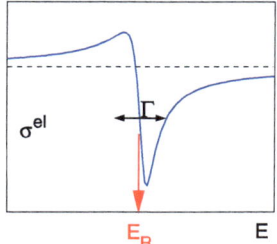

Fig. 11.5 Pure (elastic) resonance form and meaning of the total width Γ (*left*). Resonance form with interference of a pure resonance (Breit-Wigner shaped CN scattering) amplitude with a constant elastic potential-scattering amplitude. The sign of σ^{el} depends on the relative sign of the two amplitudes (*right*)

- In cross sections (as well as in other observables) as functions of energy strong excursions may be visible. Maxima in cross sections are due to the resonance denominator. For elastic s-wave scattering the maximum value of $(4\pi/k_{in}^2)\sin^2\delta_0$ occurs at $\delta = \pi/2$. Depending on whether the resonance is isolated and in a region with no background or not and whether one has a cross section or a polarization observable, the appearance of narrow resonance shows a pure Breit-Wigner shape or has *dispersion form* caused by interference. This implies that also destructive interference may occur such that the cross section is small or zero in the interference region. An example for the theoretical shape of a cross section across an isolated (elastic) resonance without background interference and with interference with a constant amplitude is shown in Fig. 11.5. Such a case is low-energy (s-wave) neutron scattering, see Sect. 4.1.1. For charged particles the interfering potential scattering ("shape elastic") amplitude is an energy-dependent Coulomb (Rutherford) amplitude (see Fig. 7.13).
- An essential feature is the increase of the resonance scattering phase over the region of the resonance, ideally by $+\pi$, with a transition through $\pi/2$ at the resonance.
- The angular distribution at the resonance is symmetric around $90°$.

11.5.2 Alternative Description of Resonances

Argand Plot If one plots $\text{Im}(f_R)$ against $\text{Re}(f_R)$ in the complex plane with the energy as parameter one expects in the ideal case of a purely elastic resonance the following behavior: The complex scattering amplitude moves counter-clockwise on a circle with the radius $\Gamma/2$, see Fig. 11.6.

Resonances as Poles in the Complex Plane In the complex energy plane (which is two-valued because of $E \propto k^2$) each resonance is a zero and has as its complex conjugate a pole in the positive energy plane. The ordinate values ($\text{Im}(E)$) are the

Fig. 11.6 Schematic Argand plot with the resonance phase η_ℓ performing a complete circle in the complex η_ℓ plane and with a potential scattering phase set to $\xi_\ell = 0$. In the case of $\xi_\ell \neq 0$ the phase angle becomes $\delta_\ell + \xi_\ell$ and the resonant circle is rotated by $2\xi_\ell$

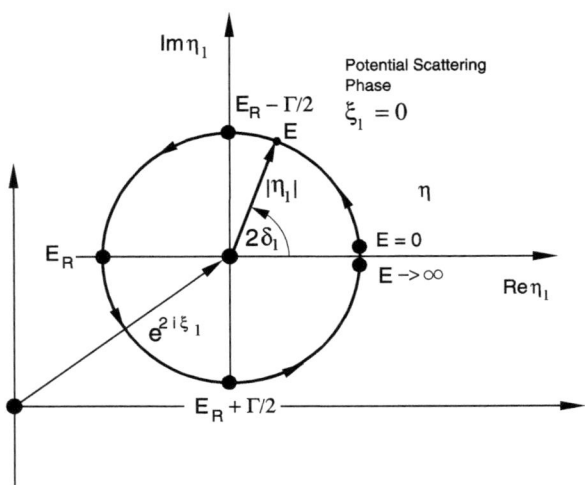

Fig. 11.7 In the complex energy plane resonances correspond to zeros and symmetrically located complex conjugate poles in the positive energy region (i.e. in the continuum). Bound and quasi-bound (excited) states are locates in the negative energy plane. Ordinates (imaginary parts of the state energies) are the total widths of the states. The connection with the scattering phase δ_ℓ is indicated for one resonance

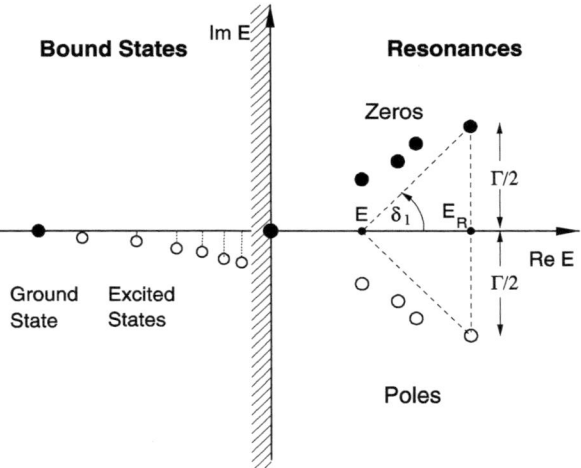

total widths of the resonances and prevent the amplitudes from becoming infinite. The scattering phase is

$$\delta_R = \arctan\left(\frac{\Gamma/2}{E_R - E}\right) \quad (11.20)$$

and passes through $\pi/2$ at resonance. Figure 11.7 explains that situation.

A good example for a measured resonance that to a large degree is isolated is the Δ^{++} resonance in π^+-p scattering. It shows only small interference with the background. Its angular distribution is symmetric about $90°$, see Figs. 11.8 and 11.9 and its Argand plot fills the unit circle completely, i.e. there is no sign of inelasticity, as shown in Fig. 11.10.

11.5 Single Resonances

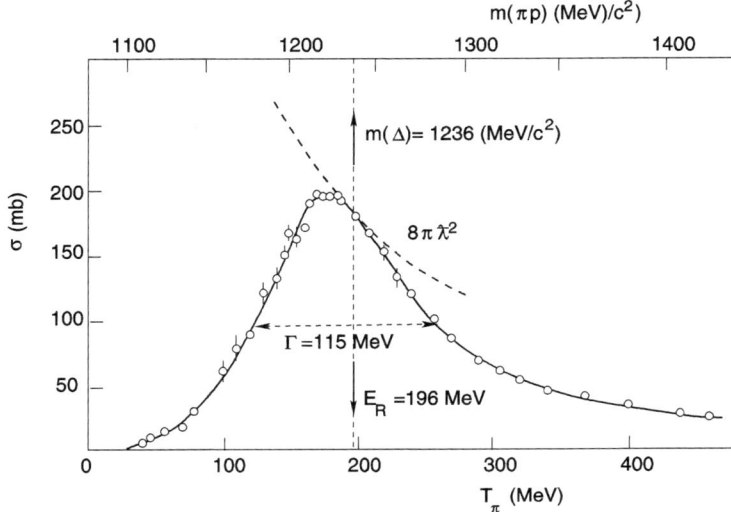

Fig. 11.8 The Δ-resonance (also called "P33") (here in the $\pi^+ - p$ scattering) is one of the most prominent resonances. It is so strong that it reaches the maximum cross section value of $8\pi\lambda^2$ allowed by the conservation of probability, which is equivalent to the unitarity of the S-matrix. Adapted from Ref. [PER83]

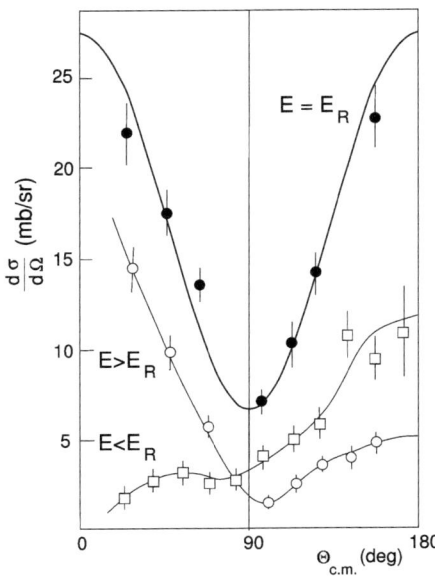

Fig. 11.9 The Δ resonance has an angular distribution symmetric about $\pi/2$. The function $\propto 1 + 3\cos^2\theta$ on resonance shows this to be a $J^\pi = 3/2^+$ resonance corresponding to an excited state of the nucleon. Adapted from Ref. [PER83]

An example for (resonance + Rutherford) scattering of protons from ^{12}C with scattering-phase analysis and Argand plot is shown in Fig. 11.11. The analysis shown was able to determine parameters of a number of strongly overlapping reso-

Fig. 11.10 The Argand plot over the Δ resonance fills the unit circle nearly perfectly. Smaller circles point to higher-energy resonances of the $\pi + p$ system. From [RPP08]

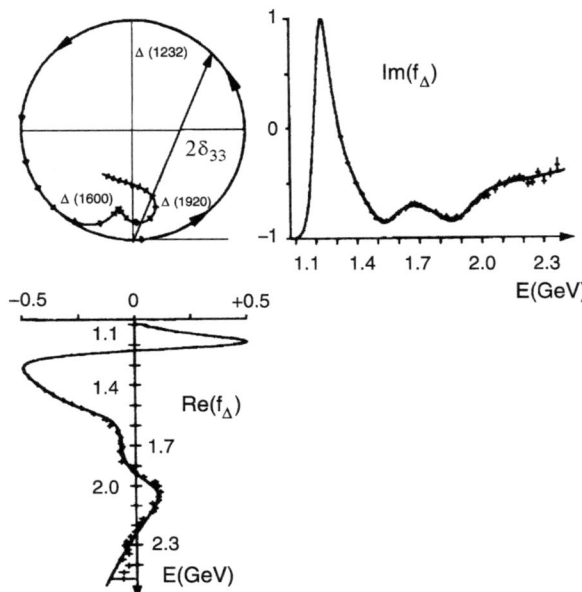

nances, also by analyzing polarization data together with unpolarized cross sections [MEY73].

11.5.3 Overlapping Resonances

The energy range, in which $\Gamma/D \gg 1$ where the compound resonances overlap is characterized by two possible types of excitation functions:

- The experimental resolution (where the energy spreading of the beams, kinematical broadening by the finite (solid) angle of the detectors etc. enter) ΔE is small: $\Delta E \ll \Gamma$. In this case the many Breit-Wigner amplitudes overlap stochastically and coherently. The cross sections fluctuate statistically with characteristic coherence widths Γ_{coh}, see Sect. 11.5.6, and they obey statistical distribution functions.
- The experimental resolution is bad: $\Delta E \gg \Gamma$. Here we have energy-averaging over the fluctuating cross sections (also over other observables such as polarization) such that their appearance is smooth. Because also the competing direct processes are smooth with energy criteria for distinguishing between both have to be developed e.g. by studying the behavior of the angular distributions.

11.5.4 Ericson Fluctuations

The phenomena of *chaotic* behavior of nuclear cross sections are described e.g. in Refs. [ERI66, VOG68, ALB71, RIC74]. The usual model assumption is that in the

11.5 Single Resonances

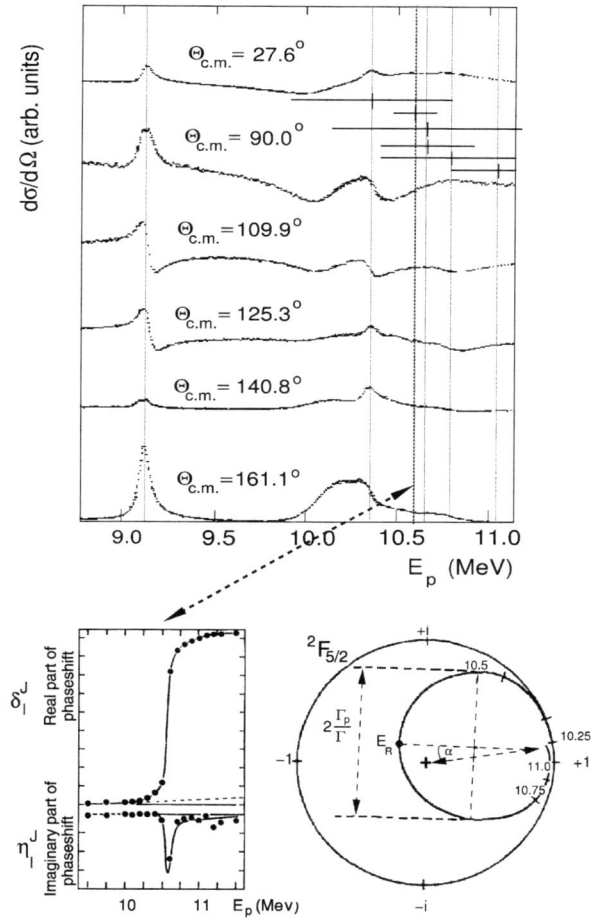

Fig. 11.11 Elastic scattering of protons from ^{12}C. *Upper figure*: Excitation functions over a number of partly overlapping resonances (states of ^{13}N, their resonance widths are indicated). *Lower figure*: Example of the passage of the $F_{5/2}$ scattering phase through $\pi/2$ and Argand plot belonging to the resonance at $E_R = 10.58$ MeV. In the analysis also polarization data were included. The analysis shows significant contributions from inelasticity, i.e. deviations from the unit circle. The (complex) phase results from a phase-shift analysis of differential cross section and analyzing-power data including a number of ^{13}N levels. Adapted from Refs. [SWI66, AJZ91, MEY73]

expressions for the S-matrix

$$S_{cc'} = S_{DI} + \sum \frac{a_{cc'}}{(E - E_R) + i\Gamma/2} \tag{11.21}$$

the partial-widths amplitudes $a_{cc'} = \Gamma_c^{1/2} \Gamma_{c'}^{1/2} \cos(\Phi)$ of the compound-nucleus contribution behave stochastically: their real and imaginary parts follow a statistically uniform distribution with the average zero:

$$\langle S_{cc'} \rangle_{cc'} = 0, \tag{11.22}$$

$$\langle S_{cc'} \rangle_{cc'}^2 \neq 0. \tag{11.23}$$

With this assumption the following features of the statistical behavior of the S-matrix elements are predicted: Their real and imaginary parts fluctuate independently and follow a Gaussian distribution around the average zero. The cross section from $S_{cc'}^2$

$$\langle S_{cc'}\rangle_{cc'}^2 \neq 0 \tag{11.24}$$

fluctuates around the smooth average cross section from $\langle S_{cc'}\rangle_{cc'}^2 \neq 0$, which e.g. may be described by the Hauser-Feshbach theory.

From the stochastic (chaotic) features of the amplitudes a number of predictions follow that can, on the one hand, be used to check the basic statistical assumptions, on the other, to obtain information about the average behavior of compound states (not about individual levels).

- The level distances D follow a Wigner distribution

$$P(D)dD = \frac{\pi D}{2\langle D\rangle^2}\exp\left(\frac{-\pi D^2}{4\langle D\rangle^2}\right) \tag{11.25}$$

about the average level distance $\langle D \rangle$. A purely stochastic distribution would be $\propto \frac{1}{\langle D \rangle}\exp(-\frac{D}{\langle D \rangle})$, i.e. would prefer small level distances. In reality levels with equal spins and equal parity repel each other, which makes the distribution more uniform.
- The level widths for one (the elastic) channel obey a Porter-Thomas distribution

$$P(x)dx = \frac{1}{\sqrt{2\pi x}}\exp(-x/2)dx \tag{11.26}$$

with $x = \Gamma/\langle\Gamma\rangle$. For more than one ($n$) channel(s) this distribution becomes a χ^2 distribution with n degrees of freedom.
- The cross sections follow a probability distribution, which in the simplest case (one spin channel) is given by

$$P(y)dy = e^{-y} \tag{11.27}$$

with

$$y = \frac{\sigma}{\langle\sigma\rangle}. \tag{11.28}$$

This case is naturally realized for spinless particles. For N spin channels:

$$P(y) = \frac{N^N}{(N-1)!}y^{N-1}e^{-Ny}. \tag{11.29}$$

- There are no channel-channel correlations. This assumption has been shown to be doubtful due to the unitarity of the collision matrix, see Ref. [MOL67].
- The correlation between the (differential) cross sections for one channel but at different angles depends on the number of independently fluctuating spin channels and on the direct-reaction contribution. Both have the effect of reducing the

11.5 Single Resonances

fluctuation "amplitude". The correlation is largest for $0°$ and $180°$ and minimal for $90°$. The number of spin channels can be calculated, e.g. for $90°$

$$N_{\text{eff,max}} = g/2 \quad \text{for } g = \text{even}, \qquad N_{\text{eff,max}} = (g+1)/2 \quad \text{for } g = \text{odd},$$

$$\text{with } g = (2S_a + 1)(2S_A + 1)(2S_b + 1)(2S_B + 1). \tag{11.30}$$

For the limiting angles $0°$ and $180°$ $N_{\text{eff,min}} < N_{\text{eff,max}}$ and depends on the spin structure of the reaction. The especially interesting case $N_{\text{eff}} = 1$ occurs at the spin structure $1/2 + 0 \to 1/2 + 0$ (in which the fluctuation is strongest). In the general case the transmission coefficients of the optical model are needed. With this, together with data from a measurement of such an angular correlation, the direct-reaction contribution can eventually be derived.

- Energy (auto) correlations: The energy correlation

$$C(\epsilon) = \frac{\langle \sigma(E) \sigma(E+\epsilon) \rangle}{\langle \sigma(E) \rangle \langle \sigma(E+\epsilon) \rangle} - 1, \tag{11.31}$$

for which under the above assumptions

$$C(\epsilon) = \frac{1}{N_{\text{eff}}} \frac{\Gamma_{\text{coh}}^2}{\Gamma_{\text{coh}}^2 + \epsilon^2} (1 - y_D^2) \tag{11.32}$$

provides the average level width (and thus the lifetime) of the CN levels. The *normalized variance*

$$C(0) = \frac{1}{N_{\text{eff}}} (1 - y_D^2) \tag{11.33}$$

may be used to determine the direct contribution y_D^2 of the reaction if the number of spin channels is known.

- In practice a correction for the finite range of data (*FRD error*) has to be applied.
- Polarization observables also are expected to fluctuate randomly under the conditions outlined above, due to statistical behavior of the T- or S-matrix elements. In the simplest cases the averaging over these results in zero average polarization (or analyzing power), see [LAM66, KUJ68]. The polarization of the particles is equivalent to reducing the number of spin channels thus increasing the fluctuation strength of the observables relative to that of unpolarized cross sections. The fluctuation analysis of polarization quantities obtained simultaneously with the cross sections provides an independent means to evaluate either Γ_{coh}, y_D, or N_{eff}.

In Fig. 11.12 examples for Ericson fluctuations in proton scattering on $N = 50$ nuclei is shown. The measurements were performed in the region around $E = 12$ MeV where in these nuclei very deep cross-section minima at extreme backward angles had been found [SCH75, BER75] that were identified as diffraction minima in the framework of the optical model, see Fig. 10.3. This small direct-interaction (DI) contribution helps to make compound-nucleus (CN) contributions, either smooth or fluctuating, visible. In these nuclei DI and CN contributions could be determined

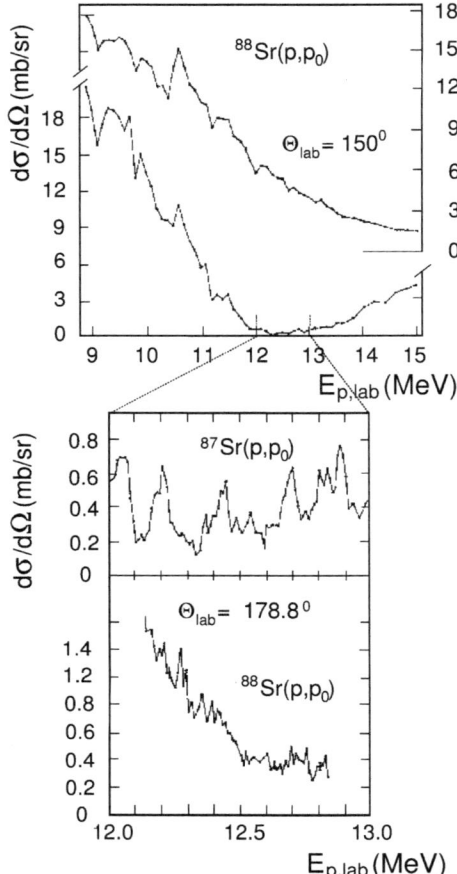

Fig. 11.12 Excitation functions at backward angles (p,p) elastic scattering on ^{87,88}Sr. Because in the *upper figure* energy steps of 100 keV were taken it shows an average behavior of the cross section that could be described by the optical model (with modifications for the very deepness of the backward minima). The *lower figures* depict data taken in 10 keV steps. The scattering from the even-even isotope ^{88}Sr still shows ≈ smooth behavior whereas the data from ^{87}Sr shows Ericson fluctuations. This is a direct consequence of the very different number N_{eff} of spin channels (2 vs. 200). Adapted from Refs. [SCH75, BER75]

quantitatively and confirmed by optical-model and Hauser-Feshbach calculations, as well as coherence widths Γ_{coh}. Similar results were obtained for Zr, Y, and Mo isotopes. It should be mentioned that indications for Ericson fluctuations of overlapping isobaric analog resonances have been seen [HGS79].

11.5.5 Analysis of Ericson Fluctuations

There are basically two established methods of analyzing statistical fluctuations and their distinction from single-resonances compound processes: correlations and Fourier analysis.

Auto- and Cross Correlations Almost by definition the formation and decay mechanism of single resonances implies that e.g. all the outgoing channels from one compound state are related, i.e. mathematically they will show correlations between the cross sections of different exit channels (*channel-channel or cross correlations*). However, it has been shown that, under certain circumstances, cross correlations and their analysis may not be sensitive to even very strong correlations of resonance parameters [KLE73].

In contrast, such correlations are absent if we have stochastically fluctuating cross sections in different exit channels that come about by squaring the contributions of many overlapping Breit-Wigner-shaped amplitudes. Interference terms between different (complex) amplitudes have statistically varying phases and, on average, cancel in the cross section. Thus, in analyzing these data, nothing can be learned about the underlying individual resonant states. However, some average quantities may be derived.

Auto-correlations between cross sections (and in suitable cases polarization observables) at different energies, separated by an energy interval ϵ reveal the average width $\langle \Gamma_{\text{coh}} \rangle$ of the underlying states and can be used for an energy systematics of this quantity. In addition the number of effective spin channels and direct-reaction contributions may be evaluated.

Angular auto-correlations of differential cross sections as function of an angular interval $\Delta\theta$ yield information about the number of contributing spin channels.

Fourier Analysis Another way of obtaining information about the average widths of the compound states is Fourier analysis. The fluctuating cross sections are treated like a mixture of frequencies in acoustics, from which the "pure" components are filtered out by Fourier analysis. In Ericson fluctuation analysis the main frequency components are again the widths $\langle \Gamma \rangle$. It is possible to obtain contributions with different widths, e.g. from overlapping intermediate structures.

11.5.6 Results for Level Widths

By studying single isolated resonances and, by fluctuation analysis of the coherence widths Γ_{coh} for overlapping resonances as discussed before, a systematic picture of the widths of nuclear levels can be gained. Figure 11.13 shows the systematics of Γ over A for one excitation energy. The widths increase with excitation energy as the number of open exit channels increases (see also Sect. 11.6.2) but decrease with mass number A for a given excitation energy E_x.

11.6 Complete Averaging over the CN States

11.6.1 Generalities

Formally the energy averaged (integrated) cross sections can be understood as an average over the S-matrix, which leads to an understanding of the apparently different

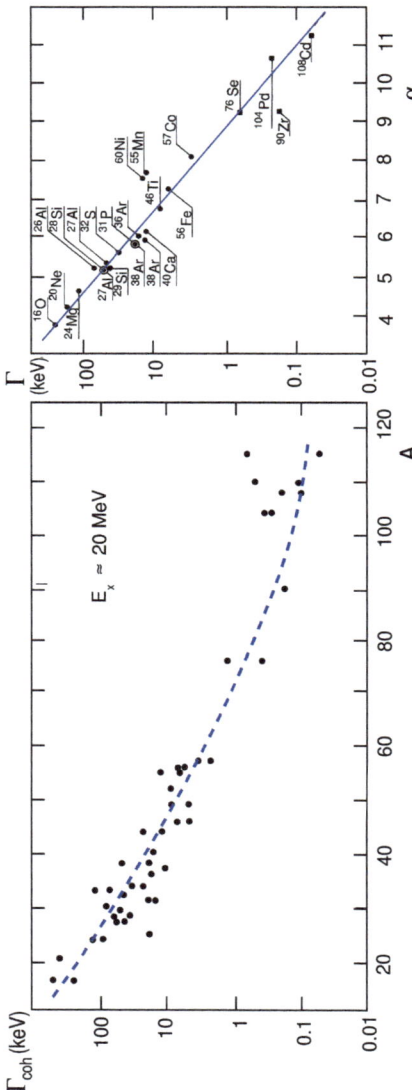

Fig. 11.13 Dependence of nuclear level widths on the mass number A for $E_x \approx 20$ MeV (*left*). On the *right* a parametrization including mass number and excitation energies is shown. The independent variable is defined as $\alpha = A^{1/2}(\frac{19}{E_x(\text{MeV})})^{1/2}$ for excitation energies from 17 to 21 MeV. For references see [ERI66, BRA70, EBE71]

11.6 Complete Averaging over the CN States

contributions.

$$\langle \sigma_{\text{el}} \rangle = \pi \lambdabar^2 \sum (2\ell + 1) \underbrace{\left(1 - \langle S_\ell \rangle - \langle S_\ell^* \rangle + \langle |S_\ell|^2 \rangle\right)}_{\langle |1 - S_\ell|^2 \rangle}$$

$$= \pi \lambdabar^2 \sum (2\ell + 1) \cdot [\underbrace{|1 - \langle S_\ell \rangle|^2}_{\to \sigma_{\text{s.e.}}(\text{O.M.!})} + \underbrace{\left(\langle |S_\ell|^2 \rangle - |\langle S_\ell \rangle|^2\right)}_{\to \langle \sigma_{\text{c.e.}} \rangle (H-F)}] \quad (11.34)$$

and

$$\langle \sigma_{\text{r}} \rangle = \pi \lambdabar^2 \sum (2\ell + 1) \Big[\underbrace{\overbrace{\left(1 - |\langle S_\ell \rangle|^2\right)}^{\langle 1 - |S_\ell|^2 \rangle}}_{\sigma_{\text{abs}}(\text{O.M.})} - \underbrace{\left(\langle |S_\ell|^2 \rangle - |\langle S_\ell \rangle|^2\right)}_{\langle \sigma_{\text{c.r.}} \rangle (H-F)}\Big]. \quad (11.35)$$

The two (averaged) cross sections σ_{el} (elastic) and σ_{r} (reaction) both consist of two different components

- *per se smooth*: $\sigma_{\text{s.e.}}$ (shape-elastic) and σ_{abs} (absorption), both described by the optical model
- in principle *fluctuating*, but which may appear smooth due to experimental averaging: $\langle \sigma_{\text{c.e.}} \rangle_{\Delta E}$ (compound-elastic) and $\langle \sigma_{\text{c.r.}} \rangle_{\Delta E}$ (compound-reaction). (ΔE is the experimental averaging interval, either determined by the finite energy resolution of the experiment or by deliberate smoothing by the experimenter). *Ericson fluctuations* are seen with sufficiently good experimental resolution.

11.6.2 Hauser-Feshbach Formalism

According to the Bohr hypothesis formation and decay of CN states are independent, which means e.g. that the integrated cross section can be factorized:

$$\sigma_{\alpha\beta} = \sigma_\alpha^{\text{CN}} \cdot P_\beta \quad (11.36)$$

with

$$P_\beta = \frac{\Gamma_\beta}{\sum_{\text{all} \beta} \Gamma_\beta} \quad (11.37)$$

and

$$\sum_\beta P_\beta = 1. \quad (11.38)$$

The formation cross section is described in the optical model (see above Sect. 10.3) via transmission coefficients T_α:

$$\sigma_\alpha^{\text{CN}} = \pi \lambdabar^2 \left(1 - |\eta_{\alpha\alpha}|^2\right) = \pi \lambdabar^2 \cdot T_\alpha. \quad (11.39)$$

Hence

$$\sigma_{\alpha\beta} = \pi \lambdabar^2 T_\alpha P_\beta. \tag{11.40}$$

From the *Principle of detailed balance* follows:

$$\lambdabar_\beta^2 \sigma_{\alpha\beta} = \lambdabar_\alpha^2 \sigma_{\beta\alpha} \tag{11.41}$$

$$T_\alpha P_\beta = T_\beta P_\alpha \tag{11.42}$$

$$\frac{P_\alpha}{T_\alpha} = \frac{P_\beta}{T_\beta} = \lambda = \text{const} \tag{11.43}$$

$$\sum P_\alpha = \lambda \sum_{\gamma = \text{all } \beta} T_\beta = 1 \tag{11.44}$$

$$\lambda = \frac{1}{\sum_\gamma T_\gamma} \tag{11.45}$$

$$P_\beta = \lambda \cdot T_\beta = \frac{T_\beta}{\sum_\gamma T_\gamma}. \tag{11.46}$$

The final result (for spinless particles) for the integrated Hauser-Feshbach cross section [HAU52] is then:

$$\sigma_{\alpha\beta}^{\text{HF}} = \langle \sigma_{\alpha\beta}^{\text{CN}} \rangle = \pi \lambdabar_\alpha^2 \cdot \frac{T_\alpha T_\beta}{\sum_\gamma T_\gamma}. \tag{11.47}$$

For reactions with particles with spins we obtain

$$\sigma_{\alpha\beta}^{\text{HF}} = \langle \sigma_{\alpha\beta}^{\text{CN}} \rangle = \pi \lambdabar_\alpha^2 \sum_{J\pi} \frac{(2J+1)}{(2i+1)(2I+1)} \cdot \frac{T_\alpha T_\beta}{\sum_\gamma T_\gamma} \tag{11.48}$$

where i, I, J, and π are the spins of the projectile, the target, and spin and parities of the compound systems, resp.

Analogously and with use of the partial-wave expansion the differential Hauser-Feshbach cross section is derived. The Racah coefficient with the symbol $Z(abcd; ef)$ is a vector-coupling coefficient used for coupling of three angular momenta and is equivalent to the 6j symbol

$$\begin{Bmatrix} a & b & e \\ d & c & f \end{Bmatrix} \tag{11.49}$$

and the Wigner coefficient $(-)^{a+b+c+d} W(abcd; ef)$, see Refs. [BRI71, BLA52, VOG68] and Sect. 22.3. Here it performs the coupling of the entrance (exit) channel spins s_a, s_A with the entrance (exit) orbital angular momenta ℓ, ℓ' to the total angular momenta J.

11.6 Complete Averaging over the CN States

$$\left(\frac{d\sigma}{d\Omega}\right)_J^{HF} = \frac{\lambdabar_\alpha^2}{4\pi(2i+1)(2I+1)} \sum_{LJ\pi}$$

$$\times \sum_{s\ell s'\ell'} Z(\ell J \ell J; sL) Z'(\ell' J' \ell' J'; s'L)(-)^{s-s'} P_L(\cos\theta)$$

$$\times \frac{T_\alpha{}^\ell T_\beta{}^{\ell'}}{\sum_\gamma T_\gamma{}^{\ell''}} \qquad (11.50)$$

where α, β, γ designate the entrance, specific exit, and all exit channels, resp., i, I, s, s' are the particle spins and channel spins, and ℓ the corresponding angular momenta.

The angular dependence of the differential cross section is governed by L, coupled from ℓ and ℓ'. The interference terms between different amplitudes, which are assumed in the model Ansatz to have statistically distributed phase relations cancel and only even-order Legendre polynomials survive leading to angular distributions symmetric around 90°. Because vector-polarization effects result from interference of different amplitudes this leads to the fact that in the statistical model no (energy-averaged) polarization effects occur (however: in the regime of Ericson fluctuations there may be fluctuations of polarization effects around zero). Both cross sections show the factorization into entrance and exit channels and are symmetric in both channels.

The calculation of the Hauser-Feshbach cross sections is performed using the transmission coefficients T_l or T_{lJ}. They are obtained by adjusting the optical-model parameters to elastic-scattering data for the entrance channel as well as the special exit channel considered and for all possible exit channels, specifically from the imaginary part of the optical potential. They must be known for all energies up to the excitation energies, which are maximally reached in the compound nucleus.

An approximation often used for heavy-ion reactions (i.e. with strong absorption) was the sharp-cutoff model, in which the smooth transmission coefficients were replaced by suitable step functions changing their values at a sharp energy from zero to one. Transitions ending in the continuum i.e. in the region of strong overlapping of final states must be treated statistically using a level-density function as weighting factor. Nuclear level densities are an important parameter that decides whether Hauser-Feshbach (statistical) assumptions are applicable. For details on nuclear-level densities see Sect. 11.4.

In summary, the salient features of the Hauser-Feshbach cross section are:

- The angular distributions—due to properties of the Legendre polynomials and the angular-momentum coupling coefficients—are symmetric around 90°. This is evident in Fig. 11.14.
- The cross sections decrease strongly with increasing energy because with the increasing number of open decay channels the transmission denominator increases strongly. Therefore one expects that in a competition with direct processes the latter will dominate at increasing energy, the compound processes, however, will die out.

Fig. 11.14 Hauser-Feshbach angular distribution, adjusted to data of the compound reaction $^{31}\text{P}(p,\alpha_0)^{28}\text{Si}$, averaged over $E_p = 8.5$ to 11.6 MeV. Adapted from Ref. [DAL68].

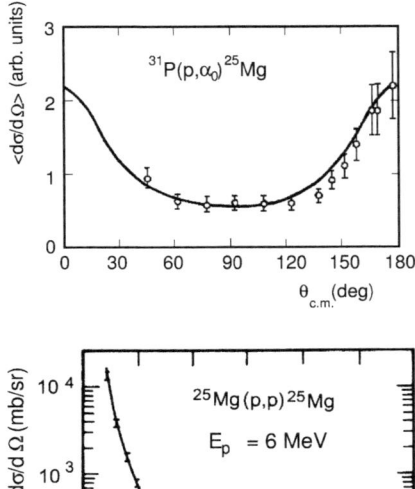

Fig. 11.15 Differential cross section of elastic proton scattering $^{25}\text{Mg}(p, p)^{25}\text{Mg}$ at $E_p = 6$ MeV with a clear disentanglement of direct (DI, shape-elastic) vs. compound (CN, compound-elastic) scattering. From Ref. [GAL66]

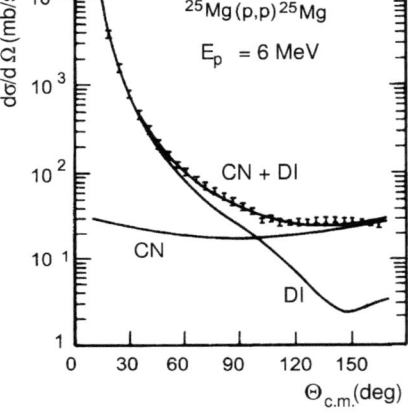

At lower energies the separation of (averaged) CN from DI processes as well as the distinction between (fine-structure) resonances and Ericson fluctuations is not always simple. One has to rely on angular distributions together with fitting of theoretical models on the one hand and correlation analyses (angular and energy autocorrelations, and cross-correlations between different channels) on the other. An example where this separation was achieved is shown in Fig. 11.15. The applicability of the statistical model for the compound-elastic scattering ensures that there is no interference between the DI and CN amplitudes and cross sections can be added. With interference the separation of the two would be more difficult.

11.7 Exercises

11.1 The Breit-Wigner resonance cross section was derived with the following assumptions:

- s-wave ($\ell = 0$) scattering only,
- No spins involved, and
- No background scattering.

11.7 Exercises

In the more general case of $\ell \geq 0$ the integrated cross section contains a term $(2\ell + 1)$ in its partial-wave sum (see Chap. 8).

(a) Show that in this case σ across a resonance contains one such term with a definite ℓ, corresponding to the resonating partial wave.

(b) If there is a (nearly) constant background (such as from Coulomb or potential scattering as for thermal-neutron elastic scattering) the partial phase shift for this ℓ is the sum of the potential (B) and the resonant (R) phase shifts.

$$\delta_\ell = \delta_\ell^B + \left(\frac{\pi(E - E_R)}{2\Gamma} + \pi/2 \right). \tag{11.51}$$

Show that the resulting integrated (total) cross section is a sum of three terms: σ^B, σ^R, and an interference term. Sketch the possible shapes of δ_ℓ and σ_ℓ as functions of E across E_R for different background contributions of $\delta_\ell^B = 0$, $\pi/4$, $\pi/2$, and $3\pi/4$.

11.2 For $\ell = 0$ (s-wave scattering) and a background from hard-sphere scattering (see Chap. 8) derive and plot the integrated cross section across a Breit-Wigner resonance. Indicate the width, the background contribution, and the maximum possible cross section as functions of E.

11.3 Discuss why the interference may be destructive, sometimes leading to zero cross section.

11.4 Single-particle states play an important role in nuclear physics (think of the (quasi-)bound states of the shell model, but also the shape resonances described by the optical model). Assume a square well as shape of the scattering potential of a nucleus A with s-wave neutrons of a potential depth of $V_0 = 50$ MeV and a potential radius of $R = 6.5$ fm.

(a) Are there bound states of the compound system $A + n$ and what are their positions on the energy axis?
(b) What happens at $E = 0$, what at $E > 0$?
(c) Why are resonances also called "states in the continuum" and what is the physical meaning of the width Γ?
(d) Verify that the internal wave function is

$$u_0 = A \sin(\kappa R) \quad \text{with } \kappa = \sqrt{2\mu(E - V_0)}/\hbar \tag{11.52}$$

(see also Sect. 8.3.2).

(e) According to Sect. 11.3 the width of an elastic resonance is related to the logarithmic derivative $L_0(E)$, i.e. follows from a Taylor expansion of $\text{Re}(L_0(E))$ after truncation after the second term

$$\Gamma_{el} = \frac{-2kR}{(dL_0(E)/dE)_{E=E_R}}. \tag{11.53}$$

Calculate

$$\frac{dL_0(E)}{dE} = \left(\frac{dL_0}{d\kappa}\right)\left(\frac{\kappa}{dE}\right). \quad (11.54)$$

Show that $\kappa a = (2n+1)\pi/2$ corresponds to "single-particle" s-wave resonances. What is the energy spacing of the sequence of s.p. resonances? With the help of Eq. (11.53) deduce the single-particle width $\Gamma_{\text{s.p.}}$ as function of the mass number A and the incident energy.

References

[AJZ91] F. Ajzenberg-Selove, Nucl. Phys. A **523**, 3 (1991)
[ALB71] J. Garg (ed.), *Proc. Int. Conf. on Statistical Properties of Nuclei*, Albany, 1971 (Plenum, New York, 1971)
[BER75] G. Berg, W. Kühn, H. Paetz gen. Schieck, K. Schulte, P. von Brentano, Nucl. Phys. A **254**, 169 (1975)
[BLA52] J.M. Blatt, L.C. Biedenharn, Rev. Mod. Phys. **24**, 258 (1952)
[BET36] H. Bethe, Phys. Rev. **50**, 332 (1936)
[BET37] H. Bethe, Rev. Mod. Phys. **9**, 69 (1937)
[BET38] H. Bethe, Phys. Rev. **53**, 675 (1938)
[BRA70] M.G. Braga Marcazzan, L. Milazzo Colli, Prog. Nucl. Phys. **11**, 145 (1970)
[BRA72] M. Brack, J. Damgaard, A.S. Jensen, H.C. Pauli, V.M. Strutinsky, C.Y. Wong, Rev. Mod. Phys. **44**, 320 (1972)
[BRE36] G. Breit, E. Wigner, Phys. Rev. **49**, 519 (1936)
[BRI71] D.M. Brink, G.R. Satchler, *Angular Momentum* (Clarendon Press, Oxford, 1971)
[CSI13] L. Csige, D.M. Filipescu, T. Glodariu, J. Gulyás, M.M. Günther, D. Habs, H.J. Karwowski, A. Krasznahorkay, G.C. Rich, M. Sin, L. Stroe, O. Tesileanu, P.G. Thirolf, arXiv:1302.3425v1 [nucl-ex] (2013)
[DAL68] P.J. Dallimore, B.W. Allardyce, Nucl. Phys. A **108**, 150 (1968)
[DIL73] W. Dilg, W. Schantl, H. Vonach, M. Uhl, Nucl. Phys. A **217**, 269 (1973)
[EBE71] K.A. Eberhard, A. Richter, in [ALB71]
[ERI66] T. Ericson, T. Mayer-Kuckuk, Annu. Rev. Nucl. Sci. **16**, 183 (1966)
[FEY77] R.P. Feynman, R.B. Leighton, M. Sands, *The Feynman Lectures in Physics I*, 6th printing edn. (Addison-Wesley, Reading, 1977)
[GAL66] A. Gallmann, P. Wagner, G. Franck, G. Wilmore, P.E. Hodgson, Nucl. Phys. **88**, 654 (1966)
[GIL65] A. Gilbert, A.G.W. Cameron, Can. J. Phys. **43**, 1446 (1965)
[HAH39] O. Hahn, F. Straßmann, Naturwissenschaften **27**, 11 (1939)
[HAHN39] O. Hahn, F. Straßmann, Naturwissenschaften **27**, 89 (1939)
[HAU52] W. Hauser, H. Feshbach, Phys. Rev. **87**, 366 (1952)
[HUI69] J.R. Huizenga, H.K. Vonach, A.A. Katsanos, A.J. Gorski, C.J. Stephan, Phys. Rev. **182**, 1149 (1969)
[HGS79] H. Paetz gen. Schieck, G. Latzel, W. Lenssen, G.P.A. Berg, Z. Phys. A **293**, 253 (1979)
[KLE73] G. Klein, H. Paetz gen. Schieck, Nucl. Phys. A **215**, 24 (1973)
[KRA12] H.J. Krappe, K. Pomorski, *Theory of Nuclear Fission*. Lecture Notes in Physics, vol. 838 (Springer, Heidelberg, 2012)
[KUJ68] E. Kujawski, T.J. Krieger, Phys. Lett. B **27**, 132 (1968)
[LAM66] M. Lambert, G. Dumazet, Nucl. Phys. **83**, 171 (1966)
[LAN58] A.M. Lane, R.G. Thomas, Rev. Mod. Phys. **30**, 145 (1958)

References

[LAN54] J.M.B. Lang, K.J. LeCouteur, Proc. Phys. Soc. Lond. A **67**, 586 (1954)
[MEY73] H.O. Meyer, G.R. Plattner, Nucl. Phys. A **199**, 413 (1973)
[MOL67] P.A. Moldauer, Phys. Rev. **157**, 907 (1967)
[PAU73] H.C. Pauli, Phys. Rep. **7**, 35 (1973)
[PER83] D.H. Perkins, *Introd. to High Energy Physics*, 2nd edn. (Addison-Wesley, Reading, 1983)
[RIC74] A. Richter, Specialized reactions, in *Nucl. Spectroscopy and Reactions*, ed. by J.B. Cerny (Academic Press, New York, 1974)
[RIPL] IAEA-TECDOC-1034, Vienna (1998). http://www-nds.iaea.org/RIPL-2/
[RPP08] C. Amsler et al. (Particle Data Group), Review of particle properties. Phys. Lett. B **667**, 1 (2008)
[SCH75] K. Schulte, G. Berg, P. von Brentano, H. Paetz gen. Schieck, Nucl. Phys. A **241**, 272 (1975)
[SWI66] J.B. Swint, A.C.L. Barnard, T.B. Clegg, J.L. Weil, Nucl. Phys. A **86**, 119 (1966)
[STR66] V.M. Strutinsky, Yad. Fiz. (USSR) **3**, 614 (1966) [Sov. J. Nucl. Phys. 3, 449 (1966)]
[VAN73] R. Vandenbosch, J.R. Huizenga, *Nuclear Fission* (Associated Press, New York, 1973)
[VOG68] E. Vogt, Adv. Nucl. Phys. **1**, 261 (1968)
[VON69] H.K. Vonach, M. Hille, Nucl. Phys. A **127**, 289 (1969)
[WIL64] L. Wilets, *Theories of Nuclear Fission* (Clarendon Press, Oxford, 1964)

Chapter 12
Intermediate Structures

Intermediate structures IS have in common that they possess fine structure, i.e. that they appear as enhancements of many fine-structure states over a certain energy region, the width of which is intermediate between that of the narrow fine-structure states and that of much wider phenomena (*coarse structure*) such as the optical-model shape resonances seen in averaged neutron elastic scattering cross sections. The term IS covers a number of quite diverse phenomena:

- Heavy-ion molecules,
- Neutron shape resonances,
- Giant resonances,
- Fission doorways, and
- Isobaric analog resonances.

A very interesting feature is the interplay of these structures and phenomena such as *doorways or hallways* and the mechanisms of *spreading* of the intermediate strengths into the fine structure measured by the *spreading width* $\Gamma^\downarrow$. The total widths then have several components, i.e. besides the spreading width there are *escape widths* $\Gamma^\uparrow$ of decays into fine-structure states and the *total width* Γ that determines the lifetime of a state is a sum of all. Early references to intermediate structures are Refs. [FES67] and [CIN73].

12.1 Heavy-Ion Scattering and Molecular Resonances

In the 1960s excitation functions of the scattering of heavy ions from heavy ions (e.g. ^{12}C on ^{12}C or ^{16}O on ^{16}O) showed regularities that were interpreted as short-lived molecular states, very much like states of atoms forming molecules like H$_2$. These "resonances" in the excitation functions are "intermediate" because their widths are larger than true compound resonances and they should have a fine structure of true compound states of e.g. ^{24}Mg, i.e. the molecular states are *doorway states* to the fine structure states, and their widths should consist of two parts: *escape width* and *spreading width*. Fourier-type or autocorrelation analyses of the

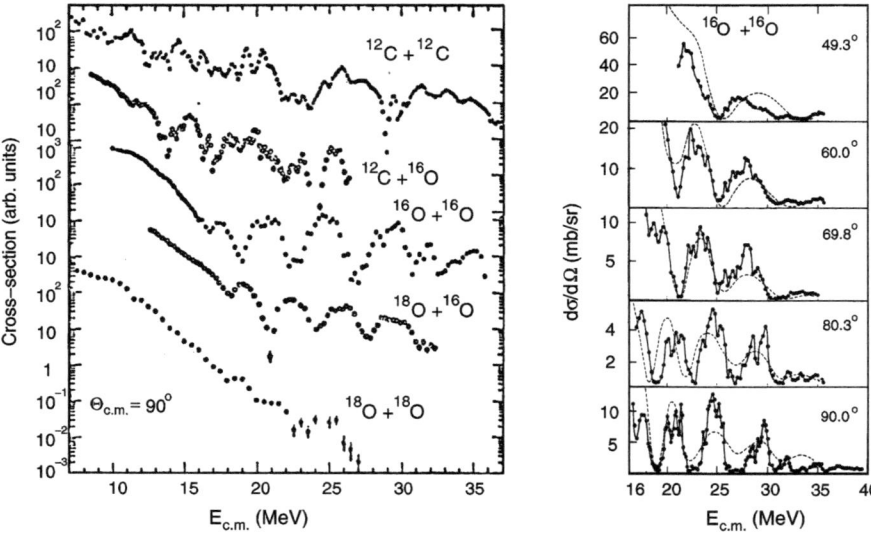

Fig. 12.1 *Left*: Intermediate-structure data in excitation functions of several heavy-ion reactions, especially prominent in the reaction $^{16}O(^{16}O, ^{16}O)^{16}O$. *Right*: CN fine structure assigned to states in ^{32}S superimposed on IS, probably molecular resonances in $^{16}O(^{16}O, ^{16}O)^{16}O$. The different angles are c.m. angles between which the data shows correlations. Adapted from Refs. [VAL81, SIE67]

resonance structures found evidence for three types of distinctly different widths: shape-elastic structures with typical widths $\Gamma_{s.e.} < 1$ MeV, intermediate structures with widths Γ_{int} of 100–200 keV, partly overlapping i.e. with $\Gamma/D \geq 1$ where $D = 1/\rho$ are the average level separation, ρ the level density, and compound fine structure resonances with typical widths $\Gamma_{c.e.} \approx 30$ keV. Very coarsely we have order-of-magnitude lifetime scales for the three cases of $10^{-21}, 10^{-20}$, and 10^{-19} s. In the classical picture of a rotating molecule this corresponds—in a semi-classical picture—to less than one complete revolution of the two nuclei forming the molecule around one another during the projectile's passage. Figure 12.1 shows a summary of older data of different molecular systems. The formation of a molecule depends on incident energy and mass of the ions and competes with complete or incomplete fusion into compound nuclei that may also lead to the same exit channels. Thus, additional criteria such as identification of rotational bands etc. are necessary. In the past also different interpretations of the IS have been put forward.

The identification of resonance-like structures in excitation functions is not easy and may be ambiguous in the presence of other, e.g. fine structure, especially when these structures have widths not very different from each other. Therefore, attempts have been undertaken to go into the complementary time regime and measure time differences from different processes directly. Direct processes have widths corresponding to the passage times ($\tau \approx 10^{-21}$ to 10^{-22} s). Similar intermediate structures have been found for different reaction channels of these same entrance-channel reactions, which is expected if the structures belong to definite quantum

12.2 Structures in Neutron Reactions

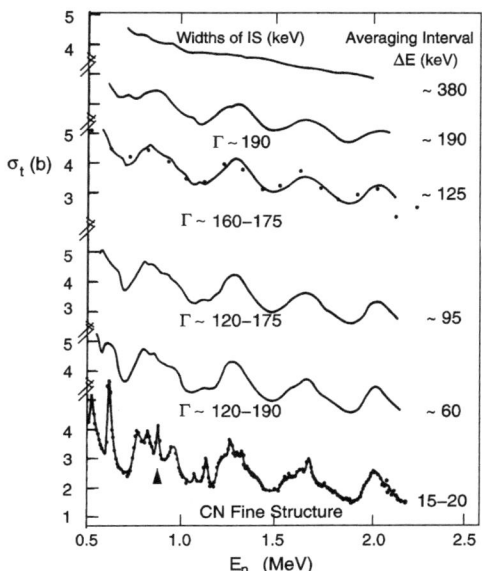

Fig. 12.2 Intermediate structures in excitation functions of neutron scattering on ^{19}F, exhibited by experimental averaging over fine structure with different energy intervals ΔE. From [MON67]

states and not to Ericson fluctuations of statistically overlapping compound states or direct processes. Typical cases are the reactions leading to α exit channels, e.g. ^{16}O(^{12}C, α_i)^{24}Mg.

12.2 Structures in Neutron Reactions

Besides the CN fine-structure resonances in neutron scattering and reactions intermediate and gross structures have been identified.

12.2.1 Neutron-Nucleus IS

After appropriate averaging neutron scattering cross sections may show intermediate structures with widths (typically around 150 keV) between the gross structures (MeV) and the CN fine-structures (eV–keV). An example is shown in Fig. 12.2.

12.2.2 Single-Particle Neutron Resonances

The fine-structure compound resonances (s-wave) found in neutron scattering at low energies exhibit—when averaged over energy with a suitable averaging interval—distributions of resonance strength (*strength function*) with resonance-like shapes,

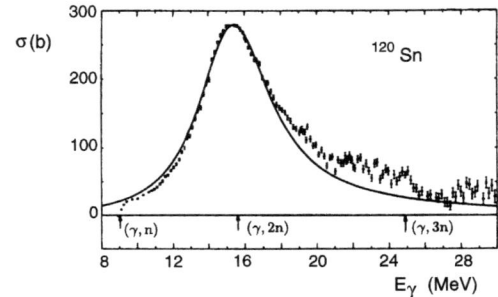

Fig. 12.3 Giant E1 electric dipole resonance in ^{120}Sn(γ, xn) with $\Delta L = 1$, and spin and isospin changes $\Delta S = 0$, and $\Delta T = 1$. From [HAR01]

but with widths of up to several MeV. Their positions on the energy scale coincide with the energies where, as functions of the mass numbers of the compound nuclei, the single-particle orbitals of the shell model become unbound, i,e, where they cross $E = 0$. This identifies these structures as *single-particle or: "shape"*, formerly also *"giant"* resonances. The single nucleon outside the core moves in the average potential of all core nucleons and therefore does not see details of nuclear structure. Also the energy and width dependence on A behave systematically.

12.3 Giant Resonances[1]

Giant Resonances are broad structures in excitation functions with large cross sections, excited by incident γ's as well as in inelastic particle reactions such as (p, p'), (p, γ), and (α, α'). They correspond to collective excitations of groups of larger numbers of nucleons moving against each other. They can be classified according to their electromagnetic modes (or multipolarities), their isospin, or their motion types. These latter are

- The *Breathing mode*: The entire nucleus "breathes" without changing shape. Electromagnetically this is an E0 mode, its isospin is 0. It is plausible that this mode is related to the nuclear compressibility.
- The E1 mode: This is the classical dipole mode, in which proton and neutron fluids move collectively against each other.
- The M1, $T = 1$ mode: This is the magnetic-dipole *Scissors mode*, in which the motions of the proton and neutron fluids have a rotational component acting like the arms of scissors.
- The E2 or quadrupole giant resonance where protons and neutrons oscillate collectively against each other such that an oscillating quadrupole moment is formed.
- Higher order resonances such as M3, E4 etc. have been found, e.g. the isoscalar octupole ($3\hbar\omega$) resonance.

---

[1]It should be noted that in the older literature the term "Giant Resonance" was also used for the total cross section behavior of neutron scattering on many nuclei in an energy region where single resonances are not resolved and which is described by the optical model, see e.g. Ref. [SAT90].

12.3 Giant Resonances

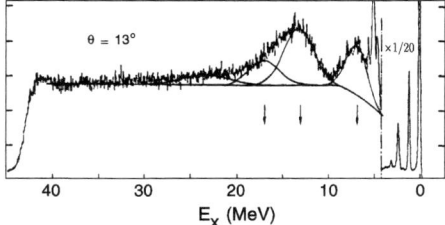

Fig. 12.4 Giant resonances, e.g. monopole, quadrupole and octupole excitations, in inelastic α scattering ^{120}Sn(α, α') with $\Delta L = 1, 2, 3$ at $E_\alpha = 152$ MeV. High background makes the disentangling of different resonances difficult and may require model assumptions. Measurements are best done at small forward angles. From [HAR01]

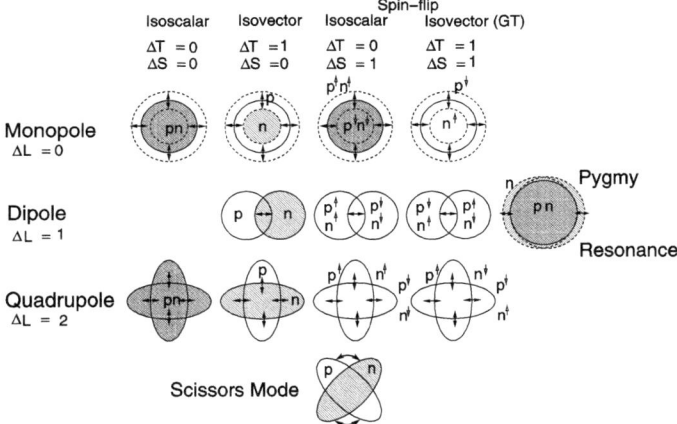

Fig. 12.5 Giant resonances classified according to their multipolarity ΔL, their spin and isospin changes ΔS, and ΔT

- Relatively new developments concern the dipole (E1) *Pygmy resonance* in nuclei with high neutron excess where—it is assumed—the neutron skin oscillates against the remaining $N = Z$ ($T = 0$) core. Radioactive ion-beam facilities will be able to investigate this phenomenon near the edges of stability in more detail.

An example for the "classical" electric E1 dipole resonance of the (γ, xn) reaction on ^{120}Sn is shown in Fig. 12.3. Besides γ interactions inelastic particle scattering, especially with α's, is an important tool for measuring giant resonances. The cross sections peak at very small forward angles and the measurements normally require the use of magnetic spectrographs, see Sect. 17.3.3. Figure 12.4 shows an example of such spectra together with the indication of the disentanglement of the different contributions. The giant resonances that have been found can be classified according to their changes of orbital angular momentum, of spin and of isospin. Figure 12.5 shows schematically the different types. According to their collective character the resonance energies and widths vary systematically and slowly with

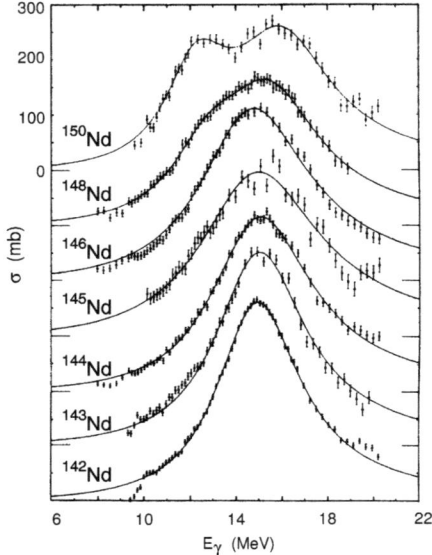

Fig. 12.6 GDR in the photoneutron (γ, n) reaction on isotopes of Nd with increasing neutron number N, showing onset of deformation. Adapted from [BER75]

the mass number A. The energies of the resonance peaks are higher than those of "normal" excitations, and the cross sections of the exciting reactions are high. This suggests the interpretation of the giant resonances as collective excitations of many (all) nucleons and—microscopically—the collective and coherent excitation of many-particle-many-hole (p-h) states across different harmonic-oscillator shells. As an example a collective $J^\pi = 1^-$ state may be constructed from excitation of $N = 5$ such p-h states from the p shell into the sd shell

$$\left|(1p_{3/2})^{-1}(1d_{5/2})\right\rangle_{1^-} \ldots \left|(1p_{1/2})^{-1}(1d_{3/2})\right\rangle_{1^-} \quad (12.1)$$

The p-h interaction acts as a perturbation that pushes one (the *coherent* state in which all amplitudes add up) upwards in energy because of the repulsive nature of the interaction and gives it nearly all strength of the N interfering states. The other $N - 1$ states interfere destructively and, therefore, are only slightly shifted in energy. (This is in analogy to the pairing phenomenon where the pairing-force is attractive and lowers the coherent ("paired") state. In both cases this is a characteristic behavior of the solution of the Schrödinger equation upon diagonalization of the perturbed Hamiltonian). In deformed nuclei the GDR peak is split into two peaks indicating different oscillation frequencies for dipole oscillations along the two principal ellipsoidal axes. This is illustrated in Fig. 12.6 where successively neutrons are added to a spherical core until deformation sets in. The number of possibilities of exciting giant resonances (excited between different harmonic-oscillator shells: $N \times \hbar\omega$, $N = 0, 1, \ldots$ which determine the multipolarities EL, ML, $L = 0, 1, 2, \ldots$, the possibility of spinflips $\Delta S = 0$ or 1, and of isospinflips $\Delta T = 0$ or 1) allow for many different giant-resonance types. Many of them have been identified although they are often difficult to disentangle due to similar energies and large widths. A comprehensive survey of giant resonances up to 1999 is given by Ref. [HAR01]. Inelastic scattering up to 1976 has been discussed in Ref. [BER75, BER76]. The more recent M1 scissors-mode giant resonance is discussed in detail in Ref. [HEY10]. Some properties of important giant resonances are collected in Table 12.1.

12.4 Fission Doorways

The phenomenon and mechanism induced nuclear fission as another form of nuclear reaction has already been discussed in Sect. 11.4.1. A compound system is formed in the entrance channel by any of a number of different reactions (e.g. neutrons or charged particles). For heavy nuclei the compound nucleus may decay by α emission or fission and in the latter case go through various stages of deformation forming potential minima at different separations of the fissioning nucleon clusters. In these minima systems of (rotational) levels may exist, which, due to different barrier heights and thicknesses, may exhibit quite different widths and lifetimes as well as level densities. In the fission channel these states will mix and show the behavior of intermediate-structure states with fine structure. A famous example is the case of ^{240}Pu $+ n$. The total (i.e. essentially the elastic (n, n)) cross section shows many fine-structure resonances, whereas in the (n, f) sub-barrier fission the cross section mediated by the same compound system displays groups of these levels that are widely separated and, if averaged, have shapes of resonances with larger widths (shorter lifetimes). Besides, the detailed spectroscopy of the states in the different minima tells about the complicated double- or triple fission barriers for which the doorway mechanism is of interest. The double-humped barrier model was confirmed by several evidences such as γ transitions between rotational levels in the first and the second minimum with different moments-of-inertia corresponding to different deformations. Also transitions between states in both minima have been observed. [SPE72]. A general survey of the fission process can be found e.g. in Ref. [SPE74].

Figure 12.7 shows the total and isomeric fission cross sections in the same energy region of neutrons.

12.5 Isobaric Analog Resonances (IAR)

Members of isobar multiplets are called *isobaric analog states IAS*. They are connected by exchanging one neutron in an unfilled shell of a nucleus by a proton (or vice versa), which corresponds to application of the isospin raising (or lowering) operator to such a state. Above a certain mass number A these $T^>$ states are raised sufficiently by the Coulomb interaction that they lie in the continuum, i.e. they can be studied as resonances (the *Isobaric Analog Resonances IAR*), see also Sect. 3.3.3 and Refs. [FOX66, TEM67, WIL69], and [AND69]. One point of interest in these IAR is due to the fact that their hadronic properties (i.e. those not due to the Coulomb interaction) are the same as those of the original $T^<$ states. These

Table 12.1 Properties of selected giant resonances

Type of resonance	Character	ΔL	ΔS	ΔT	Excitation energy E_x (MeV)	Width (MeV)	Preferred method	Relevance
ISGMR	E0	0	0	0 (IS)	$80A^{-1/3}$	3–5	(α, α')	Nuclear compressibility
IVGMR	E0	0	0	1 (IV)	$59A^{-1/6}$	10–15	Charge Exch. $(\pi^{\pm}, \pi^0)$	
ISGDR	M1	1	1	0 (IS)		4–8	Photoabsorption	Historically first (1937)
IVGDR	E1	1	1	1 (IV)	$31.2A^{-1/3} + 20A^{-1/6}$			
ISGQR	E2	2	0	0 (IS)	$64.7A^{-1/3}$	$90A^{-2/3}$	$(p, p'), (e, e')$	
IVGQR	E2	2	0	1 (IV)	$130A^{-1/3}$	5–15		
ISGOR (LEOR)	E3	3	0	0 (IS)	$41A^{-1/3}$			Low-Energy OR
ISGOR (HEOR)	E3	3	0	0 (IS)	$108A^{-1/3}$	$140A^{-2/3}$		High-Energy OR
GT	M1	0	1	1 (IV)	5–10	2–4	$(p, p'), (p, n), (n, p)$	β decay
Pygmy Res.	E1	1	0	1 (IV)	9–12		$(p, n), (^3\mathrm{He}, t)$	Neutron skin, symmetry energy
Scissors Mode	M1	1	1	1 (IV)	const ≈ 3		(e, e')	Rotational component

12.5 Isobaric Analog Resonances (IAR)

Fig. 12.7 Total (n,n) and sub-barrier (n,f) isomeric fission cross sections in the energy range from 0.5 to 3.0 keV [BNL11, MIG68]

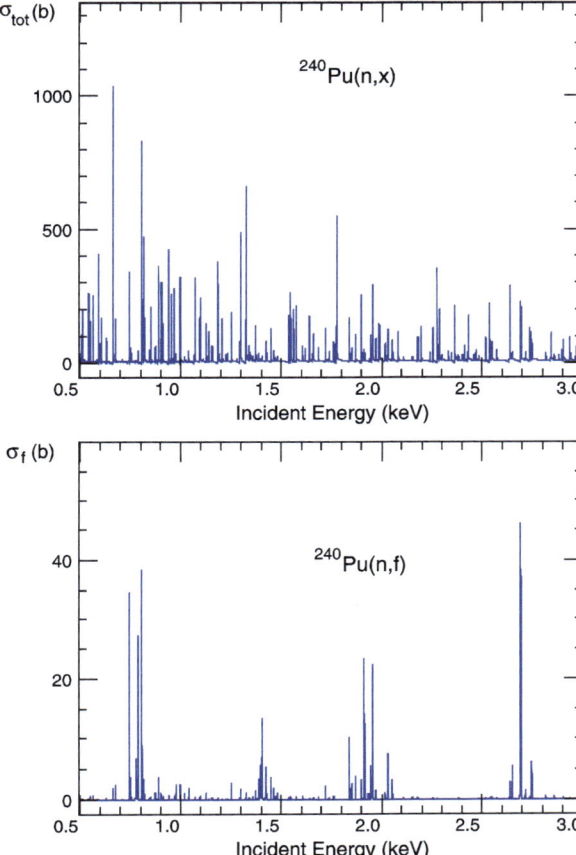

can therefore be studied in excitation functions e.g. of proton scattering and compared to the low-lying excitations of the $T^<$ states studied in transfer reactions such as (d,p). This will be explained for the case of states of $^{208}_{82}$Pb. Three neighboring nuclei have to be considered, the target nucleus, to which either a proton is added to form the compound system with the IAR, and the final nucleus of adding a neutron to the target, e.g. via the (d,p) reaction, see Fig. 12.8.

The IAR in excitation functions e.g. of proton elastic and inelastic scattering correspond to simple excitations, i.e. of single particle/single hole or particle-hole states near magic-number nuclei. This explains their typical widths. The total width of the IAR Γ consists of two parts, the spreading width of $\Gamma^\downarrow$ taking into account the decay of the IAR into the $T^<$ CN states by isospin mixing, and the escape width $\Gamma^\uparrow$ describing the proton emission into the exit channel. The IAR can be described by the classical Breit-Wigner theory, except that—due to the spreading over energy regions of varying CN-level density—the resonance shape becomes asymmetric. Figure 12.9 is a good example showing the IAR in the compound

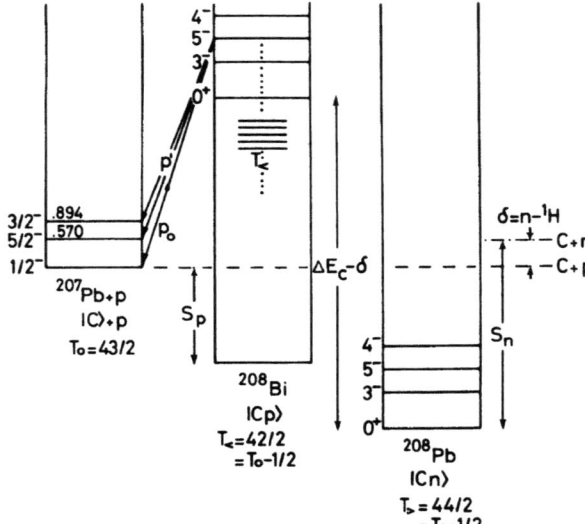

Fig. 12.8 Energy levels of the three different nuclei connected by isospin symmetry: the target nucleus $^{207}_{82}$Pb (the core state $<C>$) with isospin T_0, to which a proton or a neutron may be added, the compound (or "daughter") nucleus $^{208}_{83}$Bi, and the "parent" nucleus $^{208}_{82}$Pb. The IAR are the $T^>$ daughter states embedded in a "continuum" (in reality: a quasi-continuum) of many $T^<$ states with high level densities, with which they partly mix

nuclei $^{209}_{83}$Bi and $^{208}_{83}$Bi, the analogs to the low-energy shell model states in ^{209}Pb (single-particle states) and $^{208}_{82}$Pb (particle-hole states with different amplitude contributions (e.g. $|2g_{9/2} \otimes 3p^{-1}_{1/2}\rangle_{5^-}$ of the s.p. ground- and excited states of $^{209}_{83}$Bi coupled to single-hole states of $^{207}_{82}$Pb [LAT79, MEL85, NYG89]). The analogs are shifted into the continuum by the Coulomb energy difference $\Delta E_C \approx \frac{6}{5}\frac{Ze^2}{R}$ minus the neutron-proton mass difference $\delta_{np} = 0.139$ u. In heavy nuclei and the regions of high level densities of the $T^<$ (*parent*) states only the intermediate structures of the $T^>$ states are visible. The spectroscopic value of these studies is e.g. the spectroscopic factors (or decay amplitudes) to be compared to the spectroscopic factors obtained from corresponding (analog) (d, p) reactions to the $T^<$ isobaric analogs and the shell-model structures (shell-model state configurations), and the role of the residual interaction. The reaction mechanism reveals information about the role of isospin (breaking), see Sect. 3.3.3, the mixing of $T^>$ and $T^<$ states and the systematics of spreading widths $\Gamma^\downarrow$ (which proved to be remarkably constant over a wide mass-number range [HAR86, JAE87, NYG89]), and the systematics of Coulomb energy differences, see e.g. [JAE69]. It turns out that isospin mixing in these heavy nuclei is quite small which has been explained with shortness of interaction times.

Like in giant resonances we have a spreading of the strength into more complicated (many particle/many hole) compound states that are longer-lived, see also Sect. 3.3.3. In the excitation functions they manifest themselves as superimposed fine structure (doorway mechanism). This fine structure can be made visible, as long as the level densities are such that these CN states do not overlap and when the experimental resolution is good enough. This limits the visibility to medium-mass nuclei such as ^{40}Ca up to $A \approx 100$. Much effort has been expended to study these fine-structure states using accelerators with high energy-resolution experiments, especially at the TUNL lab. (Duke U.) at Durham, NC, USA, see

12.6 Exercises 217

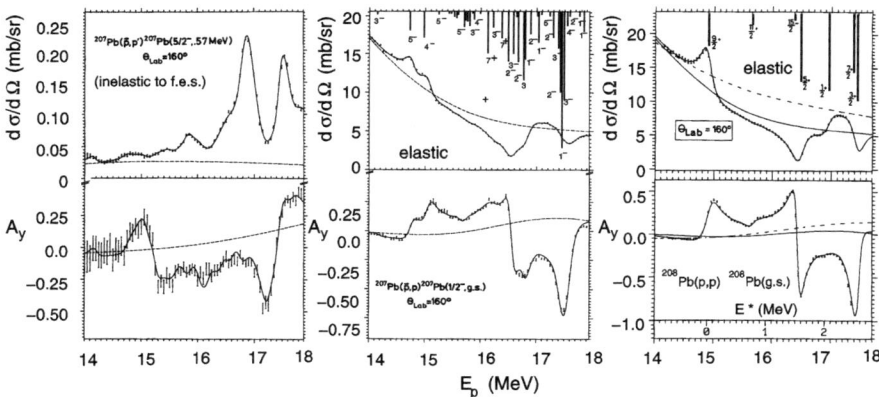

Fig. 12.9 Excitation functions of elastic and inelastic proton scattering on $^{207}_{82}$Pb (*left* and *center*) and $^{208}_{82}$Pb (*right*) across a number of low-lying IAR in $^{208}_{83}$Bi and $^{209}_{83}$Bi. The background (*dashed lines*) has been obtained in simultaneous resonance + optical model (elastic) or DWBA (inelastic) fits to the data [LAT79, MEL85]

Fig. 12.10 Excitation functions of ^{92}Mo$(p,p)^{92}$Mo at $E_p = 5.3$ MeV at two angles across a $j = 1/2$ IAR with high resolution. The heavy curves result from a single-level resonance fit with $\Gamma = 27$ keV and $\Gamma_p = 7$ keV showing interference with the smooth background. The fine structure (*thin continuous lines*, drawn through the data points) has an average width of ≈ 3 keV. The fine structure disappears outside the IAR. Adapted from [RIC64]

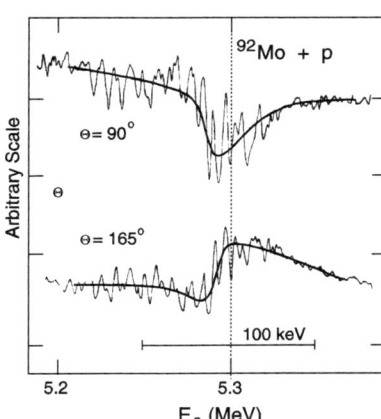

e.g. Refs. [WAT81a, WAT81b, MIT85] and references therein. The statistical distributions of these levels, their strength functions and widths etc. have been studied. Figure 12.10 shows one result of proton scattering on ^{92}Mo [RIC64].

12.6 Exercises

12.1. The Isobaric Analog Resonances in (p, p) scattering correspond to ground and excited states of the parent nuclei (accessible e.g. by neutron transfer reactions such as the (d, p) reaction, see Chap. 10), but shifted into the continuum. Parent $(T^<)$ and daughter $(T^>)$ states are assumed to have the same nuclear structure, except for possible isospin breaking. Calculate these energy shifts for the pairs of nuclei $^{209}_{82}$Pb–$^{209}_{83}$Bi, $^{139}_{56}$Ba–$^{139}_{57}$La, and $^{91}_{40}$Zr–$^{91}_{41}$Nb.

12.2. Mirror nuclei are pairs of neighboring nuclei with equal mass number A and isospin $T = 1/2$, $T_3 = \pm 1/2$ (e.g. ^{3}He and ^{3}H). Under isospin conservation they would have identical nuclear wave functions. The isospin-breaking Coulomb interaction (corresponding to an exchange of one neutron with one proton/application of isospin ladder operations) shifts the states of the $T^>$ nucleus upward in energy against the $T^<$ nucleus (see the preceding text). If—as a function of A—the energy shift exceeds the proton separation energy the $T^>$ nucleus becomes unbound, i.e. its ground state becomes a resonance in the continuum (the state acquires a width >0 and decays by particle emission). Near this region of medium-heavy nuclei the CN level densities are so low that the CN fine-structure states can be resolved and investigated. Estimate near which mass number A this transition occurs.

12.3. Nuclear matter and nuclei are weakly compressible. They share this property with macroscopic liquids, which is one evidence for a description of nuclei as liquid drops. The amount of compressibility is measured by the *compressibility* K. (In classical physics the compressibility is defined as $\kappa = -dV/V \cdot 1/dp$ describing the relative volume change per pressure change). The compressibility enables high-frequency collective oscillations, the giant monopole resonance (or breathing mode) of nuclei. An early model [BOH75] predicted a relation between the compressibility K and the frequency of the monopole excitation

$$\omega = \frac{\pi}{R}\sqrt{\frac{b_{\text{compr}}}{m_N}} \quad (12.2)$$

with $b_{\text{compr}} = \frac{1}{9}K$, R the nuclear radius, and m_N the nucleon mass. Using Table 12.1 deduce the value of K and compare with the accepted value of about 231 MeV.

References

[AND69] J.D. Anderson et al., *Conf. on Nuclear Isospin, Asilomar* (Academic Press, San Diego, 1969)
[BER75] B.L. Berman, S.C. Fultz, Rev. Mod. Phys. **47**, 713 (1975)
[BER76] F.E. Bertrand, Annu. Rev. Nucl. Part. Sci. **26**, 457 (1976)
[BNL11] ENDF/B-VII.1, Brookhaven Natl. Lab. (2011)
[BOH75] A. Bohr, B.R. Mottelson, *Nuclear Structure*, vol. 2 (Benjamin, Reading, 1975)
[CIN73] N. Cindro, P. Kulišic, T. Mayer-Kuckuk (eds.), *Proc. Eur. Study Conf. on Intermediate Processes in Nucl. Reactions*. Lecture Notes in Physics, vol. 22 (Springer, Heidelberg, 1973)
[FES67] H. Feshbach, Ann. Phys. **11**, 230 (1967)
[FOX66] J. Fox, D. Robson (eds.), *Conf. on Isobaric Spin in Nuclear Physics* (Academic Press, New York, 1966)
[HAR86] H.L. Harney, A. Richter, H.A. Weidenmüller, Rev. Mod. Phys. **58**, 607 (1986)

References

[HAR01] M.N. Harakeh, A. van der Woude, *Giant Resonances—Fundamental High-Frequency Modes of Nuclear Excitations*. Oxford Studies in Nuclear Physics, vol. 24 (Oxford Science Publ., Oxford, 2001)
[HEY10] K. Heyde, A. Richter, P. von Neumann-Cosel, Rev. Mod. Phys. **82**, 2365 (2004)
[JAE69] J. Jänecke, in *Isospin in Nuclear Physics*, ed. by D.H. Wilkinson (North Holland, Amsterdam, 1969), p. 297 ff.
[JAE87] J. Jänecke, M.N. Harakeh, S.Y. van der Werf, Nucl. Phys. A **463**, 571 (1987)
[LAT79] G. Latzel, H. Paetz gen. Schieck, Nucl. Phys. A **323**, 413 (1979)
[MEL85] R. Melzer, P. von Brentano, H. Paetz gen. Schieck, Nucl. Phys. A **432**, 363 (1985)
[MIG68] E. Migneco, J.P. Theobald, Nucl. Phys. A **112**, 603 (1968)
[MIT85] G.E. Mitchell, E.G. Bilpuch, J.F. Shriner, A.M. Lane, Phys. Rep. **117**, 1 (1985)
[MON67] J.E. Monahan, A.J. Elwyn, Phys. Rev. **153**, 1148 (1967)
[NYG89] K.R. Nyga, H. Paetz gen. Schieck, Nucl. Phys. A **501**, 513 (1989)
[RIC64] P. Richard, C.F. Moore, D. Robson, J.D. Fox, Phys. Rev. Lett. **13**, 343 (1964)
[SAT90] G.R. Satchler, *Introd. Nucl. Reactions*, 2nd edn. (McMillan, London, 1990)
[SIE67] R.H. Siemssen, J.V. Maher, A. Weidinger, D.A. Romley, Phys. Rev. Lett. **19**, 369 (1967)
[SPE72] H.J. Specht, J. Weber, E. Konecny, D. Heunemann, Phys. Lett. B **41**, 43 (1972)
[SPE74] H.J. Specht, Rev. Mod. Phys. **46**, 773 (1974)
[TEM67] G.M. Temmer, in *Fundamentals in Nuclear Theory*, ed. by A. de Shalit, C. Villi (IAEA, Vienna, 1967). Chap. 3
[WAT81a] W.A. Watson, E.G. Bilpuch, G.E. Mitchell, Phys. Rev. C **24**, 1992 (1981)
[WAT81b] W.A. Watson, E.G. Bilpuch, G.E. Mitchell, Z. Phys. A **300**, 89 (1981)
[VAL81] L. Valentin, *Subatomic Physics: Nuclei and Particles*, vol. 2 (North-Holland, Amsterdam, 1981), p. 409
[WIL69] D.H. Wilkinson (ed.), *Isospin in Nuclear Physics* (North-Holland, Amsterdam, 1969)

Chapter 13
Heavy-Ion (HI) Reactions

HI reactions are characterized by the fact that because of the higher masses of the particles—as compared to light ions—and therefore smaller de Broglie wavelengths often the phenomena can be described classically or semi-classically. Analog to light optics have been discussed. For further reading the references [BRO85, BOC81, BAS80, NOE76] are recommended.

13.1 General Characteristics of HI Interactions

In contrast to reactions between protons and nuclei interactions between heavier nuclei are characterized by strong absorption, i.e. reactions take place predominantly at or near the nuclear surfaces. In this sense, because the α particle as projectile shows already strong absorption, reactions of nuclei with $A \geq 4$ with other nuclei are considered as HI reactions. The theoretical description of HI reactions is difficult in many respects. On the other hand, HI reactions have many useful applications in nuclear physics. Their characteristics are

- The number of reaction channels is often very large.
- The number of nucleons involved is large making the systems multi-body systems that are difficult to approach with microscopic theories such as based on the NN interaction and using Faddeev and EFT methods.
- The description is therefore mostly limited to approximate reaction models.
- The many final channels make HI reactions a versatile tool for reaching many different final nuclei and their excited states for nuclear spectroscopy by fusion-evaporation reactions.
- The large atomic number Z and the strong electric fields of HIs have been widely used to excite target nuclei during elastic scattering at close distance: *Coulomb Excitation*.
- An important field for application of HI reactions is extending the periodic table as far as possible (at present to $Z = 118$) by fusion of very heavy nuclei, in competition with fast fission.

Fig. 13.1 The typical potential between two medium-mass heavy ions combined from a nuclear, mostly attractive, potential with the repulsive Coulomb and centrifugal potentials (schematically). A shallow *nuclear* potential is assumed. It is remarkable that at certain energies a shallow trough develops that, in principle, can support some (metastable) states of rotational character, suggestive of molecular resonances, see below, Sect. 13.2.3

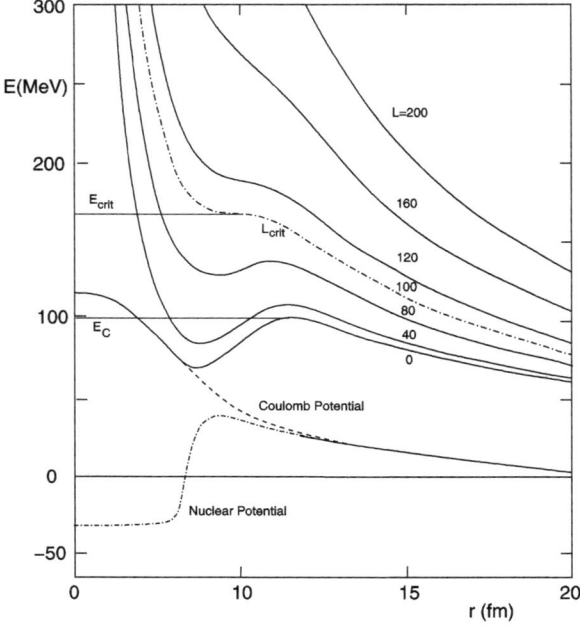

It is useful to sketch the interaction potential between two heavy ions. It consists of three parts

- At low energies, i.e. also at larger distance (large impact parameter, forward scattering) elastic scattering by a strong Coulomb potential dominates. Sub-Coulomb scattering is an important means of excitation of nuclei for spectroscopy (*Coulomb excitation*) [ALD66, ALD75]. The scattering is described by the optical model, the central part of which is often a quite shallow potential (as compared to nucleon-nucleus potentials).
- At higher energies the centrifugal potential $V_\ell \propto \hbar^2 \ell^2 / 2\mu r^2$ can reach very high values thus creating a centrifugal barrier that (classically) not only prevents the nuclei from fusing, but may cause rotational fission.
- At intermediate energies (and small impact parameters, backward scattering) above the Coulomb barrier E_C fusion into compound systems is possible. Subsequent evaporation of the highly-excited compound nuclei is a preferred method for the study of properties of final nuclei, with the interest shifting especially to exotic nuclei. Here the nuclear (hadronic) interaction is important once the nuclei overlap sufficiently.

These three components create a rather complex picture of the combined potential that is schematically depicted in Fig. 13.1. Apart from the hadronic part of the interaction, the elastic scattering between heavy ions lends itself to approximate and semi-classical phenomena and descriptions, especially because a truly microscopic theory of the interactions of systems of many nucleons is still lacking.

13.2 Semi-Classical Phenomena and Description

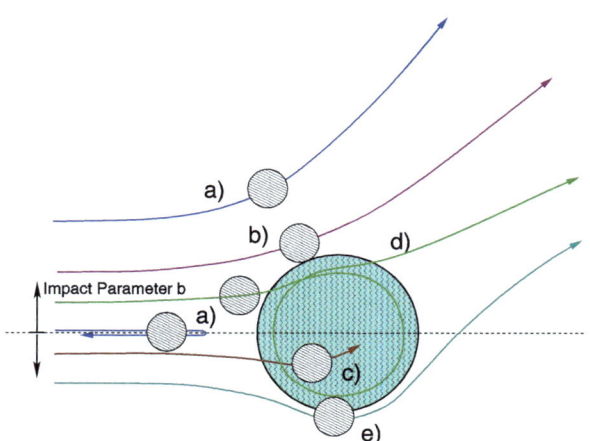

Fig. 13.2 (Semi-)classical description of different possible paths of the projectile depending on the impact parameter b (i.e. on energy and angle) of the scattering process: (**a**) Rutherford scattering, (**b**) Coulomb scattering due to the influence of the extended charge (and mass) distributions and grazing orbits, (**c**) complete fusion and CN formation, (**d**) orbiting, and (**e**) incomplete fusion and inelastic scattering

13.2 Semi-Classical Phenomena and Description

The relatively small de Broglie wavelength of the heavy ions in reactions is a condition that the projectile and target nuclei are localized and have well-defined trajectories. Especially as long as the nuclei do not merge a number of semi-classical phenomena, partly with interference between different paths have been identified. Figure 13.2 depicts a number of possible trajectories of heavy ions when passing a (heavy) target nucleus. The different modes can be understood with the deflection function, extended from the point-nucleus case, Eq. (2.5). For an arbitrary potential including the orbital-angular momentum barrier

$$V_{\text{eff}} = V(r) + \frac{L^2}{2\mu r^2} \tag{13.1}$$

the integration of the differential equation of the classical trajectory

$$d\Theta = \frac{-L}{\mu r^2}\left[\frac{2}{\mu}\left(E - V(r) - \frac{L^2}{2\mu r^2}\right)\right]^{-1/2} \tag{13.2}$$

from $r = \infty$ to $r_{\min} \equiv d$ yields

$$\Theta = \pi - 2b\sqrt{E_\infty}\int_{r_{\min}}^{\infty}\frac{dr}{r^2\sqrt{E - V_{\text{eff}}(r)}}. \tag{13.3}$$

The integrand is always >0 and $\infty \leq \Theta \leq \pi$. With the scattering angle θ we have $\Theta = \theta + 2n\pi$. Assuming a model potential that is partly repulsive (Coulomb) and partly attractive, as shown in Fig. 13.1,

$$V(r) = \frac{c}{r} - \frac{\beta}{r^2}; \quad c, \beta > 0, \tag{13.4}$$

we obtain deflection functions as in Fig. 13.3. With Eq. (2.5) the classical cross section results.

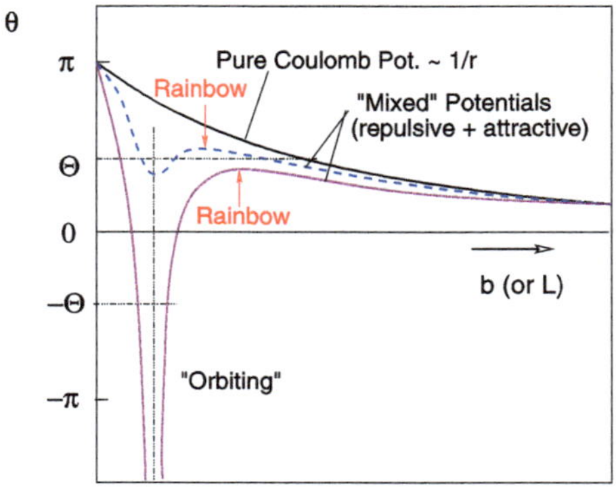

Fig. 13.3 The deflection functions $\Theta(b)$ for a pure point-Coulomb potential and for more complicated potentials with attractive parts such as the nuclear (hadronic) force and repulsive parts such as the point or extended-charge Coulomb potential show how effects such as "Orbiting" and "Rainbows" are possible

Fig. 13.4 The cross section relative to the point-Rutherford cross sections shows a typical Fresnel diffraction pattern before the strong fall-off for larger angles that is caused by the nuclear potential including strong absorption near the grazing angle θ_{gr}. Also shown (schematically) is the rainbow enhancement at the rainbow angle θ_r where classically the cross section would diverge. Adapted from [FRA72]

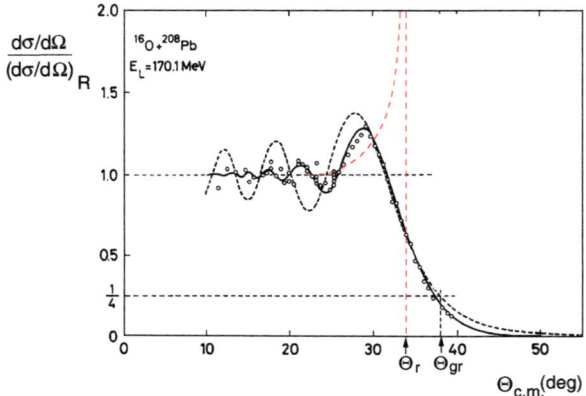

13.2.1 Elastic Scattering

For elastic scattering of heavier ions the cross sections behave very much like Fresnel diffraction in optics with an illuminated region and sharp shadow (classical trajectory picture, analogous to geometrical optics), superimposed by interference pattern due to diffraction at the (nuclear) "edge" that can be described by Airy functions. Figure 13.4 gives an example of such a semi-classical description. For lighter HI elastic scattering at higher energies where the Coulomb interaction becomes small the differential cross section assumes the form

$$\frac{d\sigma}{d\Omega} = R^2 \left[\frac{J_1(kR\theta)}{\theta} \right]^2 \qquad (13.5)$$

(with J_1 a Bessel function) of Fraunhofer scattering in light optics [NOE76], p. 122. Other analogies ("glory scattering") have been constructed. A detailed discussion of the semi-classical description of heavy-ion reactions can be found in [FRA72, SAT90, NOE76].

13.2.2 Other HI Reaction Models

Like with light ions heavy ions can be classified according to reactions times with the limiting cases of direct and complete-fusion compound-nucleus reactions. The model descriptions are similar with modifications by

- the high number of possible outgoing channels,
- the high excitation energy, and
- the high angular momentum possible.

Consequently the description e.g. by coupled-channels codes CCWBA instead of DWBA is often necessary. The complete thermodynamical equilibrium for true compound systems is often not obtained leading to effects such as *pre-equilibrium* or *incomplete fusion*. The relevant reaction mechanism is *deep-inelastic scattering*. In outgoing-particle spectra fragment masses smaller than the CN masses are found. Molecular resonances are also a consequence of systems with high angular momentum preventing complete fusion. On the other hand, the large numbers of outgoing channels, the high excitation energies and angular momenta make HI *fusion-evaporation* reactions an almost ideal tool for nuclear spectroscopy, especially of high-spin states.

13.2.3 Molecular Resonances

The interaction potential between two heavy ions is a combination of a strong repulsive Coulomb potential, a significant (repulsive) centrifugal potential, and—at sufficiently high energies—an attractive contribution from fusion (absorption) channels. Depending on energy and mass-number combinations this contribution allows either fusion into CN systems, incomplete fusion or the total potential can assume a shape with a shallow minimum at a finite separation distance > 0 in which highly excited states can exist that may be identified with short-lived molecular states (in analogy with atomic molecules). In a number of cases structures in excitation functions of "intermediate" width have been measured. For more details see Sect. 12.1.

13.2.4 Heavy-Ion Reactions and Superheavies SH

The extension of the periodic table in the direction of an assumed island of stability beyond $Z = 114$ requires beams and targets with high Z and N. Theoretical model calculations predict SH's Magic numbers $Z = 114$, 120, or 126, and $N = 184$. Thus, this new-element synthesis is a genuine field of heavy-ion physics. The production of elements up to $Z = 118$ so far teaches how to proceed further. The principles are to try paths where neither fission nor α decay are faster than the (short) lifetimes of the expected compound nuclei. Their lifetimes must be long enough to emit radiation characteristic enough to identify the new nuclides. Such were in the past α decays leading to chains with already known isotopes. For more details see Sect. 15.3.

13.3 Nuclear Spectroscopy and Nuclear Reactions

Nuclear spectroscopy, which investigates the properties of ground and excited nuclear states, has to rely in many ways on nuclear reactions and the techniques necessary (ion sources, accelerators etc.). A few examples will be discussed here:

- γ transitions in nuclei may be investigated e.g. after *Coulomb excitation*. In this case the rapidly varying electric field of an accelerated heavy ion excites the target nucleus during passage.
- In HI fusion-evaporation reactions the choice of the reaction partners and of the incident particle energies determines the range of angular momenta and of excitation energies of certain final nuclei to be investigated.
- Excitation functions of elastic and inelastic proton scattering via isobaric-analog resonances allow the detailed spectroscopy, e.g. of shell-model states, complementing the direct reactions to the low-lying final-nucleus states.
- Transfer reactions like (d, p) stripping historically served for finding and localizing of (especially single-particle) shell-model states for confirmation of the (single-particle) shell model. Most often the single-particle strength is fractionated by the *residual interaction*, i.e. the action of the other non-shell-model (or: core) nucleons. In that case a number of states is grouped around a center-of-mass (in energy), which corresponds to the shell-model state, and which has approximate Lorentz form with a width Γ connected to the strength of the residual interaction (*strength function*). From the point of view of reaction mechanism this behavior is also a doorway intermediate state with fine structure, see Chap. 12.

13.4 Heavy-Ion Reactions as Special Tools for Nuclear Spectroscopy

13.4.1 Coulomb Excitation

The high atomic numbers Z available with heavy-ion projectiles can be used for efficient excitation of many states of target nuclei by the strong Coulomb field during close-by passage, see e.g. Refs. [ALD66, ALD75]. Below the Coulomb barrier only the electromagnetic interaction acts between the nuclei. Thus, an exact description of the sub-Coulomb excitation process is possible, and, in semi-classical first-order perturbation theory the trajectories are described classically and the cross section as the Rutherford cross section times the electromagnetic excitation probability $P_{i \to f}$

$$\left(\frac{d\sigma}{d\Omega}\right) = P_{i \to f} \left(\frac{d\sigma}{d\Omega}\right)_{\text{Ruth}} \quad (13.6)$$

P is the square of the matrix element of the electromagnetic transition, expanded in multipole moments.

13.4.2 Fusion-Evaporation Reactions

The goal of nuclear spectroscopy is to elucidate as many properties of all nuclei of the nuclide chart as possible. This includes the comparison with nuclear-structure models such as the shell or collective models or their modern versions, but also the necessity of clearing all steps of the different processes of nucleosynthesis in astrophysics. It also includes detection or production of unknown nuclides at the edges of the nuclide chart. The necessity to reach these limits by fusion reactions requires many different projectile-target combinations. These fused nuclei being far from the valley of stability are unstable against decay and in general highly excited in states of high spin. The property of high mass number transfers high angular momentum, which helps in reaching these states. Figure 13.5 shows the relation between excitation energy and angular momentum ("Spin") of excited nuclei (in principle an additional variable, the level density $\rho(E_x, J)$ may be depicted in a 3D plot). The *Yrast Line* is defined as the maximum angular momentum possible with a given excitation energy (or vice versa: as the minimum excitation energy at a given angular momentum). The level density drops to zero at this line, i.e. to the right there are no levels.

At very high excitation above the particle-separation energies the states decay by particle emission, i.e. through fastest process, the strong interaction. Below the nucleon separation energy the states decay electromagnetically or by β decay. In regions of high level densities these decays occur statistically (*Evaporation*). However, at the edges of the *yrast* diagram with the maximum of angular momentum (or spin) compatible with the energy of the compound system the level density is low and well-separated γ transitions from very high-spin states are possible. Thus e.g. rotational bands have been observed up to spins of $80\hbar$. Phenomena such as *Backbending*, caused by a phase transition with change of deformation and moment-of-inertia, and *Superdeformation* of nuclei (with axis ratios of the rotation ellipsoid, assumed as nuclear shape, of 1 : 2; normal deformations in mass regions between closed shells and at lower excitation energies are much smaller) are observable in γ cascades along or near the yrast line where—due to the low level densities—single states and transitions are visible, in contrast to the statistical decays at lower angular momenta and high excitation.

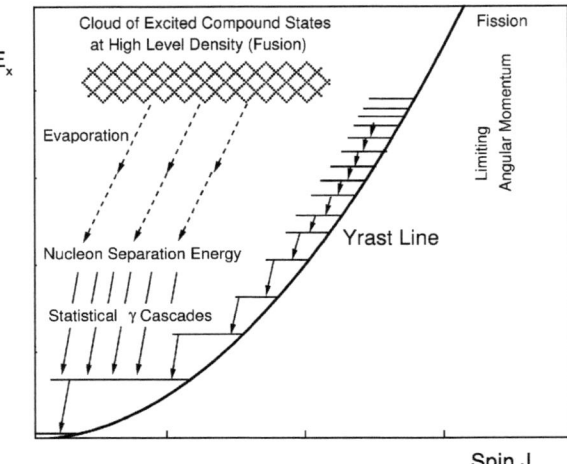

Fig. 13.5 Scheme of the possible processes after excitation by fusion reactions, and especially definition of the *Yrast Line* as the limit where the total excitation energy is rotational and beyond which the level density becomes zero

13.5 Exercises

13.1 Calculate the Coulomb energies and the Sommerfeld parameter at the Coulomb energies for the ^{12}C + ^{24}Mg, ^{127}I + ^{127}I, and ^{208}Pb + ^{208}Pb systems.

13.2 Heavy-ion reactions can often be described semi-classically. With the criteria developed in Chap. 2 test this e.g. for Ar on Th at 50 MeV above the Coulomb barrier.

13.3 Assume that transfer reactions are peripheral, i.e. take place preferentially at the surface of heavy ions. In a semi-classical description trajectories are well-defined by energy and angle. Where would you search for evidence of a transfer-reaction mechanism? Take as an example for a one-proton transfer the reaction $^{16}_{8}$O($^{103}_{45}$Rh, $^{17}_{9}$F)$^{102}_{44}$Ru at $E_{c.m.} = 135$ MeV. At which c.m. and lab. angles would you expect maximum transfer cross section?

13.4 Consider elastic Coulomb scattering of ^{16}O from gold at $E_d = 10$ MeV/nucleon. At which c.m. angle and which critical angular momentum ℓ_{crit} does $\frac{d\sigma}{d\Omega}$ start to deviate from point-Rutherford scattering, leading to a marked falloff to higher scattering angles?

13.5 Derive the classical cross section

$$\frac{d\sigma}{d\Omega} \propto \frac{1}{\sin\theta} \quad (13.7)$$

for compound-nucleus reactions with very high angular momentum (such as with heavy ions), leading not only to symmetry about 90°, but to a very marked anisotropy. An example is the angular distribution of the fission fragments of the compound nucleus ^{209}At formed in the reaction ^{12}C + ^{197}Au.

References

[ALD66] K. Alder, A. Winther, *Coulomb Excitation* (Academic Press, San Diego, 1966)
[ALD75] K. Alder, A. Winther, *Electromagnetic Excitation, Theory of Coulomb Excitation with Heavy Ions* (North-Holland, Amsterdam, 1975)
[BAS80] R. Bass, *Nucl. Reactions with Heavy Ions*. Texts and Monographs in Physics (Springer, Berlin, 1980)
[BOC81] R. Bock (ed.), *Heavy-Ion Collisions I–III* (North-Holland, Amsterdam, 1981)
[BRO85] A. Bromley (ed.), *Treatise on Heavy-Ion Science I–VIII* (Plenum, New York, 1985)
[FRA72] W.E. Frahn, Ann. Phys. (N.Y.) **72**, 524 (1972)
[NOE76] W. Nörenberg, H.A. Weidenmüller, *Introd. to the Theory of Heavy Ion Collisions*. Lecture Notes in Physics, vol. 51 (Springer, Heidelberg, 1976)
[SAT90] G.R. Satchler, *Introd. Nucl. Reactions*, 2nd edn. (McMillan, London, 1990)

Chapter 14
Nuclear Astrophysics

Because of recent developments in cosmology and particle physics ("astroparticle physics") as well as of improved methods in nuclear astrophysics the latter field has gained increased importance. In "Big-Bang" as well as later phases of nucleosynthesis the interplay between nuclear reactions and nuclear spectroscopy is strong and requires the entire apparatus and methods of nuclear physics. However, the energies required are quite low, and the description of the reaction mechanisms are simplified by the fact that often only S-waves participate. All reaction models discussed here are applicable to low-energy nuclear astrophysics. On the other hand the Coulomb barrier makes the cross sections often very small and extreme background reduction is required together with beam currents as high as possible. Accelerator facilities have therefore been placed in deep-underground laboratories. The main goal is to determine absolute cross sections and from these reaction rates of very many nuclear reactions occurring in stars and in star formation, which are partly coupled and form a complicated network of coupled rate equations. Polarization observables play a minor role here. An excellent introduction into the field of nuclear astrophysics is Ref. [ROL88].

14.1 Reaction Rates

Because of the lifetime of the neutrons that must be produced in the chain of reactions neutron-induced reactions play a major role only in rapid processes such as th r-process in later stages of element formation, and charged particle-induced reactions are the most important in earlier phases of nucleosynthesis. The Coulomb barrier

$$V_C(r) = \frac{Z_1 Z_2 e^2}{r} \tag{14.1}$$

determines the coarse behavior of the cross section whereas finer details are due to the specific reaction mechanism. The height of the Coulomb barrier even for the lightest system $p + p$ is about 500 keV (meaning that classically one proton would need $T \approx 1$ MeV in the lab. to reach the range of the nuclear interaction R_n) but tunneling makes reactions possible much below that threshold. The effect of tunneling is described by the tunneling probability

$$P = \frac{|\psi(R_n)|^2}{|\psi(R_C)|^2}, \quad (14.2)$$

where the squares of the wave function are the probabilities of finding the particles at $r = R_n$ and $r = R_C$, resp. The exact solution of the Coulomb problem [BET37] can be approximated sufficiently by

$$P = \exp(-2\pi\eta_S) \quad (14.3)$$

with η_S the Sommerfeld parameter, see Eq. (2.2). Two energy-dependent factors determine the gross behavior of the cross sections

$$\sigma(E) \propto \exp(-2\pi\eta_S) \quad \text{and} \quad \propto \frac{\pi}{k_{\text{in}}^2} \propto \frac{1}{E}. \quad (14.4)$$

Combining them the cross section may be factorized

$$\sigma(E) = \frac{S(E)}{E \cdot \exp(2\pi\eta_S)} \quad (14.5)$$

where $S(E)$ contains all information on the nuclear interaction. It is called the *astrophysical S-factor*. It is the most useful parametrization of the cross section and shows very directly the type of reaction occurring. Smooth behavior indicates direct processes, irregularities are due to resonances of the compound systems or threshold cusps, or some additional effects such as electron screening of the Coulomb field. Especially the smooth behavior of an S-factor allows the extrapolation of a cross section into energy regions not accessible experimentally, e.g. to very low energies. Figure 14.1 shows the integrated cross sections of the fusion reactions most important for primordial nucleosynthesis and for fusion energy research.

14.2 Typical S-Factor Behavior

If we take fusion reactions—which are important also for fusion-energy research—as examples, we can find the two extreme cases of the DD reactions

$$^2\text{H} + {}^2\text{H} \rightarrow \begin{cases} ^3\text{H} + p \\ ^3\text{He} + n \end{cases} \quad (14.6)$$

as direct rearrangement reactions and the

$$^2\text{H} + {}^3\text{H} \rightarrow {}^4\text{He} + n \quad (14.7)$$

14.2 Typical S-Factor Behavior

Fig. 14.1 Cross sections of a few fusion reactions relevant to primordial element synthesis and controlled fusion for energy production. Data from [NNDC]

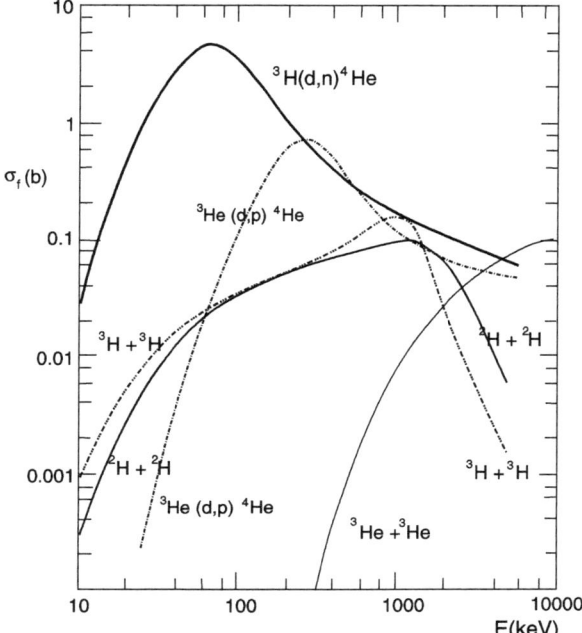

and the

$$^2\text{H} + {}^3\text{He} \rightarrow {}^4\text{He} + p \tag{14.8}$$

reactions as occurring via single resonances corresponding to excited states in ^{5}He and ^{5}Li, sitting on some direct-reaction background. Figure 14.2 shows the two cases schematically.

Electron Screening At the very low end of the energy scale the effects of screening of the Coulomb potential by the presence of electrons (either in the plasma of gaseous reaction partners or in the metallic environment of solid target materials) may modify (increase) the fusion cross sections appreciably, see e.g. [ROL95]. The present status of experiments and theory is e.g. given in Ref. [HUK08].

In the Figs. 14.3 and 14.4 [NAV11b] the effect of electronic screening is clearly visible for ^{3}He$(d, p)^4$He, not so evident for ^{3}H$(d, n)^4$He. The purely nuclear calculations [NAV11a, PIE01, EPE09] of course do not describe this part of the S-factor and a consistent theory of the screening is still missing. Precise extrapolations of the nuclear S-factor together with precise measurements of the cross sections, which are difficult due to the low energies, could help to pin down the screening details, maybe also including possible polarization effects on the screening. The figures show also the partly unsatisfactory quality of the experimental data. A very recent development is the use of high-power lasers to produce a plasma of ionized reaction partners in which no electron screening is acting and which allows the extrapolation of the purely nuclear cross sections [BAR13].

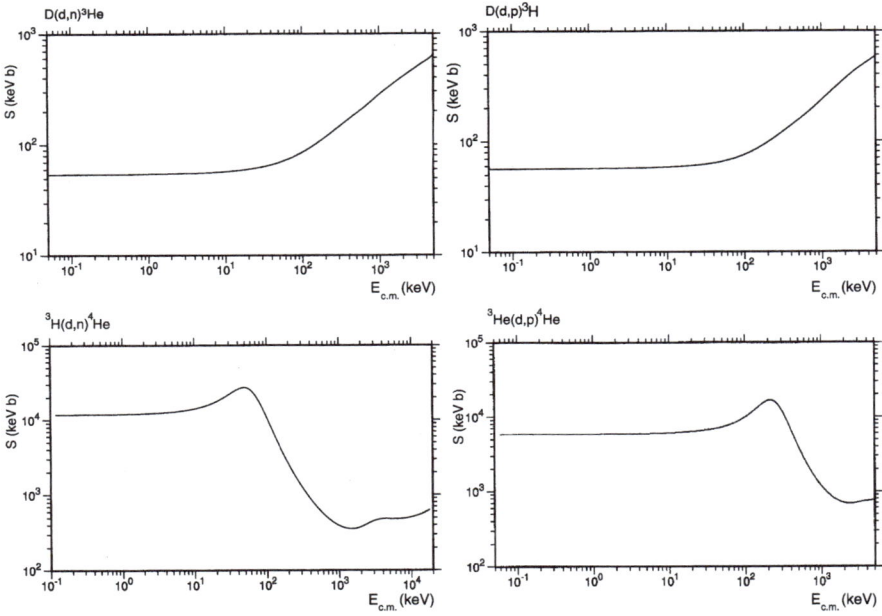

Fig. 14.2 S-factors of the direct DD reactions (*upper*) and the resonant fusion reactions $D + {}^3H$ and $D + {}^3He$ (*lower*). The nearly energy-independent parts of S indicate pure s-wave non-resonant behavior, the non-resonant increase of S towards higher energies is caused by the emergent admixtures of higher (P, D) waves. Note the logarithmic scales

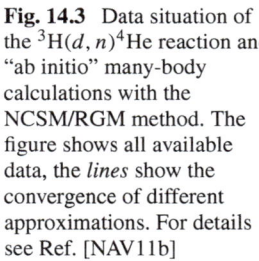

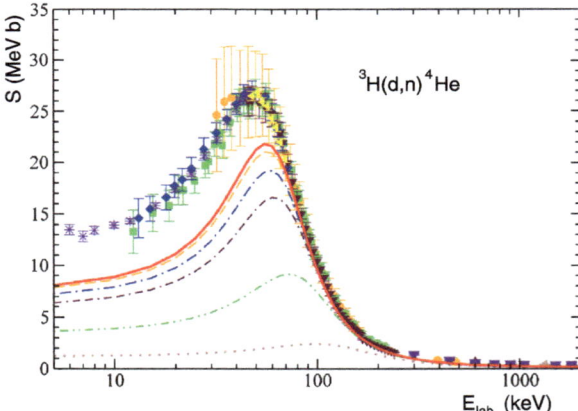

Fig. 14.3 Data situation of the ${}^3H(d,n){}^4He$ reaction and "ab initio" many-body calculations with the NCSM/RGM method. The figure shows all available data, the *lines* show the convergence of different approximations. For details see Ref. [NAV11b]

14.2.1 Calculation of Reaction Rates in Plasmas

One of the important quantities in nuclear astrophysics as well as fusion-energy questions is the rate R of reactions between at least two collision partners a and b.

14.2 Typical S-Factor Behavior

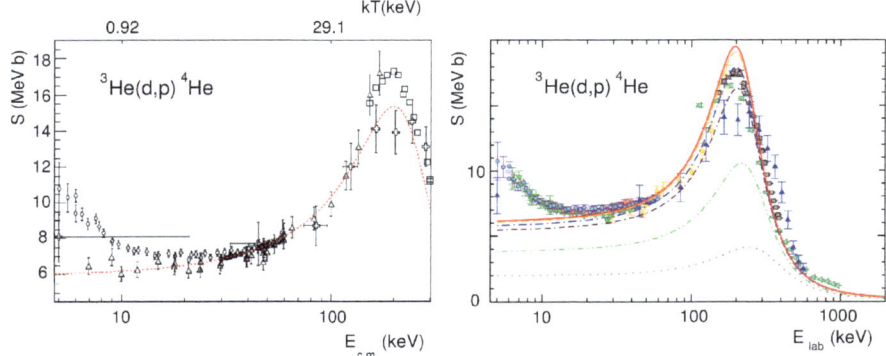

Fig. 14.4 Data situation of the $^3\mathrm{He}(d,p)^4\mathrm{He}$ reaction (*left*) and "ab initio" many-body calculations with the NCSM/RGM method (*right*). The *right figure* shows all available data, the *lines* show the convergence of different approximations. For details see Ref. [NAV11b]. The *left figure* displays the data of Refs. [KRA87] (*triangles*, data with no electron screening), [GEI99] (*squares*), [ALI01] (*diamonds*), and [LAC05] (*crosses*) and includes a recent S-factor determination using a petawatt laser beam for plasma interactions of the ionized reaction partners [BAR13] (*solid dots* and *dashed line*, no electron screening)

Different from nuclear reactions in the laboratory with one particle at rest, the other with fixed incident energy from an accelerator, in astrophysics we have a gas or plasma of at least two species of nuclei (atoms) with a relative velocity v between them. In the classical picture (see above) the total cross section is defined as the effective area A, in which collisions occur. With both partners moving this is equivalent to one partner a moving, the other b at rest, but with increased collision cross section $\sigma(v)$ times the number of target nuclei b, say N_b, per unit volume $A = \sigma(v) N_b$. All projectile nuclei a see this area and the total rate is proportional to the incident flux $j = N_a v$, thus

$$R = N_a N_b v \sigma(v) \tag{14.9}$$

where the unit of R usually is cm^{-3} s^{-1}. The velocities follow distributions $N(v)$, which are determined by the temperature T and range from 0 to ∞ such that the average product $\langle \sigma \cdot v \rangle$, the *reaction parameter* must be

$$\propto \int_0^\infty N(v) v \sigma(v) dv \tag{14.10}$$

and

$$R = N_a N_b \langle \sigma v \rangle. \tag{14.11}$$

In stars normally the gas is in thermodynamic equilibrium and the atoms (nuclei) follow the Maxwell-Boltzmann velocity distribution

$$N(v) = 4\pi v^2 \left(\frac{m}{2\pi kT}\right)^{3/2} \exp\left(-\frac{mv^2}{2kT}\right), \tag{14.12}$$

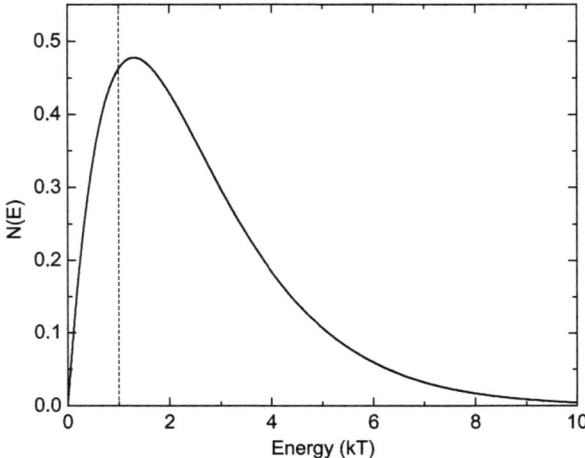

Fig. 14.5 Typical Maxwell-Boltzmann energy distribution $N(E)$ as function of the (thermal) energy of nuclei in a gas at equilibrium. The energy is measured in units of kT

see Fig. 14.5, which in terms of the energy is

$$N(E) \propto E \exp\left(-\frac{E}{kT}\right). \tag{14.13}$$

After kinematical transformation of v_i of both particles to c.m. and relative velocities v and with μ the reduced particle mass one obtains

$$\langle \sigma v \rangle = 4\pi \left(\frac{\mu}{2\pi kT}\right)^{3/2} v^3 \sigma(v) \exp\left(-\frac{\mu v^2}{2kT}\right) dv \tag{14.14}$$

or

$$\langle \sigma v \rangle = \left(\frac{8}{\pi \mu}\right)^{1/2} (kT)^{-3/2} \int_0^\infty \sigma(E) E \exp\left(-\frac{E}{kT}\right) dE. \tag{14.15}$$

Figure 14.6 shows the reaction parameters of a number of fusion reactions calculated by taking the average of the cross sections weighted over the Maxwell-Boltzmann velocity distributions as functions of temperature, measured in keV.

In the primordial phase of fusion reactions shortly after the big bang (*Big-Bang nucleosynthesis*) two chains of reactions are responsible for element synthesis: one is the pp (and pep) chain, see Fig. 14.7, the other the CNO (also "Bethe-Weizsäcker") cycle, see Fig. 14.8. Both are strongly temperature dependent ($\approx \propto T^{15}$ for CNO and $\propto T^5$ for pp). At the temperature of $16 \cdot 10^6$ K in the center of the sun (and similarly in all lower-mass, sun-like stars) the energy production of the pp chain dominates. The net reaction is

$$4p \to {}^4\text{He} + 2e^+ + 2\nu_e + \gamma, \tag{14.16}$$

and each fusion process liberates an energy of 26.4 MeV.

In stars with masses of $\gtrsim 1.5$ M$_\odot$ the CNO cycle is more important. In the latter—under participation of heavier nuclei such as ^{12}C, ^{14}N, and ^{16}O acting as

14.2 Typical S-Factor Behavior

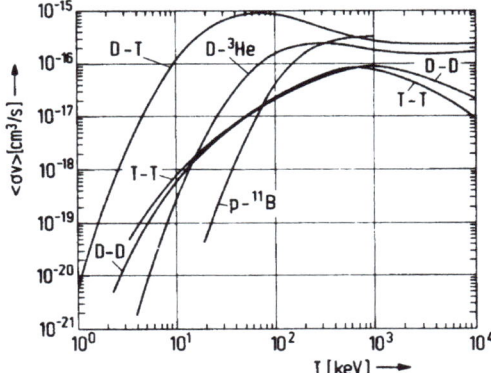

Fig. 14.6 Reaction parameters of important fusion reactions [RAE81]

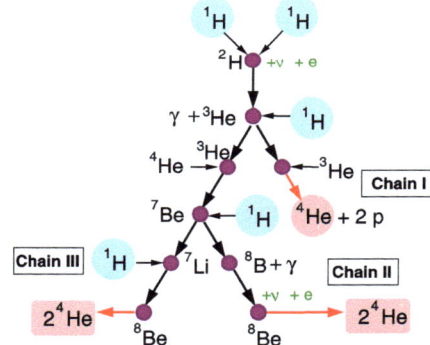

Fig. 14.7 *pp* chain of coupled reactions synthesizing ^{4}He from hydrogen. This chain is responsible for most of the energy production in the center of the sun and also for the emission of solar neutrinos

catalysts—also hydrogen is burned to ^{4}He at temperatures between about 1.5 and $3.0 \cdot 10^{15}$ K. In the lifecycle of such a star, once the hydrogen is used up, it collapses by gravitational attraction with a temperature increase to $\approx 1 \cdot 10^8$ K. This allows *He burning*, i.e. the fusion of two α's to the short-lived ($1 \cdot 10^{-16}$ s) ^{8}Be. Due to the high density in the star a certain equilibrium abundance of ^{8}Be enables α-^{8}Be and three-α collisions. Only via a 0^+ resonant state, the 2nd excited state in ^{12}C, the famous *Hoyle state* [HOY54], at an excitation energy of 7.6542 MeV and a width of only 8.5 eV is the formation of the most important element for organic life in the universe possible. The actual energy of this state seems to be extremely crucial for the formation of ^{4}He and ^{16}O by helium burning and all the following reactions.

Only recently ab initio calculations of the structure of the Hoyle state have been possible in the framework of chiral effective-field theory and Monte-Carlo lattice calculations [EPE11].

All other elements up to iron are formed by processes like (p, α), radiative-capture (p, γ), (α, n), β decay, carbon burning, (γ, α), and finally oxygen burning.

All elements beyond iron cannot be formed by fusion but need neutron-capture reactions and β decays in two different main processes, the *s* ("slow") process (with

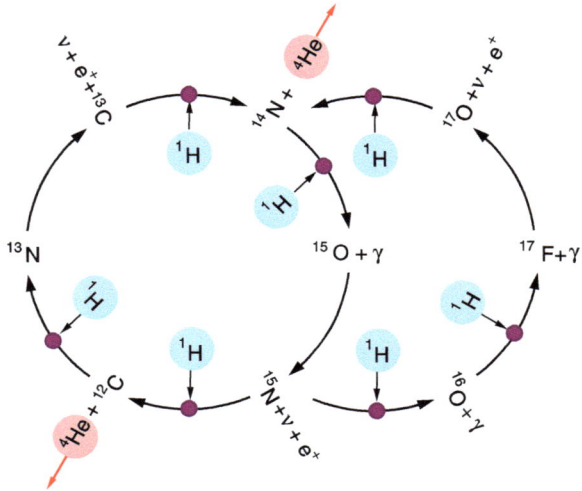

Fig. 14.8 CNO (bi-)cycle of coupled reactions synthesizing ^{4}He from four hydrogen nuclei

the neutron capture being slower than the β decay rates) and the r ("rapid") process (with fast-forming neutron-rich nuclei, typically in supernovae explosions).

Thus, nuclear astrophysics is a very rich field for the study of nuclear reactions and their mechanisms which encompass all reaction types from direct to resonant, capture reactions, charged and neutral reaction partners, heavy ions etc. It is clear that for the evaluation of such sets of coupled nuclear reactions and decays the very precise knowledge of reaction rates and cross sections is of prime importance as well as of the precise dependence on energy (temperature). Any change in cross section values and error bars requires a complete recalculation of the entire system of rate equations. Typically there are hundreds of such coupled reactions. Worldwide networks are maintained to readjust the rates with each new set of data. An example of such a network is NACRE, now updated to NACRE-II, maintained at the Université Libre de Bruxelles, Belgium, for all charged-particle thermonuclear reactions with nuclei $A < 16$, see [NAC13].

It should be noted that "primordial" (or "Big Bang") nucleosynthesis is one of three evidences for the Big-Bang model, together with the expansion of the universe and the cosmic microwave background. Abundances of ^{4}He, ^{2}H, ^{3}He, and ^{7}Li are important for the determination of the baryonic density of the universe. All reactions involving these nuclei should be measured with high precision, but they still contain some discrepancies as to Li. For a good recent survey of the status cf. e.g. Ref. [COC12]. In addition, many reactions have never been measured or the data have large systematic or statistical errors such that computer codes have to be used to produce reaction-rate data.

14.3 Exercises

14.1. Verify (using the Bethe-Weizsäcker mass formula, see e.e. textbooks on nuclear structure) that energy production in stars (or in the big bang) can only occur by nuclear fusion reactions up to medium-heavy nuclei (about to which element?).

14.2. For charged-particle reactions at astrophysical energies the Coulomb barrier determines the reaction cross sections/reaction rates. Only through the quantum-mechanical *tunneling* are there any reactions at all. Their probability is $P = |\psi(R_n)|^2/|\psi(R_C)|^2$ with R_C the classical turning point at the Coulomb barrier and R_n the nuclear radius. Solving the Schrödinger equation yields

$$P = \exp\left\{-2\kappa R_C\left[\frac{\arctan(R_C/R_n - 1)^{1/2}}{(R_C/R_n - 1)^{1/2}} - \frac{R_n}{R_C}\right]\right\} \quad (14.17)$$

with $\kappa = \sqrt{\frac{2\mu}{\hbar^2}(E_C - E)}$.

Calculate (for s waves only) P for the $p+p$ reaction as function of the c.m. energy in the range from 1 keV to 550 keV (check that this is the Coulomb barrier height). To which stellar temperatures do these two energies correspond? Discuss, why—despite P being very small at stellar temperatures—the resulting reaction rates account for stellar energy (and element) production.

14.3. The reaction rate per particle $\langle \sigma \cdot v \rangle$ must be an integral over velocities with the Maxwell-Boltzmann velocity distribution as weight. If, as is often the case, the S-factor is $\approx$ constant over E, we have

$$\langle \sigma \cdot v \rangle = \left(\frac{8}{\pi\mu}\right)^{1/2} \frac{1}{(kT)^{3/2}} S(E_0) \int_0^\infty \exp\left(-\frac{E}{kT} - \frac{b}{E^{1/2}}\right) dE \quad (14.18)$$

where the second term in parentheses describes the barrier penetrability with $b = \sqrt{2\mu}\pi e^2 \frac{Z_1 Z_2}{\hbar^2}$.

(a) Show that the integrand in $\langle \sigma \cdot v \rangle$ as a function of E has the form of a narrow peak, the *Gamow peak*.
(b) Evaluate this function for the $p + p$ and $\alpha + {}^{12}C$ reactions near $T_6 = 15$.
(c) Calculate the energy of the Gamow-peak maximum and the (total) peak width Δ.
(d) What is the (approximate) form of the integral? (Use a gaussian approximation)

$$\exp\left(-\frac{E}{kT} - \frac{b}{E^{1/2}}\right) \approx \propto \exp\left[-\left(\frac{E - E_0}{\Delta/2}\right)^2\right]. \quad (14.19)$$

Discuss the strong temperature dependences for the two reactions.

References

[ALI01] M. Aliotta et al., Nucl. Phys. A **690**, 790 (2001)
[BAR13] M. Barbui et al., Phys. Rev. Lett. **111**, 082602 (2013)
[BET37] H.A. Bethe, Rev. Mod. Phys. **9**, 69 (1937)
[COC12] A. Coc, Astrophys. J. **744**, 158 (2012)
[EPE09] E. Epelbaum, H.-W. Hammer, U.-G. Meißner, Rev. Mod. Phys. **81**, 1773 (2009)
[EPE11] E. Epelbaum, H. Krebs, D. Lee, U.-G. Meißner, Phys. Rev. Lett. **106**, 192501 (2011)
[GEI99] W.H. Geist, C.R. Brune, H.J. Karwowski, E.J. Ludwig, K.D. Veal, G.M. Hale, Phys. Rev. C **60**, 054003 (1999)
[HOY54] F. Hoyle, Astrophys. J. Suppl. **1**, 121 (1954)
[HUK08] A. Huke, K. Czerski, P. Heide, G. Ruprecht, N. Targosz, W. Żebrowski, Phys. Rev. C **78**, 015803 (2008)
[KRA87] A. Krauss, H.W. Becker, H.P. Trautvetter, C. Rolfs, K. Brand, Nucl. Phys. A **465**, 150 (1987)
[LAC05] M. La Cognata et al., Phys. Rev. C **72**, 065802 (2005)
[NAC13] Y. Xu, K. Takahashi, S. Goriely, M. Arnould, M. Ohta, H. Utsunomiya, Nucl. Phys. A **918**, 61 (2013)
[NAV11a] P. Navrátil, S. Quaglioni, Phys. Rev. C **83**, 044609 (2011)
[NAV11b] P. Navrátil, S. Quaglioni, arXiv:1110.0460v1 [nucl-th] (2011)
[NNDC] Natl. Nucl. Data Center, EANDC, Nucl. Reactions, http://www.nndc.gov
[PIE01] C. Pieper, V.R. Pandharipande, R.B. Wiringa, J. Carlson, Phys. Rev. C **64**, 014001 (2001)
[RAE81] J. Raeder et al., *Kontrollierte Kernfusion* (Teubner, Stuttgart, 1981)
[ROL88] C. Rolfs, W.S. Rodney, *Cauldrons in the Cosmos* (University of Chicago Press, Chicago, 1988)
[ROL95] C. Rolfs, E. Somorjai, Nucl. Instrum. Methods Phys. Res. B **99**, 297 (1995)

Chapter 15
Spectroscopy at the Driplines, Exotic Nuclei, and Radioactive Ion Beams (RIB)

Nuclear reactions have been a most important tool of nuclear spectroscopy. The possibility to change energies of ions easily and the use of a wide variety of beams and targets have especially made (tandem) Van-de-Graaff accelerators with their suitable energy range almost universal workhorses for this field of study. Sophisticated target manufacturing techniques as well as ion-source techniques have been developed for many nuclides, even for rare, but stable isotopes. However, with stable nuclear collision partners only a fraction of final nuclear systems and states can be reached. More than 3800 nuclides heave been experimentally observed, with 193 new entries in the list (see e.g. [KNC12]). In order to study properties of as many nuclei as possible, radioactive projectiles have to be accelerated and applied in nuclear reactions. Ref. [GEE06] gives a detailed overview of the motivations for RIB facilities.

15.1 Use of RIB

15.1.1 Nuclear Radii and Neutron vs. Proton Distributions

The almost universal $A^{1/3}$ law of nuclear radii (for non-exotic nuclei) is broken for nuclei near the two driplines, e.g. for halo nuclei such as ^{11}Li, see Sect. 2.4.2. Hadronic probes in comparison to electromagnetic ones reveal a neutron skin with a slightly larger radius than that of the protons for nuclei with large neutron excess (high isospin), see also Sect. 2.4.2. This behavior is expected for many nuclei near the driplines. The earliest such measurement on the isotopes of Na found evidence for a neutron skin increasing in thickness with increasing neutron number [SUZ95]. In the meantime especially nuclei with large neutron excess such as ^{208}Pb and ^{48}Ca that are relatively well understood in terms of the shell model have been investigated systematically as to consistency of the results obtained with different methods and projectiles.

15.1.2 Nuclear Models for Exotic Nuclei

One goal is e.g. to find the changes of the nuclear shell-model parameters and of the position and strength of the shell-model energy gaps in the $Z-N$ plane up to the extremes of the proton or neutron driplines, see e.g. [KRU11]. Also changes in parameters of mass formulae, especially of the symmetry energy and in the nuclear equation of state must be investigated. Studies of reactions with exotic nuclei may elucidate the somewhat unclear role of three-body forces (see also Sect. 9.2.5) in the nuclear equation of state and thus for astrophysical questions (e.g. the radii of neutron stars).

15.1.3 Giant Resonances of Exotic Nuclei

Collective modes of excitation of nuclei near the valley of stability (rotations, vibrations, mixed modes, and giant resonances) have been studied for a long time. One mode is the GDR (Giant Dipole Resonance) (see Sect. 12.3), in which the protons and the neutrons of a nucleus oscillate against each other and produce large and wide resonances in capture-γ reactions and their inverses as well as in charged-particle reactions, visible at extreme forward angles. The properties of these excitations change with varying neutron excess and are being studied systematically.

15.2 Production of Radioactive-Ion Beams

Two main principles of producing and accelerating radioactive beams have been developed. Figure 15.1 gives a schematic view of the main setups for producing and accelerating radioactive ion beams, ISOL (*Isotope Separator On-Line*) and IFF (*In-Flight Fragmentation*), each with the option of post-acceleration. A rather large number of such facilities has been implemented in recent years and new ones are being built worldwide.

15.2.1 The ISOL Principle

In this scheme radioactive nuclei are created in nuclear reactions in thick targets at high temperatures such that the residual nuclei diffuse out of the target into an ionization chamber. Several ionization methods may be applied and the ions are collected and formed into a (radioactive) beam that is (pre-)accelerated and sent through an electromagnetic mass separator. The ions with the selected mass is then (post-)accelerated to the desired energy required by the intended experiment with the incident radioactive projectiles. An important point of view is the lifetime of

15.2 Production of Radioactive-Ion Beams

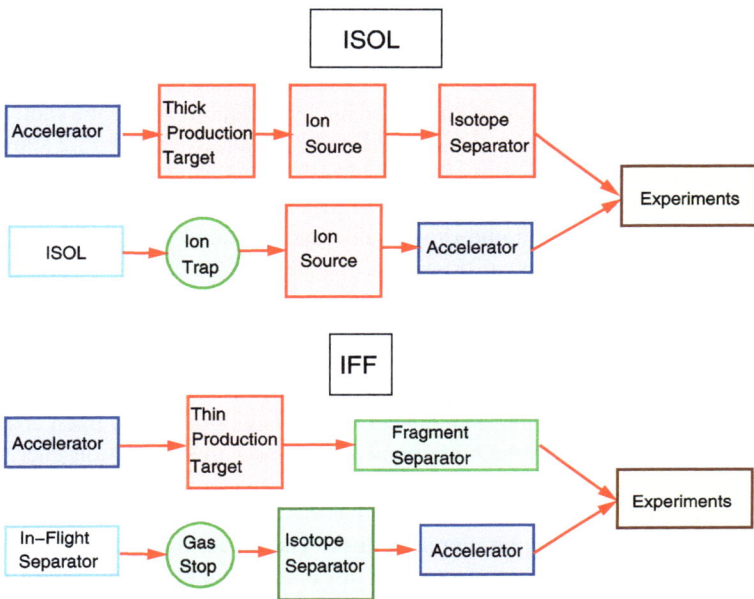

Fig. 15.1 Schemes of possible setups of radioactive-beam facilities ISOL and IFF with pre- or post-acceleration options

the radioactive species in relation to the diffusion time through the primary target, which may limit the bea intensity. Another is the ionization efficiency, which varies strongly for different elements and prevents the production of ions of refractory elements such as Zr.

15.2.2 IFF and Post-Accelerating Schemes

This scheme has the following steps: First singly charged radioactive ions are produced by collisions of an energetic beam (e.g. protons) with a target, then these ions are accumulated and "cooled" in a Penning trap, then in a bunch sent to an ion source where in collisions with a dense and energetic electron beam "charge-breeding" occurs, i.e. high charge-state ions are produced, which are extracted and post-accelerated e.g. in a LINAC. A prominent example of such a facility is REX-ISOLDE at CERN [KES00, WEN10, WEN12]. With this facility radioactive nuclei in the mass range from ^{8}Li to ^{224}Ra in charge states up to 50^+ for Ra have been accelerated.

15.3 Nuclear Reactions and the Way to Superheavies

The hope to extend the number of nuclear species beyond the known transuranium nuclei includes the idea of finding long-lived nuclei in regions where the shell model provides other energy gaps, leading to an island of stability. Figure 15.2 shows the relevant region of the chart of nuclides together with lifetime predictions (extrapolations) from nuclear-model calculations [SOB07, MOE95]. The models are generally based on collective (liquid-drop) model assumptions and shell-corrections. The elucidation of the chemical properties of the new elements is another goal of research on transfermium nuclei but requires a sufficiently long lifetime of the nuclei, presently on the order of 1 s setting a limit at $Z = 108$. It is clear that reaching further out in the (Z, N) plane requires nuclear reactions between ever heavier reaction partners, i.e. heavy-ion fusion reactions. One condition is that the fusion to new compound nuclei occurs faster than the decay into any of the many open channels. The latter is reduced when the compound system acquires excitation energies as low as possible ("cold fusion"). The selection of suitable combinations of projectile and target nuclei and incident energies is—due to uncertainties of theoretical (shell-model) predictions—largely an empirical task. The recent discoveries of nuclides with Z up to 118 were mostly the results of this approach. Depending on the atomic number Z different methods have been applied starting from neutron-capture reactions with neutrons from a reactor or from a reaction (e.g. $^2\text{H} + {}^9\text{Be} \rightarrow {}^{10}\text{Be} + n$), multiple neutron capture in nuclear-bomb explosions, and fusion of ever heavier (radioactive) nuclei with α's, ^{10}B, ^{13}C, ^{18}O up to ^{48}Ca, ^{50}Ti, and ^{54}Cr. Table 15.1 lists the most recent cases. Most results have been reproduced independently in different labs. "nC" stands for neutron capture, "CF = Cold Fusion" is the bombardment of (near-)magic nuclei such as ^{208}Pb or ^{209}Bi with medium-weight heavy ions heavier than ^{40}Ar, such that at most 2 neutrons will evaporate after fusion. "HFI = Hot Fusion I" means the bombardment of actinides ($Z = 92 - 98$) with light ions ($Z = 6 - 12$), "HFII = Hot Fusion II" has been the bombardment throughout of actinides such as ^{242,244}Pu or ^{249}Cf with the rare, stable, and neutron-rich ^{48}Ca, a rather asymmetric fusion with the aim to lower the Coulomb barrier.

As an example the discovery of the so far heaviest element ($Z = 118$) is described here. At Dubna a 1 pµA beam in the U400 cyclotron could be produced from the rare isotope ^{48}Ca that could be produced in high-flux nuclear reactors. The rare SH fusion products (cross section about 0.7 pb) were separated from a very high background in a gas-filled recoil separator (DGFRS) with a magnetic field according to mass/energy and charge state and focused into a detector system 4 m away. Time-of-flight, energy, and position of the SH recoils together with the associated decay α's and fission products were measured with almost 100 % efficiency and high suppression ratios (up to 10^{-15}) have been achieved. Experiments to synthesize element 120 are ongoing.

The island of stability concerns the half-life against spontaneous fission. In contrast to the cold-fusion method hot fusion SH's produced go more in the direction of this region and decay only by emitting α's which allows identification by decay chains through known nuclei. So far the measured half-lives follow the trend

15.3 Nuclear Reactions and the Way to Superheavies 245

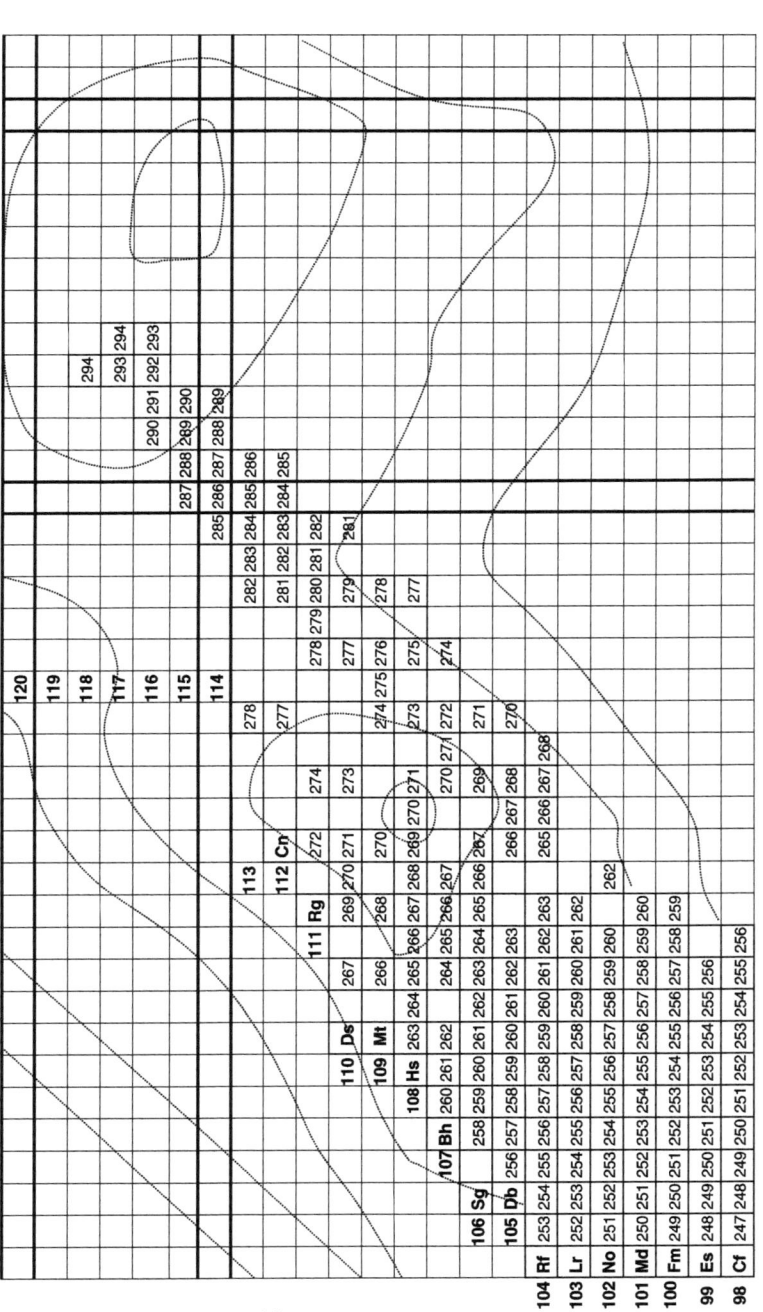

Fig. 15.2 Chart of nuclides with region relevant for superheavies production together with theoretical lifetime predictions in the region of a prospective island of stability, indicated by the *contour lines*. The possible closed-shell Z and N numbers are marked by *heavy grid lines*. Adapted from [PHI10]

Table 15.1 On the way to superheavies, method of first production, and decay properties. Only the first-discovered isotopes of a new element are listed

Element Z or isotope	Reaction or method	Decay	Exp. half life	Lab.	Year	Pres. No. of isotopes
$Z \leq 100$ (Fm)	nC	α, SF		LBL	1940	
				LANL	–	
				Argonne	1955	
$Z \leq 106$ (Sg)	HFI	α/SF	32/55 s	LBL	1957	
				Dubna	1974	
$^{262}_{107}$Bh	CF	α	84 ms	GSI	1981	11
$^{265}_{108}$Hs	CF	α	1.9 ms	GSI	1984	12
$^{266}_{109}$Mt	CF	α		GSI	1982	7
$^{269}_{110}$Ds	CF	α	179 µs	GSI	1994	8
$^{272}_{111}$Rg	CF	α	3.8 ms	GSI	1994	7
$^{277}_{112}$Cn	CF	α	0.69 ms	GSI	1996	6
^{283,284}113	CF	α	0.1/0.48 s	Dubna/LLNL	2003	6
$^{289}_{114}$Fl	HFII	α	2.6 s	Dubna/LLNL	1999	5
^{287}114	HFII	α	0.48 ms	Dubna	2004	5
^{288,291}115	HFII	α	173 ms	Dubna	2007	4
$^{290}_{116}$Lv	HFII	α	7.1 ms	Dubna/LLNL	2000	4
^{293,294}117	HFII	α	14/78 ms	Dubna/LLNL	2010	2
^{294}118	HFII	α	0.9 ms	Dubna	2007	1

predicted by modern macroscopic-microscopic models including shell corrections [MOE95, MUN03, SOB07, STR66]. The identification of the exotic nuclei is by their decay, e.g. in α-decay chains via already known nuclei and ending in spontaneously fissioning nuclei in a region with lowered fission barrier heights. For more details see the recent Refs. [HOF10, OGA10, HOF00, MOR04, OGA07]. For recent predictions up to $Z = 120$ see also Refs. [SOB10, SOB11].

It is clear that dedicated facilities to produce and accelerate the heavy-ion species with sufficient intensities had to be developed. There are four main laboratories that have advanced these technologies: Berkeley, Dubna, RIKEN, and GSI(Darmstadt). Some more details are described in Chaps. 16 and 13. A number of references describe the development of the field, especially the technically refined apparatus e.g. of the separators SHIP at GSI and DGFRS at Dubna in more detail, see [SEA94, ZAG06, BLO10, HOF00, MOR07, OGA07, HOF12, OGA13, MOE09].

References

[BLO10] M. Block et al., Nature **463**, 785 (2010)
[GEE06] D.F. Geesaman, C.K. Gelbke, R.F.V. Janssens, B.M. Sherrill, Annu. Rev. Nucl. Part. Sci. **56**, 53 (2006)

References

[HOF00] S. Hofmann, C. Münzenberg, Rev. Mod. Phys. **72**, 733 (2000)
[HOF10] S. Hofmann, In "viewpoint". Physics **3**, 31 (2010)
[HOF12] S. Hofmann et al., Eur. Phys. J. A **48**, 62 (2012)
[KES00] O. Kester et al., Hyperfine Interact. **129**, 43 (2000)
[KNC12] Karlsruhe Nuclide Chart, 8th edn., Nucleonica (2012)
[KRU11] R. Krücken, Contemp. Phys. **52**(2), 101 (2011)
[MOE95] P. Möller et al., At. Data Nucl. Data Tables **59**, 185 (1995)
[MOE09] P. Möller et al., Phys. Rev. C **79**, 064304 (2009)
[MOR04] K. Morita et al., J. Phys. Soc. Jpn. **73**, 2593 (2004)
[MOR07] K. Morita et al., J. Phys. Soc. Jpn. **76**, 45001 (2007)
[MUN03] I. Muntian, Z. Patyk, A. Sobiczewski, Acta Phys. Pol. B **34**, 2073 (2003)
[OGA07] Y.T. Oganessian, J. Phys. G, Nucl. Part. Phys. **34**, R165 (2007)
[OGA10] Y.T. Oganessian, Phys. Rev. Lett. **104**, 142502 (2010)
[OGA13] Y.T. Oganessian, Nucl. Phys. News **23**, 15 (2013)
[PHI10] S. Hofmann, Isotopenkarte, in *Physik in unserer Zeit 1/2010, Supplementary Material* (Wiley, New York, 2010)
[SEA94] G. Seaborg, *Modern Alchemy* (World Scientific, Singapore, 1994)
[SOB07] A. Sobiczewski, K. Pomorski, Prog. Part. Nucl. Phys. **58**, 292 (2007)
[SOB10] A. Sobiczewski, Acta Phys. Pol. B **41**, 157 (2010)
[SOB11] A. Sobiczewski, Acta Phys. Pol. B **42**, 1871 (2011)
[STR66] V.M. Strutinsky, Yad. Fiz. (USSR) **3**, 614 (1966). [Sov. J. Nucl. Phys. **3**, 449 (1966)]
[SUZ95] T. Suzuki et al., Phys. Rev. Lett. **75**, 3241 (1995)
[ZAG06] V.I. Zagrabaev et al., Phys. Rev. C **73**, 31602 (2006)
[WEN10] F. Wenander, J. Instrum. **5**, C1004 (2010)
[WEN12] F. Wenander, K. Riisager, CERN Cour. **52**, 33 (2012)

Part II
Tools of Nuclear Reactions

Chapter 16
Accelerators

The energy dependence of the cross sections and polarization observables of nuclear reactions is—besides angular distributions that reveal basic information about angular momenta involved—the most prominent feature. Salient examples are the existence of resonances in excitation functions, which correspond to excited states in the continuum of compound nuclei but reach into the region of highest energies where "new" particles such as the Δ, the J/Ψ or the Higgs boson show up as resonances. Thus, the wealth of nearly all important knowledge about nuclear and particle interactions is due to the use of ever larger and more sophisticated accelerators. In addition, and only partly noticed by the public, applications of accelerators have multiplied in recent years; not the least important uses are in nuclear medicine and cultural sciences such as archaeology. Here, the basic principles of the very different accelerating schemes will be described. No completeness on technical details or many different designs and installations can be aspired here.

16.1 Electrostatic Accelerators

16.1.1 The Cockroft-Walton Accelerator

Though not the first accelerator generally, the *Cockroft-Walton* accelerator was the first to be used for initiating a nuclear reaction (the reaction $^7\text{Li}(p,\alpha)^4\text{He}$ with a proton energy of 600 keV) in 1932 [COC32, COW32]. It used the voltage-multiplication circuit of Schenkel, Delon, and Greinacher and is also known as *Cascade generator*. One of the principles still in use is the subdivision of the total voltage across several accelerating gaps in order to even out the electric field strength. The high voltage is limited to about 2 MV by the dielectric strength of the surrounding air. Thus, by enclosing the accelerator by a pressure tank filled with insulating gas at higher pressure enabled voltages up to 6 MV. Because of its reliable and stable operation the Cockroft-Walton accelerator is still widely used as first accelerator stage of the large accelerator installations such as CERN, BNL, ANL, GSI etc. Figure 16.1 shows the scheme of a typical Cockroft-Walton machine.

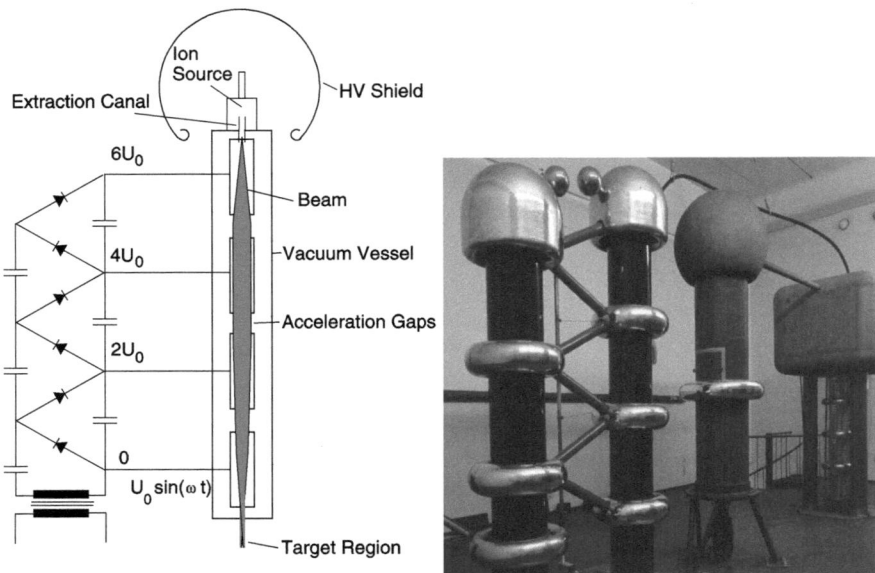

Fig. 16.1 Typical setup and picture of a classical low-energy Cockroft-Walton accelerator. *On the left* in both pictures the voltage multiplier, *on the right* the accelerating tube in vacuum, an ion source *on the top*, lenses for beam focusing, and possibly a target at ground potential (or an efficient beam forming system of lenses when used as an injector)

16.1.2 The Van-de-Graaff (VdG) Accelerator

Almost simultaneously R.J. Van de Graaff developed another method to produce high voltages [VDG31]: An endless insulating belt running between ground potential and the inside of a high-voltage terminal transports charge that builds up the desired terminal voltage of up to 15 MV. This value requires a pressure tank with insulating gases such as CO_2, N_2, or SF_6 at pressures up to 12 bar. The total electric field is homogenized by a modular design of the entire structure including the accelerating tube and the mechanical structures with the electrodes kept at constant voltage steps (typically ≈40 keV/inch) by a resistor chain. These electrodes of aluminum, stainless steel, or titanium are glued or fused together with insulating ceramic or glass elements. The voltage-holding capability of these structures, the dielectric strength of the gas, and the resulting necessary size of the tank limit the maximum voltage. The single-ended VdG machine requires an ion source in the terminal with remote controls etc. Figure 16.2 shows the principle of a single-ended VdG accelerator. Modern designs place the entire system in a pressure tank and use modular acceleration tubes consisting of many flat electrode disks glued to ceramic or glass spacing rings with high vacuum inside.

The VdG principle has been extended into a design, the *Tandem Van-de-Graaff* accelerator, that avoids this problem, and at the same time has a number of important advantages: the high voltage is put to multiple use by injecting first negative

16.1 Electrostatic Accelerators

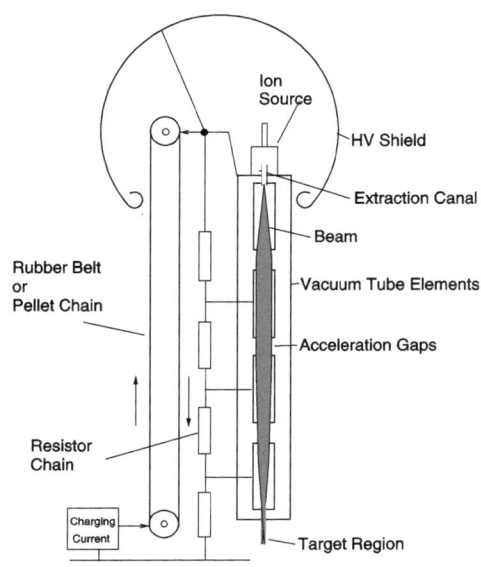

Fig. 16.2 Typical setup scheme of a classical low-energy single-ended VdG accelerator. *On the left* the moving belt or chain charging system *on the right* the accelerating tube with vacuum, a positive ion source *on the top*, lenses for beam focusing, and possibly a target at ground potential (or an efficient beam forming system of lenses and deflection magnets when used as an injector)

ions, have them stripped in the terminal (by a gas in a stripper tube, but mostly by very thin carbon foils) into a number of charge states Z of positive ions that are accelerated back to ground potential thus acquiring kinetic energies

$$T = (1 + Z)eU_{\text{terminal}}. \tag{16.1}$$

Not only the higher energies, especially for highly-charged heavy ions, but also a number of other features made these machines the "workhorse" of many low-energy nuclear laboratories. These are:

- Feedback circuits, coupled to sensing the beam position after a deflection magnet (measuring the momentum of the particles) and sending a controlled electron current from a corona-needle set to the terminal provide a fine and fast adjustment of the terminal voltage, resulting in a good energy resolution (and stabilization) on the order of ≈0.1%.
- The need to produce negative ions imposes the restriction that not all atoms form negative ions of sufficient lifetime (e.g. Ne or N). On the other hand, especially with modern sputter ion sources, the number of different negative ions from many isotopes, which can be produced is immense. The changes from one isotope to another can be done very quickly.
- The simple methods of charging the belt or chain by controlling the belt-charge current make energy changes very quick and easy, requiring sometimes some high-voltage "conditioning" of the accelerator parts.

Thus, the Tandem VdG machine is a very versatile tool for many applications. Figure 16.3 shows a typical tandem Van-de-Graaff accelerator scheme. The actual accelerators follow horizontal (HVEC) as well as vertical (NEC) designs, have tank

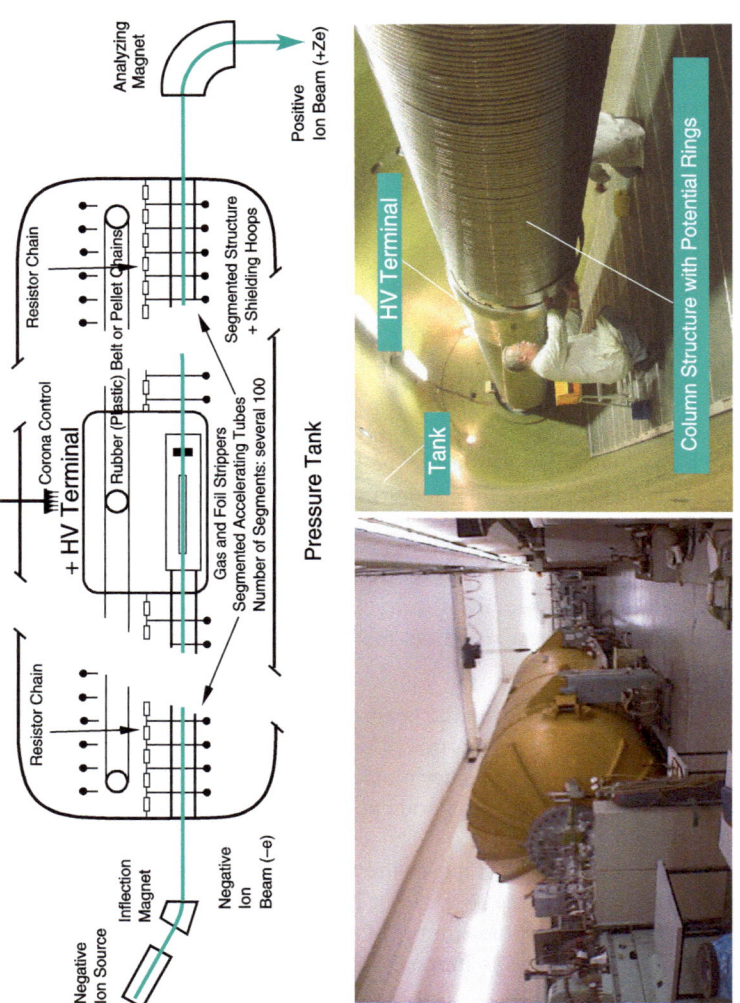

Fig. 16.3 Typical setup of a modern tandem VdG accelerator. It shows the charging elements of the VdG principle. The highly segmented and electrically insulated construction is self-supporting under the pressure of a large spring, in vertical machines under its own weight. The breaks in the graphics indicate that the length of the tank and internal structure is such as to accommodate several hundred segments on each side, typically with lengths of 1 inch. The *lower pictures* show the outside appearance of the tandem accelerator and the inside during maintenance work

lengths up to 20 m, diameters up to 12 m and terminal voltages up to 20 MV. Thus, for protons energies of several tens of MeV, for heavy ions such as e.g. ^{16}O in higher charge states >100 MeV are possible. A unique application of the tandem VdG principle is Accelerator Mass Spectroscopy (AMS), see Sect. 20.1.

16.2 RF Accelerators

16.2.1 The Linear Accelerator (LINAC)

The very first accelerator design was a LINAC prototype by R. Wideröe at Aachen Technical University in 1928 [WID28] after an idea of G. Ising [ISI24]. The idea was, because the maximum voltage across an accelerating gap was limited, to use such a given voltage more than once, i.e. by accelerating ions by the same phase of an RF voltage while shielding them from the opposite phase. This led naturally to the drift-tube design with increasing lengths of the drift tubes according to increasing ion velocity. Naturally, for relativistic particles, especially electrons, the lengths are constant. For higher energies a different design with less RF losses was developed by L.W. Alvarez in 1946 [ALV46] by mounting the drift tubes in an RF cavity with a standing-wave electric field such that the travelling ions always see an accelerating field. In both cases the synchronization of the particle motion requires that the drift-tube lengths increase with $\propto \sqrt{U_0}$ where U_0 is the peak value of the RF voltage. Figure 16.4 shows both designs schematically. Electron LINACs are simpler, consisting of cavities with equally-spaced disk-loaded electrodes. Typical modern applications of hadron LINACs are injectors for high-energy accelerators such as the UNILAC at GSI (Darmstadt) or high-current medium-energy installations such as pion factories (LANL) or spallation-neutron sources. Electron (positron) LINACs such as at SLAC (Stanford) allow the highest possible energies due to lower synchrotron radiation losses than for circular accelerators (synchrotrons). Consequently, the planned next-generation electron-positron collider, intended for the study of Higgs physics, will be of LINAC design.

16.2.2 The Cyclotron

The idea of a proton beam being bent on circles by a homogeneous magnetic field B while being accelerated by an RF field across a gap led E.O. Lawrence to invent the cyclotron in 1931 [LAW31, LAW32]. To do this using an electric RF field with fixed frequency and to keep the particle motion synchronized with it the path length in B the radii of the circles had to increase accordingly. With

$$R = \frac{mv}{qB}, \tag{16.2}$$

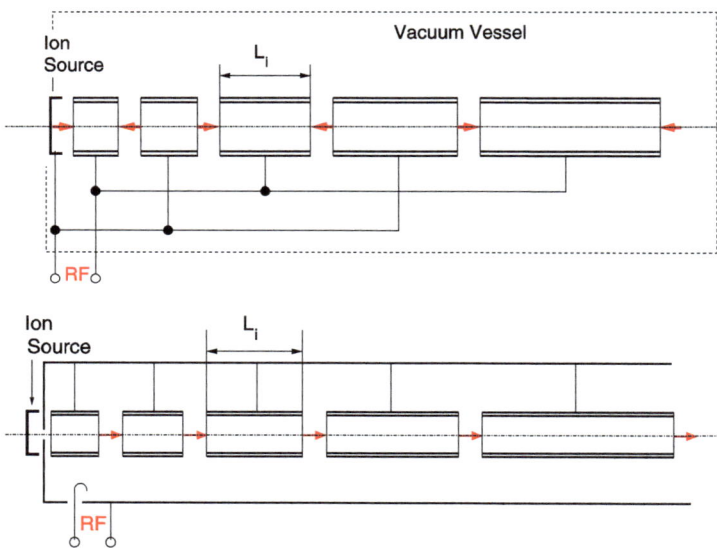

Fig. 16.4 Schemes of the Wideröe (*upper picture*) and the Alvarez (*lower picture*) designs of LINACS

where m and e are the particle's (relativistic) mass and charge and v its velocity, and in addition $v = R\omega$ (ω the angular velocity), we have the condition

$$\omega = \frac{q}{m}. \qquad (16.3)$$

Thus ω is independent of v. If ω is interpreted as the RF's circular frequency the particle will always arrive at an accelerating gap with the same (accelerating) RF phase. Figure 16.5 shows very schematically the essential structure of a classical cyclotron. The classical cyclotron can accelerate all particles with about the same ratio q/m. The relativistic energy limit for protons is about 50 MeV where m begins to deviate substantially from m_0, the rest mass.

$$m = \frac{m_0}{\sqrt{1 - v^2/c^2}}. \qquad (16.4)$$

In addition, the size of the magnetic region, i.e. the maximum trajectory radius $R_{\max}$ and the value of B limit the maximum energy T ($T \propto R_{\max}^2 B^2$). Thus, the practical limits for deuterons, or $^4\text{He}^{2+}$ are about 100 and 200 MeV, resp. The largest such cyclotron was the 184-inch machine at Berkeley.

The idea to increase B in step with increasing m leads to the difficulty that such a magnetic field defocuses the particle beam. Another idea is to decrease the RF frequency during acceleration, which necessarily allows only pulsed operation. A way out has been to decrease the field B with the beam radius R while making use of another (and stronger) additional focusing principle: *Strong Focusing*. It relies on the

16.2 RF Accelerators

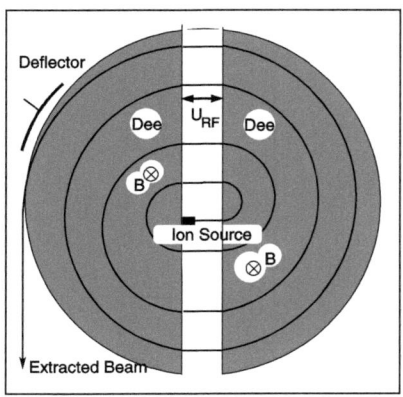 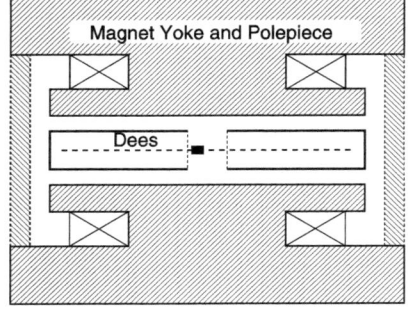

Fig. 16.5 Scheme of the classical cyclotron. B is a magnetic field produced by an electromagnet. The acceleration structure is in a vacuum chamber containing the "Dees", so called after their shape, which provide field-free space for the particles during the decelerating phase of the RF voltage on the dees. In reality the radii increase as $\sqrt{T}$ and the trajectories get accordingly closer to each other with increasing energy

Fig. 16.6 Schemes of the sector-focusing (*left*) and spiral-sector (*right*) cyclotrons. High (H)- and low (L)-field regions are produced by ridges and valleys in the pole surfaces

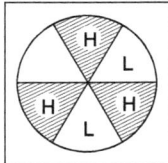 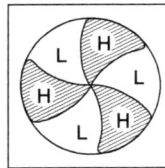 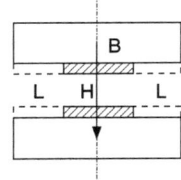

fact that a combination of two equally strong optical elements, one focusing (F), the other defocusing D), with a certain drift space of length d between them (O), is always focusing (FODO or DOFO structures), see Sect. 16.3. This is achieved in cyclotrons by *Sector Focusing SF*, i.e. by shaping the magnetic field in radial sectors such that high- and low-field regions alternate periodically: SF cyclotrons. In addition, giving the sectors a spiral shape helps to compensate for the relativistic mass increase: spiral-sector cyclotrons, see Fig. 16.6. The modern spiral-sector cyclotrons are also called *Isochronous Cyclotrons* and are in use at many installations, e.g. as 50 MeV proton injector to the COSY cooler synchrotron at Jülich. At PSI (Villigen, Switzerland) a smaller isochronous cyclotron injects protons into another such cyclotron that, however, has completely separated magnetic sectors, delivering up to 590 MeV protons. At Michigan State, at Groningen and LNS Catania superconducting cyclotrons have been operating. Figure 16.7 opens a view into the spiral-ridge structure of the superconducting spiral cyclotron at LNS Catania [LAT12].

Fig. 16.7 View into the spiral-shaped superconducting polefaces of the LNS SC cyclotron, that have a diameter of 1.8 m, magnetic fields between 2.2–4.8 T, accelerating fully stripped light ions such as ^{12}C up to 80 MeV/u

16.2.3 Betatron and Synchrotron

The betatron principle and beam stability criteria were also published first by Widerøe in 1923 (actually he obtained a patent on it) and practically realized first by Kerst (with theoretical help from Serber) in 1940. Betatrons have mainly been used for electron and bremsstrahlung irradiation in medicine, but have now been superseded by LINACs. Only for portable X-ray applications (e.g. materials testing, art investigations etc.) are betatrons still in use.

The principle of particle acceleration in time-varying magnetic fields was also applied to *synchrotrons* that combine several principles such as guiding the beam independent of energy on a fixed radius thus allowing for smaller magnets than in cyclotrons, but requiring ramping up magnetic and RF accelerating fields with energy. The principle of AG or Strong Focusing in addition made smaller magnet gaps possible, see Sect. 16.3.3. Instead extra focusing multipole elements between the bending dipole magnets have also been used. Synchrotron energies in principle are not limited. However, size of the accelerator ring, its cost, and its complexity set practical limits (the project of a superconducting super-collider (SSC) was therefore abandoned midway). Superconducting magnets with higher magnetic fields as well as SC RF cavities provide higher energies for a given accelerator circumference. At CERN, Geneva, the presently largest synchrotron-type collider LHC (*Large Hadron Collider*) is designed to provide collision energies for protons on protons of 14 TeV. At half this energy the long-sought Higgs boson (the keystone of the Standard Model) was found recently.

A special type of synchrotrons with relevance to nuclear reactions are medium-energy *storage rings*, with the capacity of beam *cooling*. In a storage ring the beam circulates many times and crosses an internal target region not once but e.g. a million times. This requires excellent ultra-high vacuum, carefully designed beam optics, and excellent beam stability. Targets may be internal, i.e. in the accelerator beam, or the beam may be extracted. Of course, an internal target can only be very thin,

16.2 RF Accelerators

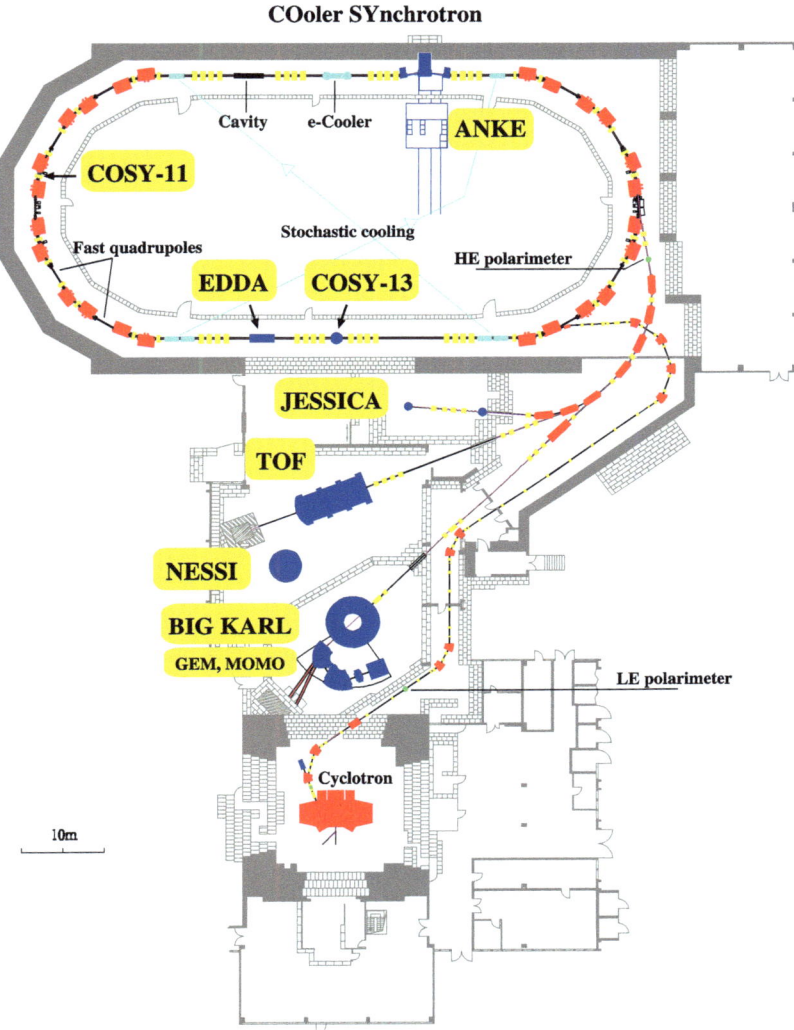

Fig. 16.8 View of the COSY cooler ring with a number of installed experimental areas at positions with suitable optical properties

which favors the use of jet-like beam targets such as polarized beams from ABS (see Sect. 16.5.1), liquid-cluster targets, or targets consisting of thin filaments only. A successful example of such a facility is COSY = *Cooler Synchrotron* at the Jülich center. Figure 16.8 shows a top view of COSY-Jülich. Another feature successfully applied in medium-energy facilities is the principle of phase-space "cooling" of circulating beams, i.e. basically a reduction of the phase-space volume (seemingly in contradiction to Liouville's Theorem, see below). The acceleration of polarized particles in such a ring requires special care, because the spin not only precesses in the many magnetic fields but at certain energies there are resonances, which, after small perturbations, can lead to strong oscillations and therefore depolarization. Several methods to either avoid depolarizing resonances such as by using *"Siberian Snakes"* or letting the polarization vector completely invert have been designed and successfully applied. In a circular accelerator with dipole magnets the polarization vector precesses around the magnetic field direction (see Sect. 16.8.1) except when both are exactly parallel. Thus, in order to have a fixed polarization direction at an experimental station any precession must be compensated e.g. by a spin-rotating device such as a solenoid. A Siberian Snake in the original version is a sequence of dipole magnets acting in two dimensions such that the polarization direction is restored in each cycle. Resonances that flip the spins can be overcome by kicker devices that provide a fast crossing of the resonance energies during ramping of the energy of a synchrotron.

16.3 Beam Forming and Guiding Elements

Inevitably experimenters require sufficiently intense beams on target, especially when using rare-ion or polarized beams in order to get "good statistics" (small statistical errors in reasonable time). In addition, nuclear experiments rely on "good" geometries, i.e. well-defined collimated incident beams with low divergence in order to have well-defined scattering angles. Thus, a beam from the source to the target has to be guided, deflected, and focused by lenses and deflectors, both electric or magnetic. Besides technical questions concerning the design of these devices the governing physical principle is the *Liouville Theorem*: Under the action of conservative forces the $6N$-dimensional phase-space volume of an ensemble of N particles (such as a beam) remains invariant under these operations. For one particle:

$$\Delta \vec{p} \cdot \Delta \vec{x} = \text{const.} \tag{16.5}$$

$\Delta \vec{p}$ and $\Delta \vec{x}$ are the spread in momentum (i.e. also velocity) and space of a given beam at a given location along a beam trajectory. The phase-space volume of a beam from an ion source is always finite and is therefore a limiting factor of the quality of a beam at the target. Consequences of the Liouville theorem are:

- Focusing to a smaller spot is always connected to an increase in divergence and vice versa. The special requirements of the experiments determine the beam-formation details.
- Acceleration of a beam, which corresponds to an increase in the longitudinal part of phase space ($\propto U^{1/2}$) will lead to a decrease in the lateral four-dimensional phase volume (for one particle).
- We measure the "quality" of a beam by defining its *emittance*. For a given energy the boundaries of the phase-space volume filled by beam is used for this discussion; with z fixed and $p_z \approx =$ const the momenta p_x/p_z and p_y/p_z define

16.3 Beam Forming and Guiding Elements

the "divergences" p'_x and p'_y. If, for simplicity, we assume rotationally symmetric beams or beams, for which the two transverse phase spaces are decoupled ($x = y = r$ and $p'_x = p'_y = p'_r$) the emittance is

$$\epsilon = r \cdot p'_r \cdot eU^{1/2}. \tag{16.6}$$

- To ensure maximum transmission of a beam through the entire system a similar quantity *acceptance* can be defined for each device along the beam path (diaphragms, lenses, deflectors, vacuum chambers, accelerator tubes, stripping canals etc.), which is the maximum phase space volume that will be transmitted through the device. Transmission must be optimized by *matching* emittances to acceptances at each location z. Matching in two dimensions can be visualized by geometrical figures in the r, p_r plane. Emittance figures, which often are represented by phase ellipses of constant area are sheared by lenses as well as by drift spaces. Matrix methods are used for transporting the figures along the beam path and for doing the matching. In the simplest approximation the transport of beams (and its representation by transport matrices) are *linear*, which is fulfilled for *paraxial* rays or beams, i.e. when the radii and divergences are small as compared to system dimensions. Then (again for one spatial dimension)

$$\begin{pmatrix} x' \\ p'_x \end{pmatrix} = \begin{pmatrix} a & b \\ c & d \end{pmatrix} \begin{pmatrix} x \\ p_x \end{pmatrix} \tag{16.7}$$

with

$$\mathbf{M} = \begin{pmatrix} a & b \\ c & d \end{pmatrix} \tag{16.8}$$

the *transfer matrix* of the optical element. Its elements may be identified with properties of real systems. The element c is connected with fcal properties, the element b is the length of a drift space. For discussion of the six-dimensional phase space the dimension of **M** may be 4 × 4 or even 6 × 6 where, with no coupling between two dimensions (no angular momentum), there are two (or three) independent submatrices.

- The Liouville theorem seems to set an absolute limit to improving the beam quality. However, e.g. the action of diaphragms will change the phase volume, but at the expense of intensity. Better methods are: *Stochastic Cooling* where large transverse components of the phase space of beam particles are sensed and electrically corrected (like by a Maxwell's demon) and *Electron Cooling* where such large transverse components are exchanged with small components of an electron beam of equal velocity by Coulomb interaction. Both methods are successfully used in circular accelerators such as COSY-Jülich, allowing for high-precision experiments with high beam and target densities and well-defined interaction geometries.

Electric or magnetic fields can act as lenses for ion beams.

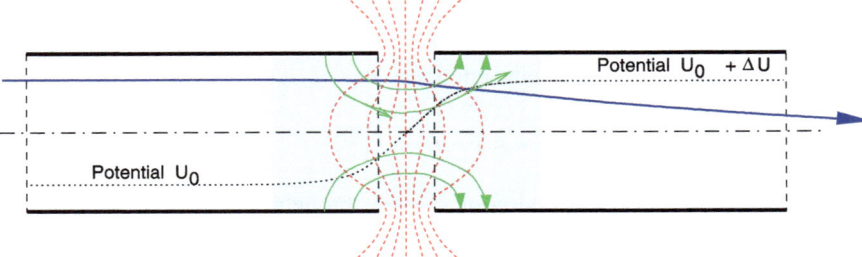

Fig. 16.9 Electric field and potential distributions in an *immersion lens* that, when run with accelerating polarity, focuses a beam, conversely defocuses it. The (rotationally symmetric) fields seem to compensate each other but the focusing force, which ideally goes $\propto r$ at the entrance is stronger than the defocusing force at the exit. The potential on the axis is also shown (*dotted line*). The *shaded area* indicates the approximate length (about one tube diameter) of the field region

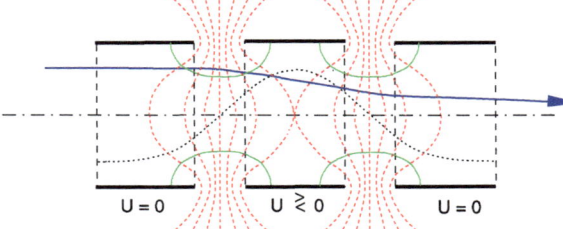

Fig. 16.10 The two successive gaps of an einzellens are always focusing due to the principle of *strong focusing*, see below, whether in accelerating or retarding mode. However, the optical properties, e.g. the maximum beam diameter depend on the mode

16.3.1 Electrostatic Lenses

The prototype of an electrostatic lens is a rotationally symmetric gap, across which an electric field $\vec{E}$ exists, i.e. two conducting tubes at different potential. Depending on polarity this field may accelerate or decelerate a beam with energy qU_0. The force component qE_r is directed either inward or outward, and the lens changes the velocity of the beam. Figure 16.9 shows the electric field and potential distributions, and the shape of the central potential $\Phi(z)$ on axis. By combining two successive gaps with different polarities at a distance d one obtains the widely used *einzellens* that does not change the beam velocity (i.e. the functions of focusing and velocity change are decoupled). Such a lens is always focusing (except when equipped with a grid in the center plane). This can be understood with the very important principle of *strong focusing*, explained below: The combination of a focusing and a defocusing element at a certain distance d is always focusing, which is already known from light optics. Figure 16.10 depicts the geometry and fields in a schematic einzellens. In reality the electrodes have to be in high vacuum.

16.3 Beam Forming and Guiding Elements

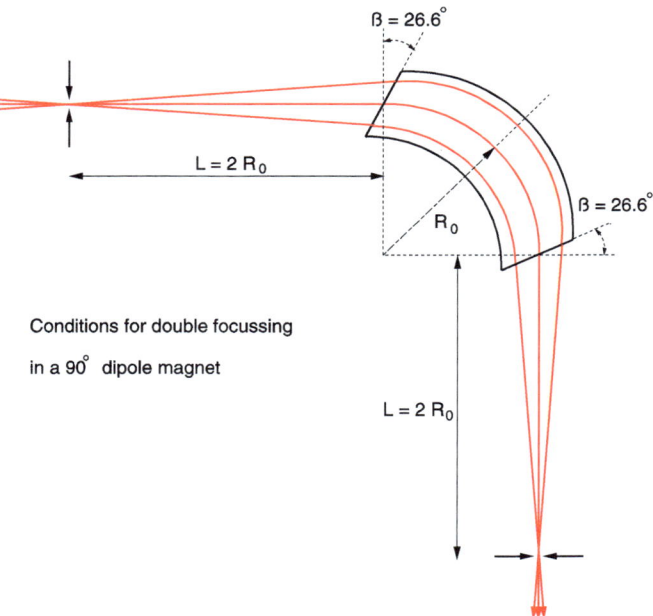

Fig. 16.11 Typical double-focusing dipole magnet used for momentum analysis and feedback control of a tandem VdG accelerator via a difference signal from the exit slit electrodes

16.3.2 Magnetic Lenses

The simplest magnetic lens is just a solenoid in beam direction. More common are magnetic multipole lenses, especially quadrupole lenses.

Already magnetic dipole fields B do not only bend beams but have focusing effects. Particles entering at angles relative to the "reference" central beam have longer or shorter paths through B and are therefore more or less deflected leading to a more or less well defined crossover for an initially divergent beam. The magnet can be made approximately double-focusing, i.e. focusing in two dimensions x and y, by shaping the entrance and exit polefaces such that the central beam does not enter and exit the polepiece perpendicularly, but at 26.6°, and the object and image distances are just $2R$ with R the curvature radius of the dipole magnet. Such dipoles are widely used e.g. in tandem laboratories where they serve also for energy control by selecting a small momentum interval across a pair of exit slits while keeping the beam focused in two dimensions. Figure 16.11 shows a typical scheme of a double-focusing analyzing magnet at a VdG laboratory.

Electrostatic lenses (and also electrostatic deflectors) become inefficient at higher beam energies (their action goes as $\Delta U/U_0$ with limited ΔU, the voltage across the gap and U_0 the equivalent voltage of the beam energy qZ_0). Thus magnetic devices are predominantly used. An ion beam in a magnetic quadrupole field "sees" in

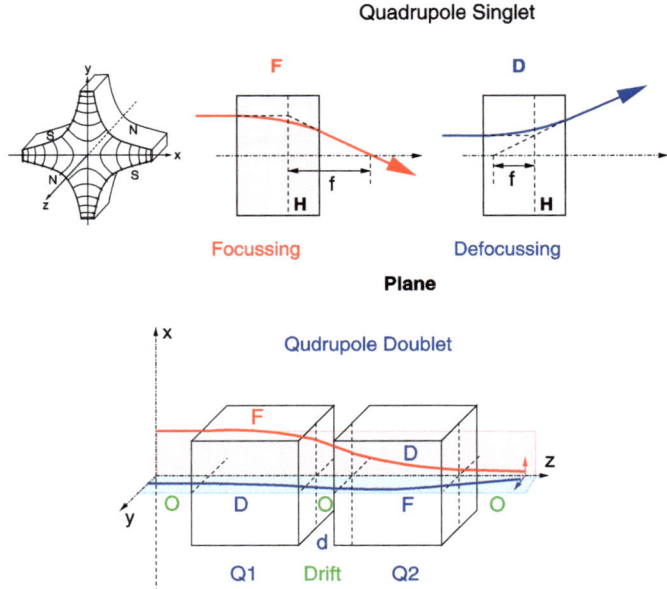

Fig. 16.12 Magnetic field distribution in a quadrupole magnet. The beam direction is assumed along the z axis. The *upper right figure* explains the action of a quadrupole singlet in two planes, the *lower* the double-focusing of a quadrupole doublet. This is an example of "strong" (or AGS) focusing in two dimensions x and y, see next subsection

one direction an harmonic restoring force with trajectories being harmonic ($\sin z$ or $\cos z$) functions leading to beam crossover at a certain distance (a focus for an incident beam parallel to z). Likewise in the perpendicular direction there is a diverging force following hyperbolic functions ($\sinh z$ and $\cosh z$). The combined action of such a quadrupole singlet is *stigmatic focusing*, i.e. the image of a point (or small circular) source becomes a line in the plane perpendicular to the focusing direction. For a singlet with x and y fields interchanged we get the line focused in the perpendicular direction. Combining the two singlets with a certain distance d between them, forming a *quadrupole doublet*, we obtain focusing in two dimensions, i.e. with a point (or small circular) image of a point (or small circle). Figure 16.12 illustrates the action on a beam of charged particles. The equation of motion with the Lorentz force acting in its simplest form is that of a harmonic oscillator

$$x'' + kx = 0, \qquad (16.9)$$

where $k \gtrless 0$ depending on whether we have a focusing/defocusing quadrupole. k is the gradient of the quadrupole field, with ℓ its length, and by defining $\phi = \sqrt{k} \cdot \ell$, we have the focusing/defocusing transfer matrices

$$\mathbf{M}_F = \begin{pmatrix} \cos\phi & \frac{1}{\sqrt{k}}\sin\phi \\ -\sqrt{k}\sin\phi & \cos\phi \end{pmatrix} \quad \text{and} \quad \mathbf{M}_D = \begin{pmatrix} \cosh\phi & \frac{1}{\sqrt{k}}\sinh\phi \\ \sqrt{k}\sinh\phi & \cosh\phi \end{pmatrix}. \quad (16.10)$$

16.3.3 Strong Focusing

The principle of "strong focusing" has been very fruitful for different development of and around accelerators. It was independently invented by Christofilos [CHR50] and Courant, Livingston and Snyder [COU52]. It can be stated as: The combination of two optical elements behind each other, one focusing (F), one defocusing (D), often with equal refractive power, at a distance $d > 0$ from each other, results in an element that is always focusing. The matrix representations of thin lenses (in one spatial dimension) and drift space (O) in phase space are

$$\mathbf{F} = \begin{pmatrix} 1 & 0 \\ \frac{1}{f_F} & 1 \end{pmatrix} \quad \mathbf{D} = \begin{pmatrix} 1 & 0 \\ \frac{1}{f_D} & 1 \end{pmatrix} \quad \mathbf{O} = \begin{pmatrix} 1 & d \\ 0 & 1 \end{pmatrix} \quad (16.11)$$

where $f_F < 0$ and $f_D > 0$ are the focal lengths of the F and D elements, and $d > 0$ the length of the drift space between them. The combination of the three elements (FOD) is a product of the three matrices

$$\mathbf{F} \cdot \mathbf{O} \cdot \mathbf{D} = \begin{pmatrix} 1 + d/f_D & d \\ \frac{d}{f_F f_D} & 1 + d/f_F \end{pmatrix}$$

$$\rightarrow \begin{pmatrix} 1 + d/f & d \\ -\frac{d}{f^2} & 1 + d/f \end{pmatrix} \quad \text{for the case of } f = f_D = -f_F. \quad (16.12)$$

The focal length of the ensemble is $\frac{f_F f_D}{d}$, which is always <0 if f_F and f_D have different signs, i.e. the element is always focusing.

Besides applications such as the electrostatic einzellens and the quadrupole doublet this principle has been used in circular accelerators and enabled to go to the highest energies. The development of the classical cyclotron into modern isochronous cyclotrons required some focusing element because the fringe fields of the magnet near the outer parts of the beam path act defocusing in the direction of the central magnetic field. Shaping the magnet such that the pole gap becomes narrower at the edges defocuses the beam in the radial direction. The solution were sections with stronger B fields alternating with weaker ones by having wider and narrower gap sections (see Sect. 16.2.2). Thus, focusing in axial and radial directions was provided.

In synchrotrons with the beam being guided by separated dipole magnets this principle was applied in different ways. By tapering the polepieces of the dipoles, i.e. giving the magnetic field a gradient the Lorentz force gets a component that focuses in one direction and defocuses in the perpendicular one. By alternating dipoles with gradients outward and inward we get again double focusing (i.e. in radial and axial) direction. This principle is therefore also called *Alternating-Gradient Focusing* and gave one of the first such accelerators its name AGS (Alternating-Gradient Synchrotron at Brookhaven National Laboratory) and contributed practically simultaneously to the successes of the first CERN PS (proton synchrotron) with $T = 25$ GeV. One of the main advantages was that the magnet gaps could be much smaller ("strong" focusing) and therefore much higher energies could be reached than with straight magnets (ZGS: zero-gradient synchrotron at Argonne). The focusing can also be obtained with straight dipoles plus extra focusing elements such as quadrupoles and sextupoles. For details see a number of references such as [HIN97, WIL01, LEE11, CHA13, WIE93].

16.4 Ion Sources

Ion sources for accelerators in nuclear physics are essential tools. Many different designs have been developed but over time some standard configurations have emerged. For nuclear-reaction work some special features are required or advantageous:

- Versatility: e.g. the possibility of producing many different kinds of beams (different ion species) and the ability to switch quickly from one to the other.
- High beam currents.
- Long running times and at the same time easy and quick maintenance.
- High brightness, i.e. high intensities into a small phase space.

To optimize these with given sources often resembles an art more than a science (one might speak of "so(u)rcery").

16.4.1 Unpolarized Beams

The different designs of current ion sources for accelerators can be classified especially according to the polarity of the ions. Tandem VdG's and some cyclotrons need negative ions. They can be produced in two ways: either from an intense positive ion source such as a duoplasmatron with following charge exchange or in a direct-extraction negative-ion source such as Cs sputter sources.

The RF Ion Source In a radiofrequency (RF) discharge at reduced pressure (i.e. sufficiently long path lengths for electrons and gas atoms/molecules) collisions between electrons and heavy particles lead to ionization and a plasma is formed, from which ions may be extracted. The excitation may be either electric where the RF

Fig. 16.13 Cut view of a standard RF positive-ion source

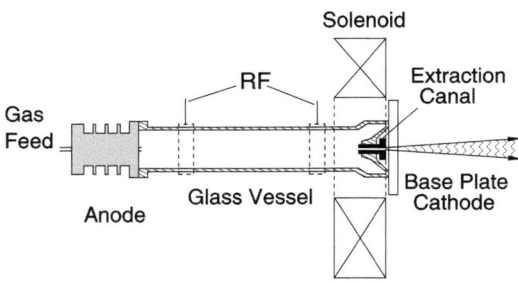

is applied by electrodes (see Fig. 16.13) or magnetic by a coil around the vacuum vessel.

Typically a few mA of protons, deuterons and He isotope-ions can be extracted.

The Duoplasmatron Source Invented by M. von Ardenne in 1948, the duoplasmatron has been a true workhorse for positive-ion work, i.e. for single-ended VdG's, mainly for protons and deuterons, and as primary source for charge exchange to negative ions of H, D, ^{3,4}He, and ^{6,7}Li. Figure 16.14 shows a scheme of a duoplasmatron. The basic principle is to produce a dense plasma region by ionization by an electron beam in a strong magnetic field, in which the electrons spiral and from which the positive ions can be extracted. Like in all ion sources the size and shape of the extraction canal and optics determines the phase space, i.e. the optical quality of the beam.

Direct-Extraction Negative-Ion Sources From the off-axis region of a duoplasmatron negative ions can be extracted. More powerful and versatile, however, is the sputter ion source with Cs^+ as sputtering beam. Sputtering from all surfaces by ions is a common phenomenon and is e.g. used in industry for surface modifications. The efficiency of the sputtering process depends on the material and its structure, but the atomic weight of the sputtering ion should be as high as possible. When using a Cs beam, which can be produced conveniently and efficiently by surface ionization on a hot tungsten surface, its electro-negativity causes the sputtered atoms, molecules, or clusters to attach an electron. Thus, negative ions can be extracted directly. There is a large choice of the material to be sputtered, and provided negative ions can be formed (i.e. live long enough, which is not the case e.g. for ^{20}Ne and ^{14}N), intense beams of negative ions can be extracted and injected into a tandem VdG. Figure 16.15 shows a design that has become a standard. The quick change of ion species can be effectuated by a wheel-type or other design of sputter-target reservoir.

For the special importance of tandem VdGs and their ion sources for accelerator mass spectroscopy see Sect. 20.1.

Charge-Exchange Negative-Ion Source Especially for negative ^{3,4}He and ^{6,7}Li ions at tandem VdG accelerators the combination of a duoplasmatron and a charge-exchange section with Li vapor is used. Figure 16.16 shows such a system schematically.

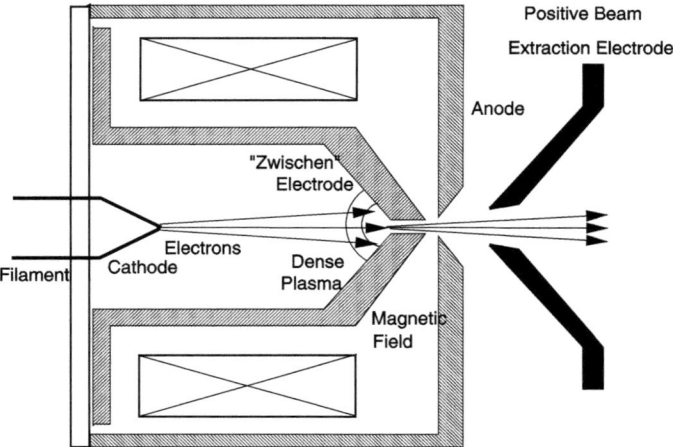

Fig. 16.14 Cut view of a standard duoplasmatron positive-ion source

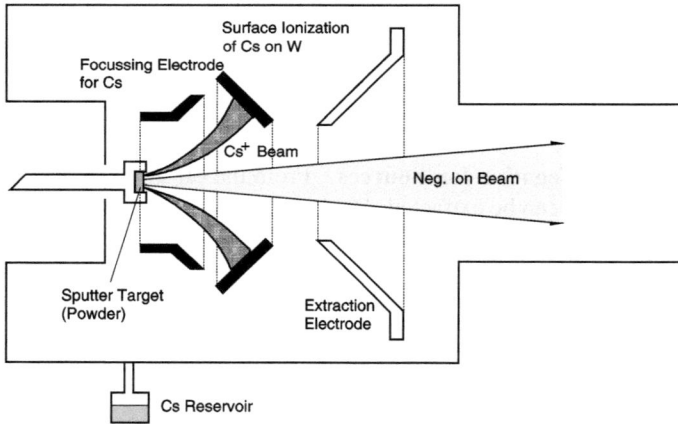

Fig. 16.15 Cut view of a standard negative-ion sputter source. The Cs vapor from the Cs oven when touching the ring-shaped hot W ribbon is ionized. The electric fields are such that the positive Cs ions are accelerated and focused to a fine and intense spot on the sputter-material pill and, at the same time, the negative ions formed are extracted and focussed into a negative-ion beam

There have been other types of ion sources, which will not be discussed here: the ECR (electron-cyclotron resonance) ion source, different types of Penning ion sources and the special sources used in cyclotrons. For fusion research intense neutral beams are required for injection into fusion reactors with strong magnetic fields, which are also produced by charge exchange from intense charged-particle sources.

16.4 Ion Sources

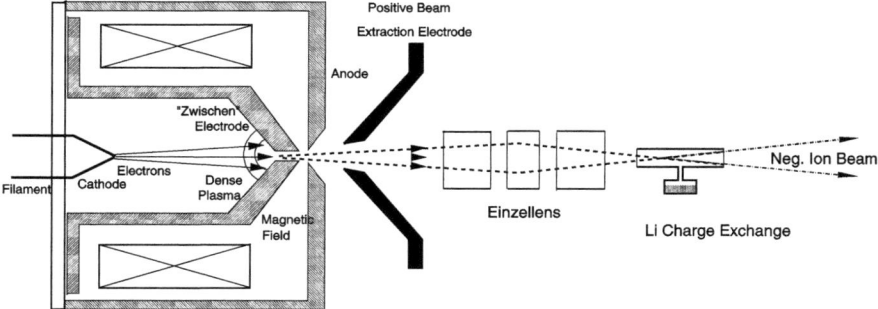

Fig. 16.16 Cut view of a negative-ion source where the negative ions are produced in a charge-exchange region with (mostly used) Li vapor. The positive ions come from a standard duoplasmatron source

16.4.2 Polarized Beams and Targets

The rather complex sources of spin-polarized ions for accelerators have had a long development history since the first publication of the principle of such sources [CLA56, CLA59] and the first nuclear reaction with polarized particles at Basel, the ^{3}H$(\vec{d}, n)^4$He reaction on resonance at $E_d = 107$ keV [RUD60]. They are based on spin-state separation of the atomic electrons and transfer of the polarization to the nuclei via the hyperfine interaction, because the magnetic moments of nuclei are about 1800 times smaller than the Bohr magneton. There are essentially two types of such sources for isotopes of hydrogen (and alkali atoms; for polarized ^{3}He beams of sufficient intensity no viable scheme has been developed, in contrast to polarized targets).

- The ground-state atomic beam source (ABS) with different methods of ionization that is now the superior method concerning intensity.
- The Lamb-shift polarized-ion source (LSS), which uses the tiny Lamb shift to polarize ions.

The scope of this book allows only a short summary of these sources. For some more detail, especially also about details of the description of polarization see Chap. 5 and [HGS12] and the relevant references therein.

The Atomic Beam Polarized-Ion Source (ABS) The separation of spin states (actually: hyperfine states) in an ABS, like in the famous Stern-Gerlach experiment [GER22], is achieved by the different forces on magnetic moments in an inhomogeneous magnetic field. Due to the small nuclear magnetic moments, as compared to the electron's, the coarse separation occurs according to electron spin projections (in sextupole fields this means: focusing or defocusing). The nuclear spin contributes only small corrections, and nuclear polarization needs additional measures. The force of the inhomogeneous magnetic field acting on such an "effective magnetic

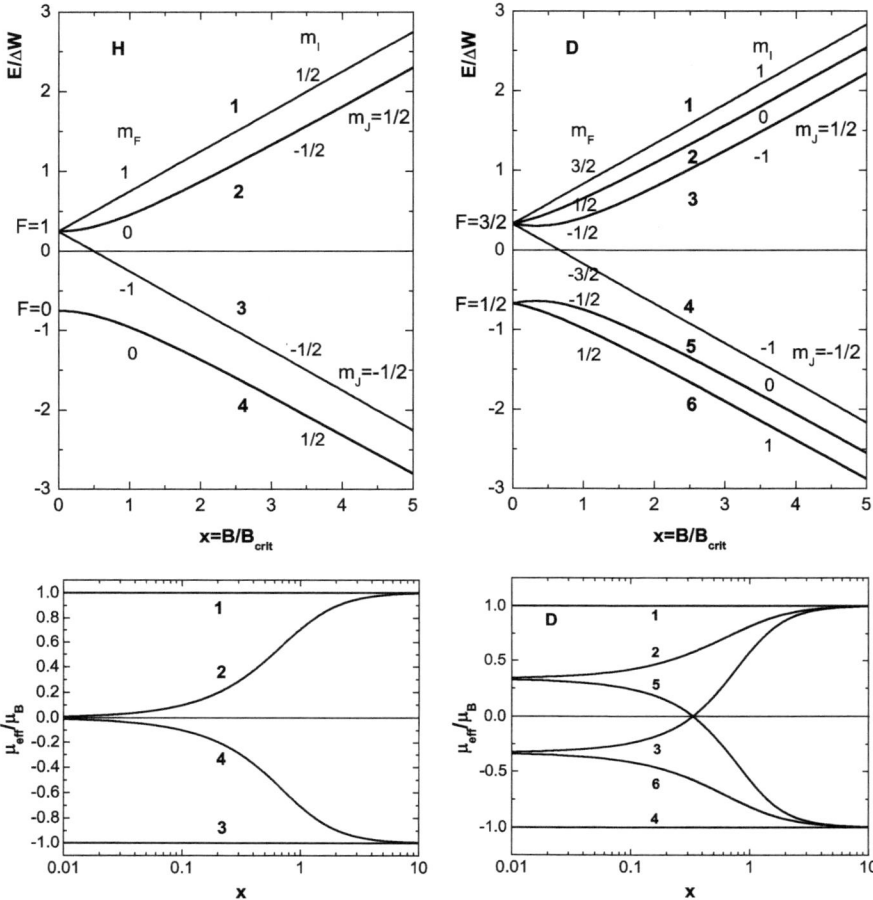

Fig. 16.17 Hyperfine Zeeman energy levels (Breit-Rabi diagrams) and "effective" magnetic moments in units of μ_B ($\propto$ the forces) of H and D in an inhomogeneous magnetic field with $x = B/B_{\mathrm{crit}}$. The deflecting force in an inhomogeneous magnetic field B is $\mu_{\mathrm{eff}} \cdot \nabla |\vec{B}|$

moment" that is defined as derivative of the energy $W(B)$, given by the Breit-Rabi formula (see below), after the field strength B is

$$\vec{F} = -\vec{\nabla} W_{F,m_F} = -\frac{\partial W}{\partial B}\vec{\nabla} B = \mu_{\mathrm{eff}}\vec{\nabla} B. \qquad (16.13)$$

Only for the "pure" components is $\mu_{\mathrm{eff}} = \mu_B$. Therefore, the separation according to the m_J works only for large B. Figure 16.17 shows the Breit-Rabi diagrams together with the effective magnetic moments of the Zeeman components as functions of the magnetic field. The principles of ABS polarized-ion source, e.g. for use on accelerators and thus obtaining much higher intensities at high beam quality and complete control over the polarization parameters, as compared to using nuclear reactions as primary source of polarized particles, were first formulated by Clausnitzer et al. [CLA56, CLA59].

The radial dependence of the force on magnetic moments is given by r^{L-2} (L = multipole order of the magnetic field). Thus, the force in a quadrupole field is constant and that in a sextupole is linear in r. In a sextupole there is a lens-like focusing action on one spin component whereas the other is being defocused. Therefore, an "optics" for spin-magnetic moments with features like beam transport, phase space, emittance, acceptance in analogy to the optics of charged particles in electric fields can be defined and be used to optimize a Stern-Gerlach system ("matching"). Note, however, that the sextupole provides no state separation on the axis, whereas for the quadrupole the state separation force is uniform with r, but there is no focusing. For several reasons (among these better pumping, the requirement of leaving space for intermediate radiofrequency transitions, and higher flexibility to optimize the atomic-beam optics) modern ground-state atomic-beam polarized ion sources (ABS) use not one, but a number of spin-separation magnets. It is suggestive to use a quadrupole magnet as first magnet leading to a better spin-state separation and somewhat higher polarization. The main quantities characterizing an ABS are the polarization p, the beam intensity I, but also the beam quality ("brightness" = intensity per two-dimensional transverse emittance). From the point of view of minimizing the measurement time for a given statistical error in experiments the figure of merit is $p^2 I$, which is valid for vector and tensor polarization components. When ionization of the neutral beam takes place in a strong magnetic field the ion beam acquires transverse momentum thus increasing the transverse phase space, i.e. the emittance. Therefore the usual atomic-beam sources (ABS) have emittances (typically 2 cm rad $(\text{eV})^{1/2}$) about twice those of Lamb-shift (LSS) and colliding-beams (CBS) sources (typically <1 cm rad $(\text{eV})^{1/2}$).

With the above outlines the principles of common types of polarized-ion sources can be understood. These are:

- Ground-state atomic-beam sources (ABS). They differ in the way the atomic beam is ionized:
  - Electron-bombardment and ECR ionizers
  - Ionizers with colliding beams of Cs^0, H, or D
  - Optically-pumped ion sources.
- Lamb-shift polarized-ion sources (LSS).

16.5 Physics and Techniques of the Ground-State Atomic Beam Sources ABS

16.5.1 Production of H and D Ground-State Atomic Beams

In order to produce atomic beams of H/D dissociators of different designs, all based on radiofrequency (RF) excitation, are used. The atomic beam intensity depends on

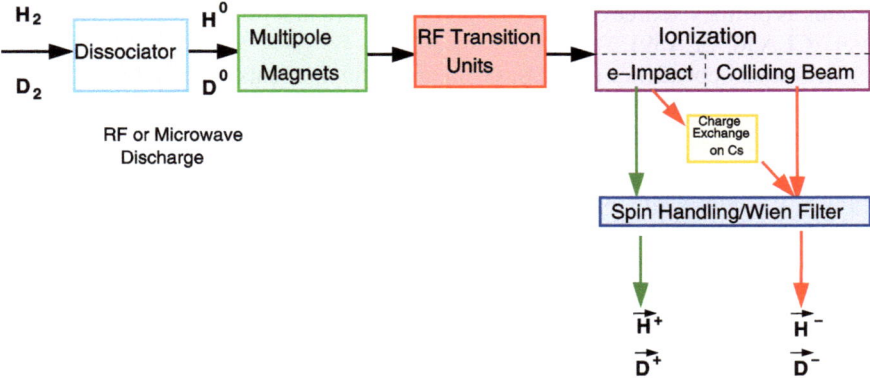

Fig. 16.18 Principle of the ABS. Depending on the type of ionization positive ions (electron-impact ionizer) or negative ions (directly: colliding-beams source, CBS, indirectly by charge exchange from positive ions) are produced. The different components are discussed briefly in the text

a number of parameters: gas pressure and gas flux, RF power, recombination rate on surfaces and their temperatures, and intra-beam scattering processes. After many years of development (since 1965) optimal design schemes have evolved that will be described here. Figure 16.18 shows schematically the main components of a modern atomic-beam polarized-ion source.

16.5.2 Dissociators, Beam Formation and Accommodation

Dissociators Two types of dissociators have evolved, RF discharge and microwave dissociators. In both a gas discharge is excited by an RF field in a cylindrical Pyrex glass or quartz vessel producing atoms from H_2 or D_2 being fed in from one end and H or D atoms streaming out at the other.

The discharge is maintained by a coil around the glass bottle (magnetic coupling) and normally runs at about 13 MHz with an RF power up to 200 W. The proper matching of the discharge assembly to a power oscillator is achieved by a matching circuit. The atomic beams are formed by a nozzle, typically from aluminum with a canal about 8–20 mm long and 2–3 mm wide. They have a velocity distribution somewhat narrower than Maxwellian due to the action of the nozzle. The nozzle is cooled, which is essential for several reasons: Slower atoms are accepted more easily by the first separation magnet, they are deflected more strongly, the chromaticity of the magnet system is reduced, and the beam has a higher density $\rho = j/v$ in a following ionizer volume (j is the particle number density of the beam, v the particle velocity). The intensity of the beam is determined by a number of parameters: The discharge is burning best in a certain pressure range, within which the gas feed

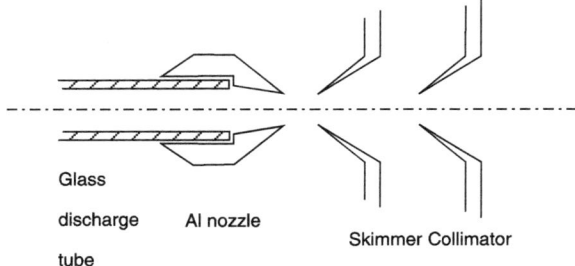

Fig. 16.19 Typical nozzle-skimmer-collimator arrangement of ABS sources. Typical diameters: nozzle 1 mm, skimmer 1.5 mm

should be as high as possible. The limit is, however, set by the pumping speed, with which the space after the nozzle can be maintained at such a low pressure that the mean free path is long enough to avoid or at least minimize intra-beam scattering of the atoms. These conditions are constrained by the small available space for pumping thus limiting the conductance, i.e. the effective pumping speed in this space. Differential pumping has to be applied and better beam quality is achieved by a skimmer. The collimator is necessary for differential pumping and facilitates maintaining high vacuum in the separation-magnet regions. A typical setup is shown in Fig. 16.19. As nozzle materials copper and aluminum have been used because of their heat conductivity but due to layers formed on the inner surface the choice is not critical for recombination. For the reasons discussed above, over the years, a saturation of the polarized atomic-beam intensities achieved at slightly above 10^{17} atoms/s in the relevant region behind the last separation magnet is observed. The role of cooling the dissociator arrangement is twofold: first the dissociator vessel must be cooled to prevent the glass from heating up (causing background residual gas and increasing recombination), secondly the nozzle must be cooled to make the atoms slower. The effect of this is that the acceptance of the entrance to the separation magnet system is increased, the separation power of the magnets improves, and the ionization yield of an ionizer increases because the beam density $\rho = j/v$ is higher (j = beam particle current density, v = particle velocity). Some modelling showed that the sum of these effects scales as $\propto T^{-3/2}$.

16.5.3 State-Separation Magnets—Classical and Modern Designs

Historically starting from the Stern-Gerlach spin-state separation magnet working in one dimension only much better intensity can be achieved with rotationally-symmetric magnetic fields provided by quadrupole and sextupole magnets acting in two dimensions. Permanent magnets as well as electromagnets have been used where the latter could be turned off for an unpolarized beam. However, the advantages of permanent magnets of modern design ("Halbach" magnets) are such that almost all sources use them.

Multipole Fields   The properties of magnetic multipole fields are:

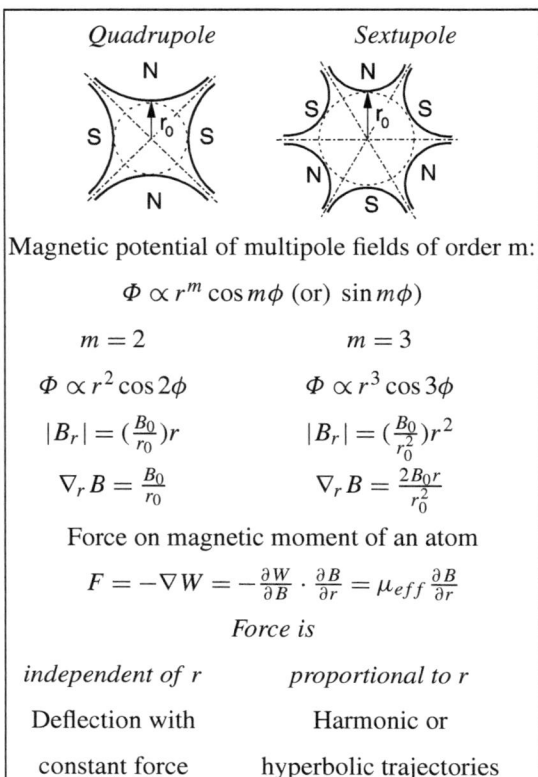

Only sextupole fields focus the atoms like an optical lens. Atoms in the opposite spin states are defocused. It is obvious that on the axis (in an infinitesimal volume) in a sextupole there is no spin-state separation whereas in a quadrupole the separation force is constant over the entire volume. At least in principle this should guarantee a somewhat higher beam polarization than from a sextupole. The rather wide velocity distribution of the atoms leads to a strong chromatic aberration of multipole magnets. This can be partly offset by tapering the magnetic fields along the z axis.

Since about 1980 with the advent of improved magnetic materials sextupole magnets follow a new design [HAL80]. Figure 16.20 [VAS00] shows the setup of such a typical Halbach sextupole and a measured field distribution. With a modification of the Halbach design, see Fig. 16.21 a poletip field of 1.8 T has been achieved. In modern ABS an arrangement of several (typically four to six) separated short sextupoles is used. The field strengths and locations of the magnets are determined by numerical trajectory calculations taking into account other requirements such as optimum pumping and insertion of RF transition units. Figure 16.22 shows the results from one such calculation.

16.5 Physics and Techniques of the Ground-State Atomic Beam Sources ABS

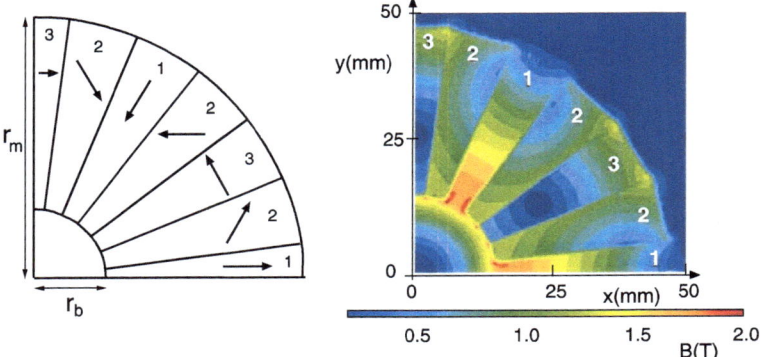

Fig. 16.20 Segment scheme and field distribution of a typical Halbach sextupole magnet [VAS00]

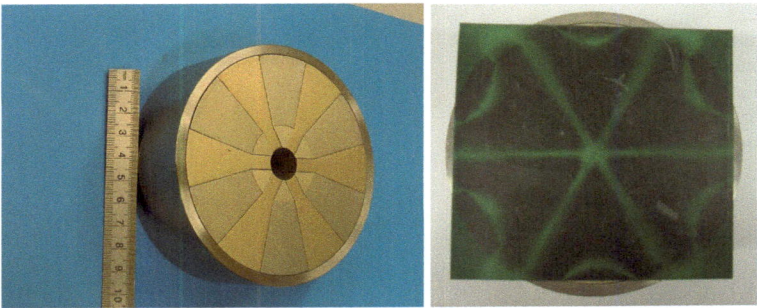

Fig. 16.21 View of the modified Halbach ("hybrid") sextupole magnet developed for SAPIS-Cologne with a measured maximum field value of 1.8 T. *On the right* a field image is shown obtained with "Magna View Film"

16.5.4 RF Transitions

After a Stern-Gerlach device and in a strong magnetic field (e.g. in an electron-bombardment ionizer) the particles in a beam are highly polarized with respect to the spin of their valence electron, but nearly unpolarized in nuclear spin. For higher intensities ionization in a strong magnetic field is necessary (see below). Therefore high nuclear polarizations can only be achieved by RF transitions between hyperfine states. The adiabatic-fast passage method proposed by Abragam and Winter [ABR58] as is explained in Fig. 16.23 is used throughout. Three field regions for RF hyperfine transitions may be classified according to the value of the static magnetic field B_0, e.g. in relation to the critical field B_{crit}: weak-field (WFT, $B_0 \ll B_{\text{crit}}$, transition frequencies typically 5–15 MHz), medium-field (MFT, $B_0 < B_{\text{crit}}$), and strong-field (SFT, $B_0 \geq B_{\text{crit}}$, transition frequencies typically several hundred MHz to GHz) transitions. Another classification refers to the change of quantum numbers by the transitions. Transitions within one F multiplet ($\Delta F = 0$, $\Delta m_F = \pm 1$) are π

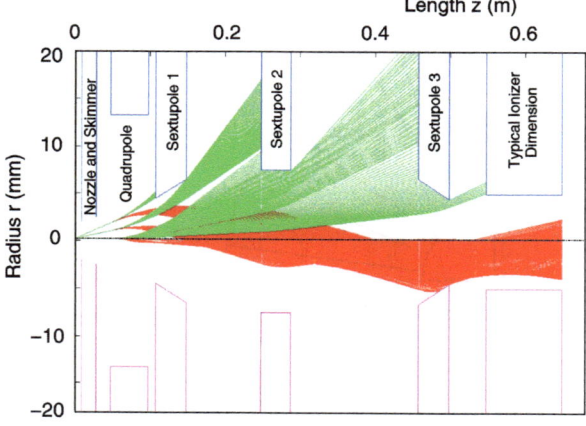

Fig. 16.22 Result of optimized trajectory calculations for a deuterium beam through one quadrupole and three sextupole magnets

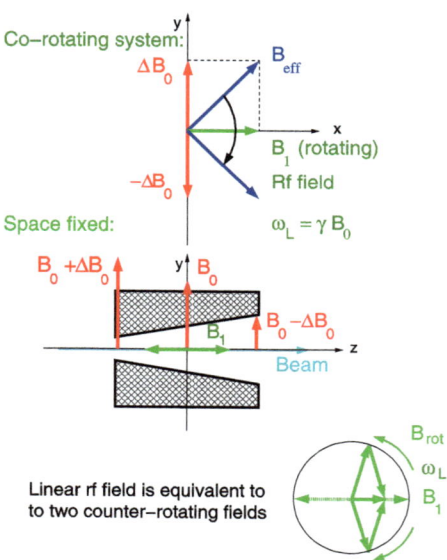

Fig. 16.23 Semi-classical explanation of the "adiabatic-fast passage" method of producing a spin flip during passage through an inhomogeneous magnetic field

transitions and they are induced by the RF field $B_1 \perp B_0$. Transitions between different F multiplets ($\Delta F = \pm 1$, $\Delta m_F = 0, \pm 1$) are σ transitions, and the two fields are parallel to each other.

In the practice of polarized-ion sources the WFT and MFT used are low-B_0 π transitions. The WFT occur in the Zeeman region of the HFS where the m_F states belonging to one F are nearly equidistant, leading to multi-quantum transitions within the F multiplets. The MFT are similar π transitions at somewhat higher B_0 and RF frequencies such that the energies of single-photon transitions in one F multiplet are sufficiently separated, i.e. with a field region short enough that only single transitions do occur. SFT, however, are σ or π single-quantum transitions at

16.5 Physics and Techniques of the Ground-State Atomic Beam Sources ABS

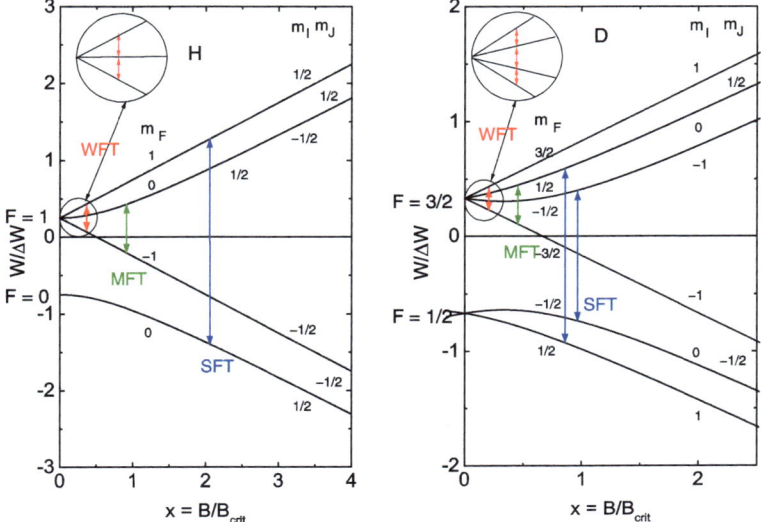

Fig. 16.24 RF transitions and transition types as functions of the field parameter x

still higher B_0 between single HFS states. Figure 16.24 illustrates the three types of transitions.

Quantum-Mechanical Treatment The quantum-mechanical transformation of the two-spin system with time-varying magnetic fields (the "static" field B_0 slowly varying due to the particle motion in its field gradient $2\Delta B_0/\Delta z$ and the (fast-changing) RF field $B_1(\omega t)$) into a co-rotating system is equivalent to unitary transformations of the Hamiltonian of the system. If the transformation is appropriately chosen (i.e. such that the rotation occurs with the Larmor frequency ω) the slowly- and the fast-varying parts can be separately diagonalized.

The "quasi-stationary" slow-solution part of the Schrödinger equation leads just to a different picture of the Breit-Rabi energy eigenstates (as functions of B_0 or, equivalently, x, or t), which follow a linear dependence, making the states cross at just the field corresponding to the particular transition frequency. With the RF field B_1 switched on, the q.m. calculation including this perturbation leads to (in this representation)

- up and down shifts of the eigenvalues ("level repulsion") and to
- non-diagonal terms of the Hamiltonian matrix, i.e. mixing of states and therefore transitions between them.

Figure 16.25 shows this schematically. These effects are strongest at the crossing points, and the efficiency of the transitions is governed by the degree of adiabaticity, i.e. whether the occupation of the initial states involved will stay on its original levels (non-adiabatic or "diabatic" transitions, see also Sect. 16.7.3) or undergo a more or less complete transition to the other level. The strength of the RF field B_1

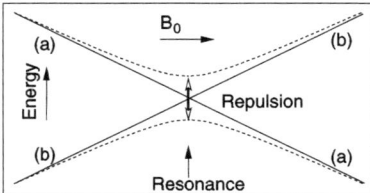

Fig. 16.25 Schematic q.m. representation of two Zeeman hyperfine states (*a*) and (*b*) as functions of the external static magnetic gradient field B_0 without perturbing RF field B_1: states crossing (*solid lines*) and with perturbation: mixing, repulsion of states, and more or less transition between them (*dashed lines*)

is one determining factor. In order to quantify this degree an adiabaticity parameter was derived from the adiabaticity condition

$$\kappa = \frac{\mu_J B_1^2}{2\hbar \dot{B}_0}, \tag{16.14}$$

leading to the transition probability

$$P = \exp(-\pi \kappa). \tag{16.15}$$

The formalism is applicable to two-state strong-field as well as medium-field transitions.

Detailed information about the RF transitions is obtained by the solution of the time-dependent Schrödinger equation, in the case of multiple transitions like the WFT by a set of $2m_F + 1$ coupled, time-dependent equations. Figure 16.26 shows an example.

Experimentally as well as in theoretical studies differences in occupation numbers and therefore polarizations have been found to depend on the sign of the gradient of the static field B because Zeeman HFS states are not exactly equidistant and therefore the several transitions between Zeeman-HFS states did not occur simultaneously but sequentially, but in different order.

For the WFT being π transitions the direction of the RF field B_1 is along the beam axis (z direction) and perpendicular to the static field B_0. Due to the low frequency required it is realized by a coil with a small number of windings (e.g. about 5–10 for frequencies of 8 to 12 MHz).

Medium-Field Transitions (MFT) Like the SFT the MFT are transitions between single states. They are, however, π transitions occurring at rather low B_0. Typical transitions are between states $1 \leftrightarrow 2$ and $2 \leftrightarrow 3$ for H and $1 \leftrightarrow 2$, $2 \leftrightarrow 3$, $3 \leftrightarrow 4$, and $5 \leftrightarrow 6$ for D.

Strong-Field Transitions (SFT) Strong-field transitions take place at values of $x \approx 1$. There, at fixed magnetic field and frequency only transitions between single Zeeman states are possible. The transitions are π or σ transitions. Typical transitions

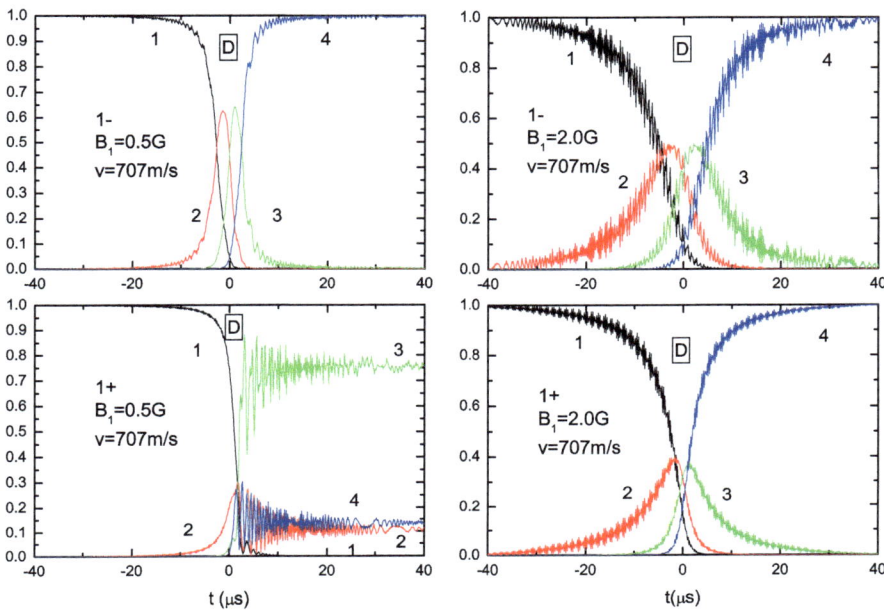

Fig. 16.26 Plot of the change of occupation numbers of Zeeman hyperfine states of deuterium starting with state 1 occupied only, through the WFT region. *Upper plots* are for negative, *lower* for positive field gradient. *Left*: $|B_1| = 0.05$ mT; *right*: $|B_1| = 0.2$ mT

are $1 \leftrightarrow 4$ for H, $2 \leftrightarrow 6$, and $3 \leftrightarrow 5$ for D. For these, being π transitions, the field direction of B_1 is parallel to the static field B_0. The higher frequencies require single-loop or, more modern, RF-cavity designs.

16.6 Ionizers

In order to convert polarized atomic beams into polarized ion beams a number of different schemes have been developed:

- Electron-bombardment ionizers
- ECR ionizers
- Colliding-beams ionizers

16.6.1 Ionizers—Electron-Bombardment and Colliding-Beams Designs

Electron Bombardment Ionizers The cross section for ionization of hydrogen by electron impact has a maximum near $E_e = 70$ eV, see Refs. [FIT58, KIE66].

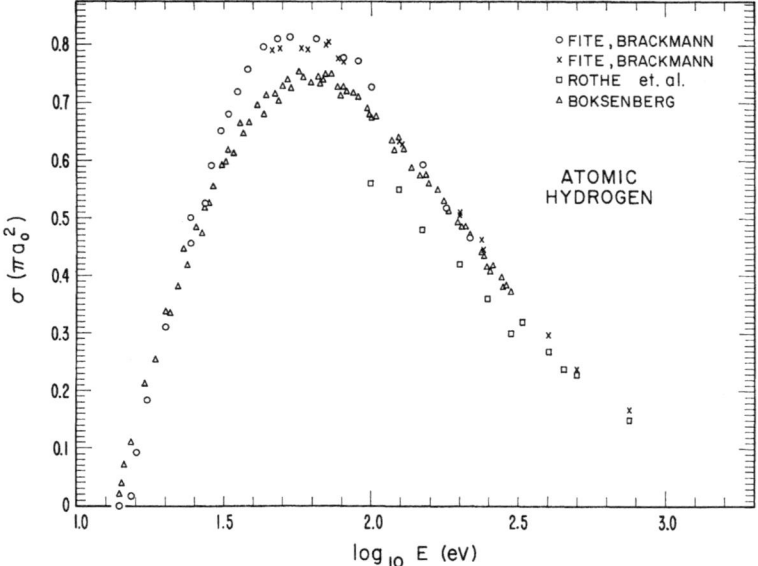

Fig. 16.27 Cross section of ionization by electron impact of atomic hydrogen with a maximum near 70 eV

Figure 16.27 shows the results of several authors. The classical strong-field ionizer makes use of a long ionization path by three measures:

- Spiraling of the electrons, emitted by a cathode wire and accelerated by a positive grid or ring electrode, along a strong magnetic field of >0.1 T.
- Long ionization volume.
- Multiple use of electrons by reflection from a repulsive electric field at the end of the volume, serving at the same time as extraction field for the ions.

The electron space-charge depression has to be compensated by injecting the electrons at voltages much higher than 70 eV. The prototype of this ionizer was developed by Glavish [GLA68] and has been used in many positive-ion sources, see Fig. 16.28. When negative ions were required—such as for tandem Van-de-Graaff accelerators—an additional charge exchange in alkali vapor had to follow.

ECR Ionizers The high ionization efficiency of electron-cyclotron-resonance ionizers was exploited in some polarized-ion sources. The ECR principle is to ionize the polarized beam by electrons accelerated in a plasma created by an intense RF discharge in a strong magnetic field. The RF frequency corresponds to the electron-cyclotron resonance and is therefore coupled to the magnetic field. At the magnetic field of >100 mT optimal RF frequencies around 3.8 GHz are necessary with an RF power up to several 100 W. The fields are shaped to confine the electrons to the ionizing region and to extract the ions efficiently and without depolarization. The plasma discharge has to be maintained stably at rather low pressures $<1 \cdot 10^{-6}$ mbar,

16.6 Ionizers

Fig. 16.28 Scheme of a Glavish-type electron-bombardment ionizer

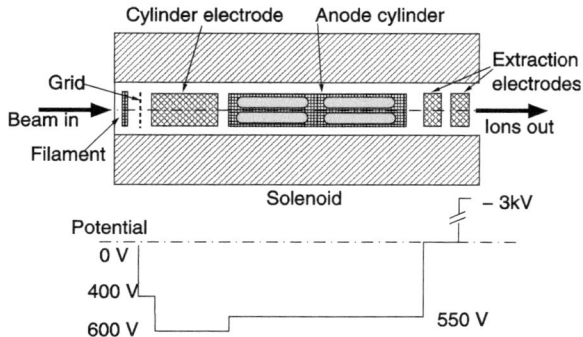

Fig. 16.29 Cross sections for ionization of H or D into positive or negative ions by charged H (or D), neutral Cs, and electron beams

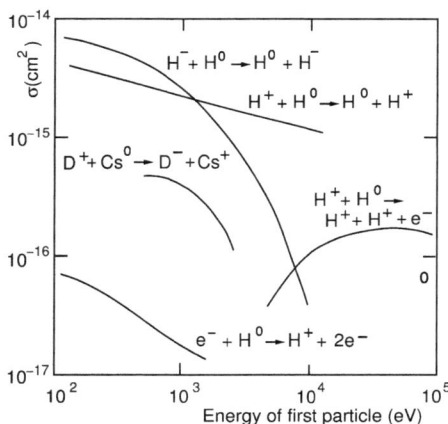

which has been achieved by bleeding inert gas like N_2 into the discharge volume. One advantage of ECR ionizers is their small beam emittance. Ionization efficiencies of up to $6 \cdot 10^{-3}$ have been reported.

Colliding-Beams Ionizers, CBS This type of source was proposed and realized by Haeberli et al. [HAE68, HAE82]. The very high cross sections for ionization of atomic hydrogen/deuterium into negative ions in collisions with neutral Cs beams appeared very attractive. Even more attractive is the ionization of the polarized thermal atomic beams by intense colliding beams of negative or positive unpolarized ions. This is because the cross sections are larger by about two orders of magnitude at very low energies due to resonant charge exchange. In Fig. 16.29 the relevant cross sections are compared. The CBS with Cs requires an energetic Cs beam of about 45 keV, as shown in Fig. 16.30 with the charge-exchange cross section into negative ions as function of Cs-beam and relative energies. Cs^+ ion beams with currents of many mA (up to 15 mA) are extracted from a hot tungsten surface ionizer button with about 45 keV energy, then neutralized efficiently in Cs vapor. Due to the high current density nearly complete space-charge neutralization takes place

Fig. 16.30 Cross section for ionization of H or D into negative ions by a neutral Cs beam. From Ref. [GIE86]

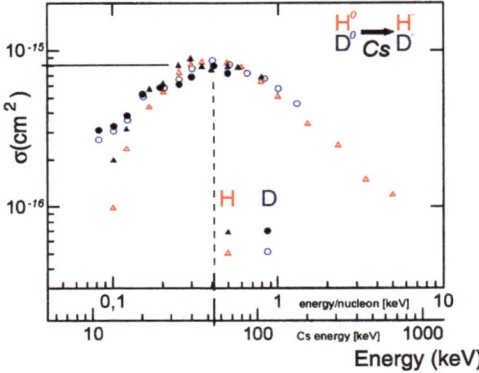

Fig. 16.31 Example of a 1 mA Cs^+ beam emitted from the hot tungsten cathode *on the right* and exciting residual-gas atoms to fluorescence while exiting *to the left*

thus allowing beams of very high brilliance to reach the charge-exchange cell, see e.g. Fig. 16.31. Such colliding-beams sources were successfully built at Madison, Brookhaven, Seattle, and, finally, for COSY-Jülich. Figure 16.32 shows the principal design of a CBS. Essential features of the CBS are high polarization, reliability and long-time running capability.

The use of resonance ionization by low-energy, but high-intensity beams of H^-, D^-, H^+, or D^+ meets the difficulty of high space charge that so far restricts the realization only as pulsed systems with very short pulses (μs). Plasma ionizers with very high ionization efficiencies have been developed. Though very high peak pulse currents (up to 50 mA) have been reached the average number of polarized particles per unit time remains relatively small. The CBS with Cs is in principle a DC source, but in connection with pulsed accelerators such as COSY/Jülich with long (20 ms) pulses the performance is much improved by pulsing the entire source, i.e. the polarized atomic as well as the Cs beam synchronously.

16.6 Ionizers

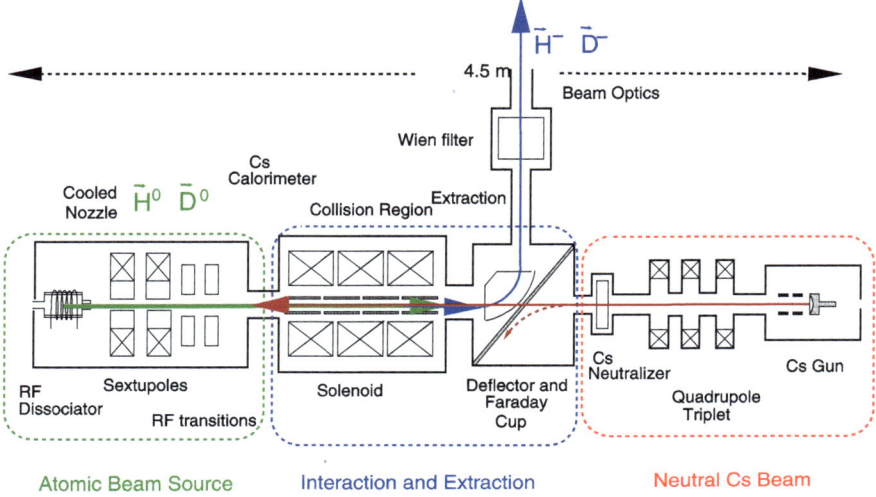

Fig. 16.32 Scheme of a colliding-beams polarized ion source

16.6.2 Sources for Polarized ^{6,7}Li and ^{23}Na Beams

In the past atomic-beam polarized ion sources for $\overrightarrow{^{6,7}\text{Li}}$ ($I = 1$ and $3/2$) and $\overrightarrow{^{23}\text{Na}}$ ($I = 3/2$) beams have been developed. In the Spin-3/2 cases the complete description of the polarization requires tensor moments t_{kq} up to rank $k = 3$.

The techniques of producing the atomic beams are different from the hydrogen case: atoms are evaporated from an oven and ionization can be done by surface ionization on heated W metal. In the first such sources Stern-Gerlach separation magnets have been used for spin-state separation.

16.6.3 Optically Pumped Polarized Ion Sources (OPPIS)

The principle used here is the same as that of polarized targets applying the optical pumping of alkali vapors (especially rubidium) and transfer of the high electronic polarization by collisions to the ground-state atoms and nuclei of H or D (spin-exchange method SEOP). The relevant wavelength (795 nm) is in the near infrared and pumping can be done by different lasers (e.g. Ti:Sapphire), but high-power laser diodes (diode arrays) have recently become available and are most convenient. Figure 16.33 shows a schematic of the principal functions of an OPPIS. Different functions such as neutralization of the injected H^+ beam, optical pumping of Rb, and spin-exchange collisions occur in an integral vessel in a common magnetic field region.

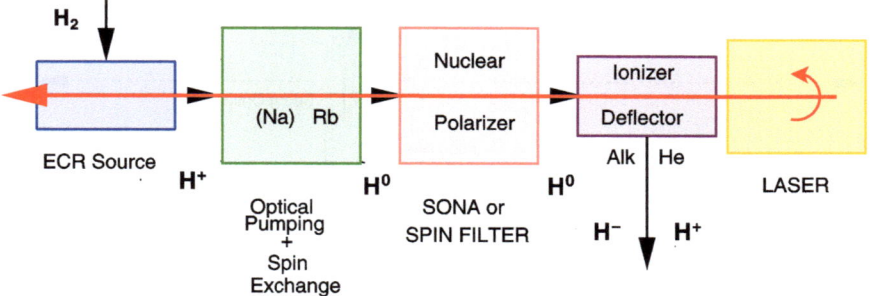

Fig. 16.33 Scheme showing the principles of optically pumped polarized-ion sources

A source of this type, using the charge-exchange reaction

$$H^0 + He \rightarrow H^+ + He + e \qquad (16.16)$$

for positive ion output, developed originally at TRIUMF/Vanvouver, is being used at the RHIC accelerator at BNL/Brookhaven. It has superior properties: ionization efficiency up to 0.8, DC $\vec{H}^-$ currents of up to 15 mA (pulsed about 25 mA) at high polarizations.

The polarized jet target used at RHIC is in principle a classical ABS. It has the highest polarized atomic-beam intensity achieved so far (≈ 1.2 s^{-1}) [ZEL05, WIS06]. Another recent polarized jet target based on an ABS is the ANKE target at COSY-Jülich (for a detailed account see Ref. [MIK13]).

16.7 Physics of the Lambshift Source LSS

16.7.1 The Lamb Shift

Although Lamb-shift sources are not at the forefront any more they have been used for decades and rely on interesting physics. The *Lambshift*, relevant here, is the energy difference between the $2S_{\frac{1}{2}}$ and the $2P_{\frac{1}{2}}$ states [LAM50] of H or D atoms and is explained only by quantum electrodynamics. For hydrogen this shift (without a magnetic field) is about 1057 MHz or $4.38 \cdot 10^{-6}$ eV. The lifetime for the transition $2S_{\frac{1}{2}} - 2P_{\frac{1}{2}}$, due to the very small energy difference, is about 20 years. A dipole transition (E1) to the $1S_{1/2}$ ground state is forbidden ($I = 0 \rightarrow I = 0$), as is the corresponding quadrupole transition (E2) ($J = 0 \rightarrow J = 0$). A magnetic dipole transition (M1) is allowed and its lifetime was calculated to be about two days (Breit and Teller [BRE40]). The main contribution comes from a two-quantum transition with $\tau \approx \frac{1}{7} s$. An electric field reduces the lifetime of the $2S_{\frac{1}{2}}$ and increases that of the $2P_{\frac{1}{2}}$ state via the Stark effect, which mixes states of different parity (i.e. here the

16.7 Physics of the Lambshift Source LSS

parity is not a good quantum number). Following Lamb and Retherford [LAM50] the lifetime of the $2S$ state is

$$\tau_S \cong \tau_P \frac{\hbar^2(\omega^2 + \frac{\gamma^2}{4})}{|V|^2} \tag{16.17}$$

with

$$\tau_P = \text{lifetime of the } P \text{ state} = 1.595 \cdot 10^{-9} \text{ s} \tag{16.18}$$

$$\hbar\omega = \Delta E$$

$$= \text{energy separation between } S \text{ and } P \text{ state (field dependent).} \tag{16.19}$$

$$\gamma = 1/\tau_P \tag{16.20}$$

$$V = \langle \varphi_S | eEr | \varphi_P \rangle$$

$$= \text{matrix element of the dipole transition} \tag{16.21}$$

which, for small electric fields, (≤ 100 V/cm) is approximately

$$\tau_S = \tau_P (475/E)^2. \tag{16.22}$$

16.7.2 Level Crossings and Quench Effect

In the picture of the fine structure (FS) the Stark effect mixes states with $\Delta m_J = 1$, $\Delta \pi = +$, i.e. (in the historical nomenclature of Lamb and Retherford [LAM50]) the states α and f, β and e, respectively. Because the states β and e cross at a magnetic field of about 57.5 mT, the transition probability there becomes maximal. The lifetime of β becomes shorter with smaller ΔE (the transition probability (perturbation calculation!) contains $(\Delta E)^2$ in the denominator). $\tau_S(\alpha)$ increases with B because of increasing state separation, while $\tau_S(\beta)$ has a minimum near 57.5 mT. The lifetime of the S state is empirically given by the formula:

$$\tau_S = \frac{1.13}{E^2} \left[(574 \pm B)^2 + 716 \right] \text{ ns.} \tag{16.23}$$

For $E = 15$ V/cm one obtains e.g. $\tau_S(\alpha)/\tau_S(\beta) = 1850$. For a hydrogen beam with 500 eV (or $3.1 \cdot 10^7$ cm/s) practically all atoms have decayed into the state β after 6.5 cm, but only 3.5 % of the α states (see Fig. 16.34). In this way an atomic beam is obtained that is about 96 % polarized in the electronic spin. The HFS Zeeman splitting leads to four (or nine, resp.) crossings around 57.5 mT, of which two (for H) or three (for D) can undergo Stark-effect quenching (see also Fig. 16.41). If the beam is (adiabatically) transported and ionized in a weak magnetic field a nuclear polarization of half of the theoretical value of the electronic polarization (and a correspondingly polarized proton or deuteron beam) results.

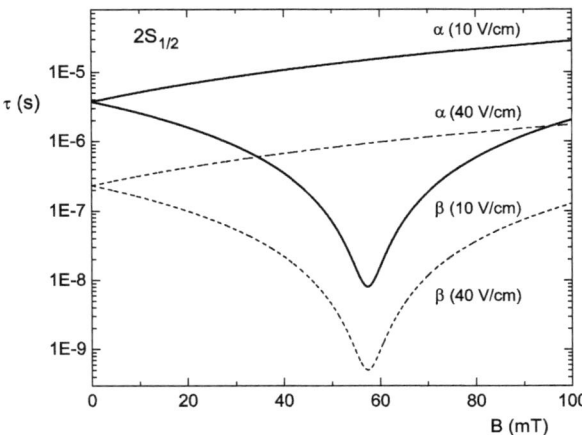

Fig. 16.34 Lifetimes of the $n = 2$ Zeeman states as functions of the magnetic field for two electric quenching field strengths

16.7.3 Enhancement of Polarization

There are two ways to enhance and change the nuclear polarization in the metastable beam. One is the use of a non-adiabatic (fast) transition with a change of the occupation of the Zeeman states, the SONA transition scheme [SON67]. The other is the use of a spinfilter [KIB68], in which a combination of a longitudinal magnetic field, a transverse static electric field and an RF field allow the selective transmission of single HFS states. These methods result in polarization values close to the theoretical maxima.

Figure 16.35 depicts the Breit-Rabi diagrams for H and D with a sudden field reversal via a zero-crossing. Depending on the degree of adiabaticity of the crossing, the occupation of the Zeeman states follows different "trajectories" on the Zeeman levels (see Sect. 16.5.4). The practical realization of the LSS will be addressed below. The LSS with SONA transitions can be run in several different modes, e.g. with one or two quenching processes (the latter for deuterium). The resulting polarizations of a beam can be calculated from the occupation numbers of the remaining Zeeman states and depend on the magnetic field. Figure 16.36 shows different modes of operation of the LSS with one (for vector polarization of protons or deuterons) or two quenching processes (for deuteron tensor polarization). Figures 16.37 and 16.38 show the dependence of the proton or deuteron polarizations on the magnetic field at the ionizer location.

Production of the Beam of Metastables H and D atoms in the $2S$ state may be produced by electron impact, but the charge-exchange reaction

$$\mathrm{H}^+ + \mathrm{Cs}^0 \rightarrow \mathrm{H}^0(1S, 2S) + \mathrm{Cs}^+ + Q \tag{16.24}$$

16.7 Physics of the Lambshift Source LSS

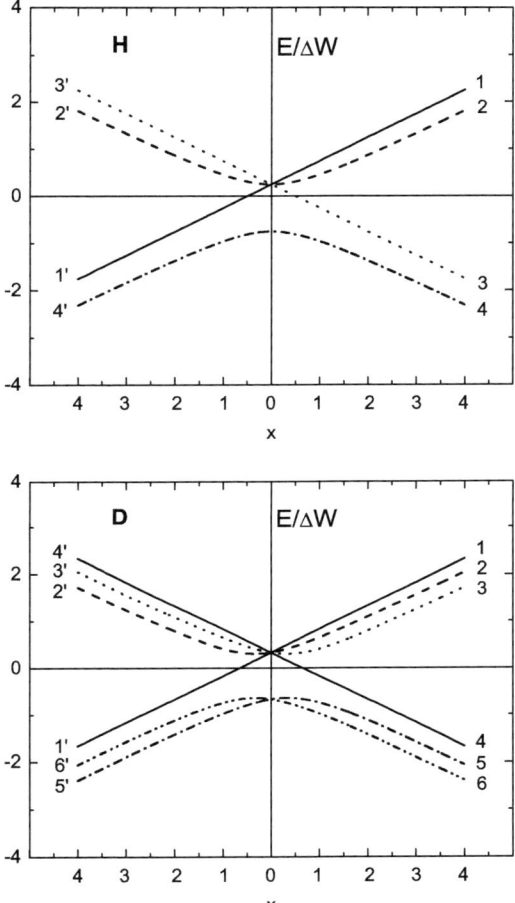

Fig. 16.35 Zeeman levels of hydrogen and deuterium with (non-adiabatic) zero-field crossing/field inversion

is much more efficient. The primary positive ion beam is produced in a standard ion-source system such as an RF discharge, a duoplasmatron or even ECR ion source, see Sect. 16.4.1. The required fixed energy of the ions of 500 eV for H^+ or 1 keV for D^+ and therefore relatively high space-charge limit the beam current that can be injected into a charge-exchange cell containing Cs vapor of appropriate density. Figure 16.39 shows the charge-exchange cross sections to the neutral ground and to the metastable $2S$ states as functions of the energy, and the relative yields as functions of the Cs target thickness. For Cs $Q = 0.50$ eV, and the ionization energy of 3.89 eV is very small. The yield is 10–15 % at 500 eV for a target thickness of $5 \cdot 10^{-3}$ Torr cm. The measured fraction of metastables in the full beam of neutral particles ($1S, 2S$) amounted to $f_{max} = 0.430 \pm 0.03$. It is serendipitous that the cross sections for metastable production and selective ionization in argon (see below) have maxima at about the same beam velocity. The positive beam into the Cs-vapor cell

Fig. 16.36 Different SONA modes of operation of the LSS for H (*top*) and D (*bottom*)

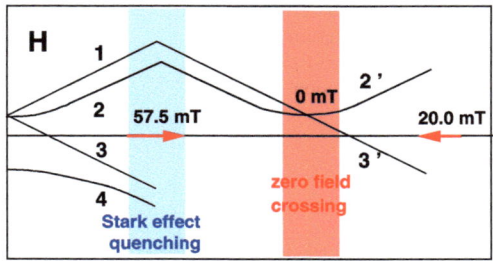
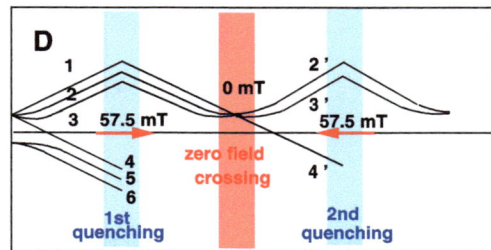

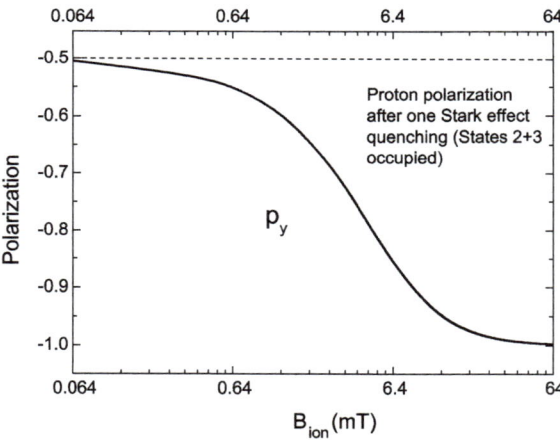

Fig. 16.37 Proton polarization as function of the ionizer field after one quench

is produced either by a conventional RF ion source or by a duoplasmatron, see the preceding Sect. 16.4.1.

16.7.4 Production and Maximization of the Beam Polarization

In order to obtain maximum values of the polarization with a LSS, in analogy to the ABS transitions between hyperfine states are induced. However, because the (metastable) beam, in comparison to the ground-state atomic beam, is "fast", none

16.7 Physics of the Lambshift Source LSS

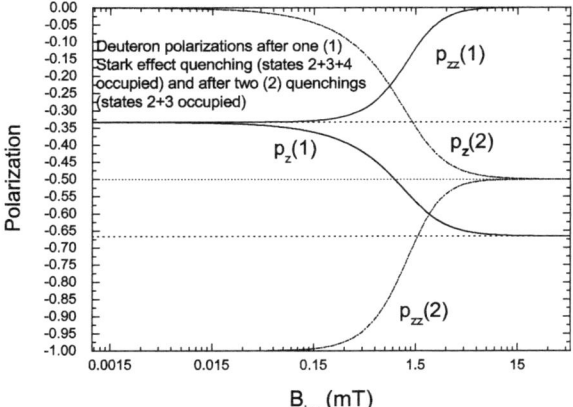

Fig. 16.38 Deuteron vector and tensor polarization as functions of the ionizer field with one (*1*) or two (*2*) quench processes

of the usual adiabatic RF transitions can be used, but either non-adiabatic transitions (SONA transitions) or a SPINFILTER. This leads to two possible schemes for the construction of the LSS, depicted in Fig. 16.40.

SONA Transition The idea is to perform a non-adiabatic transition in a "rapidly" sign-changing magnetic field (zero crossing) [SON67], i.e. the field changes in time intervals short against $1/\nu_L = 2\pi/\omega_L$, from which the condition

$$1/B(dB/dt) \gg \omega_L/2\pi = (\gamma/2\pi)B, \tag{16.25}$$

i.e.

$$dB/dt \gg (\gamma/2\pi)B^2 \tag{16.26}$$

is derived. In this case the atoms stay in their respective Zeeman HFS states while the field is reversed, see also Fig. 16.25 and the discussion there. Thus the original state 1 becomes state $1' \equiv 4$, leading to a theoretical nuclear polarization of 100 % instead of 50 %. There is a critical volume: The non-adiabaticity condition is always fulfilled for $B = 0$, i.e. on the beam axis as long as the field has a gradient at all. Away from the beam axis the field can only be $\neq 0$. Therefore, there is a critical beam radius, beyond which this condition is not fulfilled.

Spin Filter The theory of the spin filter [OHL67, KIB68] is somewhat complicated because its function rests on the simultaneous interaction of three states:

- The $1S_0$ ground state,
- the metastable $2S_{1/2}$ state, and
- the short-lived $2P_{1/2}$ state.

The Breit-Rabi diagram Fig. 16.41 illustrates (for H) the simultaneous interactions. Near the level crossings the β states are quenched, i.e. decay rapidly into the $1S$ ground state. The RF transition depopulates the substate $\alpha 2$ while the substate $\alpha 1$

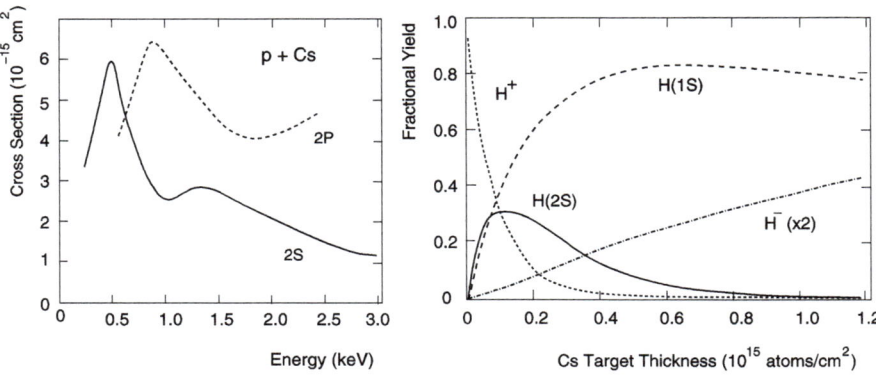

Fig. 16.39 Charge-exchange cross sections of protons in Cs vapor into the metastable $2S$ state and the $2P$ state as functions of the energy (*left*) and relative contributions in the beam after passage through Cs vapor as functions of the areal thickness of Cs (*right*). Adapted from Ref. [PRA74]

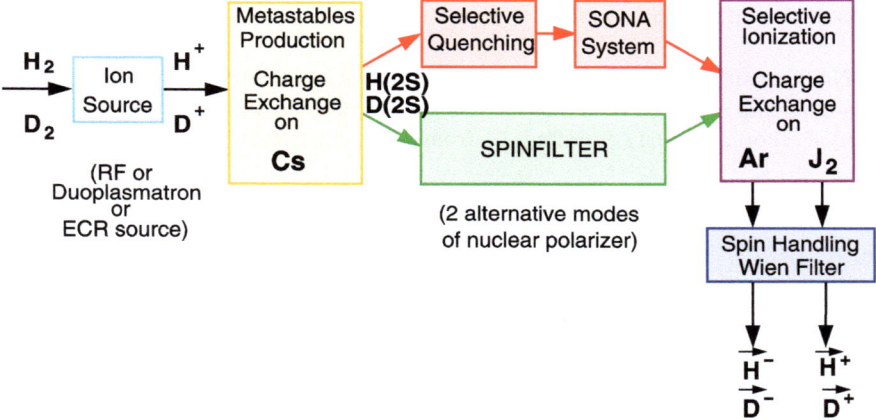

Fig. 16.40 Realization of the LSS with two principles: SONA method (*upper*) and SPINFILTER (*lower*). Negative as well as positive ions can be obtained

is constantly repopulated from one β substate. After exiting the spin filter only one hyperfine substate remains populated. The choice of the magnetic field value at fixed RF frequency (or vice versa) determines, which state is being transmitted. For deuterium the interactions are analogous. The interactions are realized by the static longitudinal magnetic field of a solenoid of about 57.5 mT, which must be quite homogeneous, a static electric quenching field realized by segmenting the RF cavity into quadrants and applying a DC voltage to an opposing quadrant pair, and an electric RF field with a frequency of $\nu = 1.60975$ GHz in the TM_{010} mode in a resonator cavity. The spin-filter setup is illustrated in Fig. 16.42. The spin-filter principle has advantages over the SONA principle, at least for deuterium. They derive from the

16.7 Physics of the Lambshift Source LSS

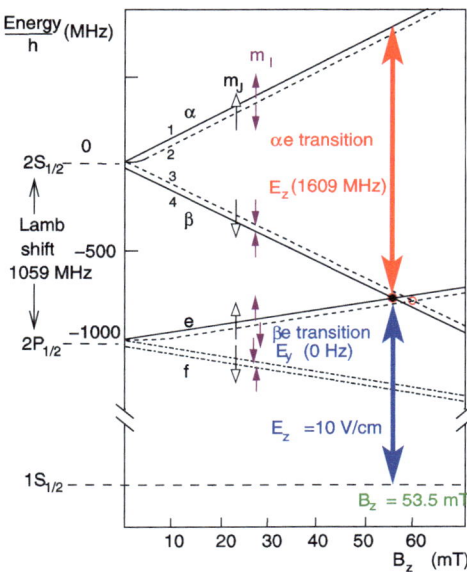

Fig. 16.41 $2S/2P$ hydrogen hyperfine-Zeeman states diagram with transitions caused by the static and the RF electric fields

fact that single hyperfine components can be selected and transmitted whereas the usual SONA scheme transmits two states. Therefore, only with the spin filter the theoretical maximum values of the polarization between $p_{ZZ} = +1$ and -2 can be obtained together with the possibility to change the sign of the polarization. On the other hand the intensity is reduced, as compared to the SONA scheme. Therefore, the figure of merit $p^2 \cdot I$ has to be evaluated for each scheme, and in general, the use of a spin filter may not be useful for protons, also in view of the simpler operation of the SONA scheme, whereas for the deuteron tensor polarization a doubling of the figure of merit was proven experimentally.

The function of a spin filter is illustrated by Fig. 16.43, which shows the polarization and the transmitted intensity (current) of the deuterons as functions of the spin-filter magnetic field keeping the E field and the RF frequency constant.

Selective Ionization of the (Polarized) Beam of Metastables This is achieved by a quasi-resonant charge-exchange process

$$H(2S) + Ar \rightarrow H^- + Ar^+. \tag{16.27}$$

A similar charge exchange leads to positive polarized ions [KNU70, BRU70]

$$H(2S) + I_2 \rightarrow H^+ + I_2^-. \tag{16.28}$$

Figure 16.44 shows the high value and weakly resonant behavior of the cross section σ_{2S-} for negative-ion formation from metastables as compared to σ_{1S-} from ground-state atoms (left). The right part of the figure shows the strongly energy-dependent (relative) H^- ionization yields of these processes, especially the high

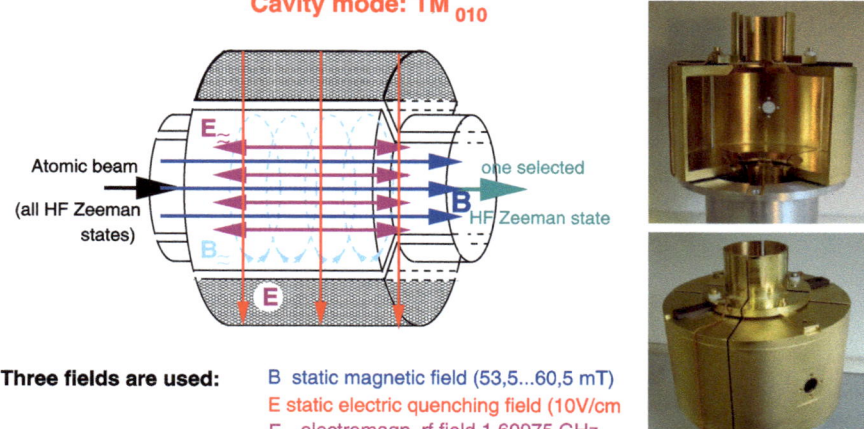

Fig. 16.42 Scheme of a spin filter with the relevant fields and photographs

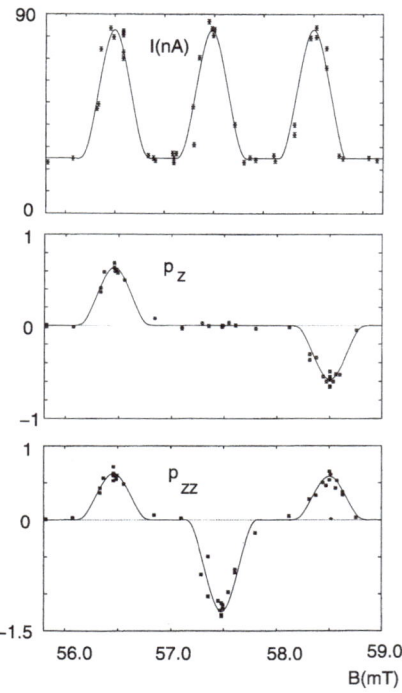

Fig. 16.43 Transmitted current and polarization of the deuteron hyperfine components appearing at three different values of the spin-filter magnetic field B. Data measured with the Cologne LSS and the FN tandem VdG accelerator

selectivity of the metastable relative to the ground state. For the $2S$ state an ionization energy of at least $10.19 + 0.75 = 10.94$ eV (i.e. the excitation energy of the $2S$ state plus the binding energy B.E. of the electron in H$^-$) is required. Argon has

16.8 Spin Rotation in Beamlines and Precession in a Wien Filter

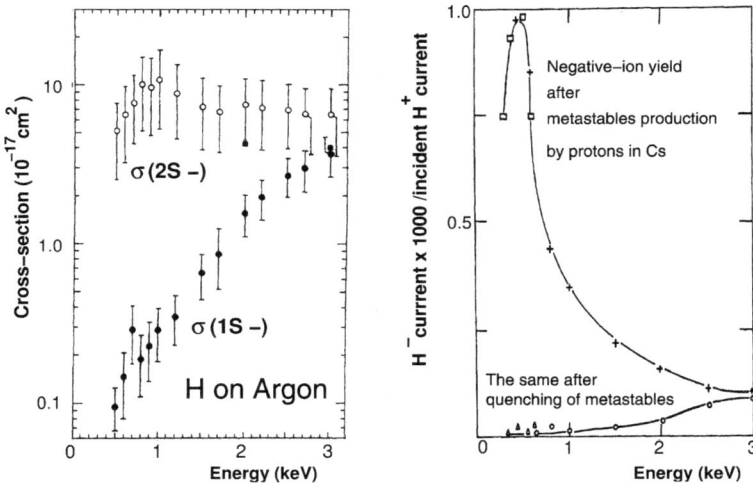

Fig. 16.44 Electron capture cross sections σ_{1S-} and σ_{2S-} on argon for ground-state and metastable H atoms as functions of the energy, adapted from Roussel [ROU77] (*left*) and $H^+ \to H^-$ charge-exchange yield (through the entire system of the Lambshift source) of ground-state and metastable H atoms on argon, adapted from Donnally and Sawyer [DON65] (*right*)

an ionization energy of 15.8 eV and is therefore especially suited. The selectivity $2S/1S$ is almost 100 %, the ionization yield is near 1 %. In practice, like always in polarization work, all source parameters must be optimized with respect to the maximum figure of merit $p^2 \cdot I$.

The final limitations of the LSS are at least twofold. They consist in the necessity of working at given low energies for the production of metastables, which limits the current of H^+/D^+ ions that can be injected into the Cs charge-exchange region due to space charge.

Another limiting factor is the relativistic quenching by the electric field $\vec{E} = \gamma \vec{v} \times \vec{B}$ generated by the fast motion of the metastables in magnetic fields such as the SONA fields.

16.8 Spin Rotation in Beamlines and Precession in a Wien Filter

Each polarization facility must—in addition to simple beam management—take care of the spin behavior during beam transport. The polarization at the source and/or in the beamline must be prepared such that the absolute value of the polarization as well as its orientation in space can be optimized, the latter freely chosen. The direction of the polarization vector coincides with the direction of the principal axis of the polarization tensor (for deuterons).

16.8.1 Spin Rotation in Beamlines

Normally the beamlines contain deflection magnets and electric deflection fields. The latter do not influence the direction of the polarization in a space-fixed coordinate system, but of course the angle between the polarization vector and the direction of motion of the beam may change. In magnetic fields (such as from analyzing and switching magnets) the spin polarization precesses except when $\vec{p}$ is parallel to $\vec{B}$.

- The precession of a nuclear spin $\vec{I}$ in a magnetic field can be described by the classical relation between the torque $\vec{M}$ and angular momentum $\vec{M} = \vec{\mu} \times \vec{B} = g_I \mu_N \frac{1}{\hbar}(\vec{I} \times \vec{B})$ and $\vec{M} = \frac{d\vec{I}}{dt}$:

$$\frac{d\vec{I}}{dt} = \frac{q}{2m} g_I \cdot (\vec{I} \times \vec{B}) = g_I \cdot \mu_N \frac{1}{\hbar} \cdot (\vec{I} \times \vec{B}). \tag{16.29}$$

For a particle with spin $\vec{I}$, mass m, and charge q (m_p = proton mass)

$$g_I = \frac{m}{m_p} \cdot \frac{e}{q} \cdot g_{\text{Landé}}. \tag{16.30}$$

(The $g_{\text{Landé}}$ factors are fundamental constants of particles and measure the deviation from point-particle behavior, i.e. hint at internal structures of nuclei and nucleons or at QED corrections to the interaction of leptons with external fields). $d\vec{I}/dt$ is oriented perpendicular to $\vec{I}$, i.e. only the direction but not the absolute value of $\vec{I}$ changes. Likewise, $d\vec{I}/dt$ is perpendicular to $\vec{B}$ and the spin precesses around $\vec{B}$. The Larmor precession period T is given by

$$\frac{2\pi}{T} = \left|\frac{d\vec{I}}{dt}\right| \frac{1}{|\vec{I}|\sin\phi} = g_I \cdot \mu_N \frac{1}{\hbar} |\vec{I}||\vec{B}| \sin\phi \frac{1}{|\vec{I}|\sin\phi}. \tag{16.31}$$

The rate of precession is thus independent of the angle between the spin and the magnetic field and occurs with the circular *Larmor frequency*

$$\omega_L = \frac{g_I \mu_N B}{\hbar} \tag{16.32}$$

with the nuclear magneton

$$\mu_N = \frac{e\hbar}{2m_p} = 5.05 \cdot 10^{-27} \text{ J/T}. \tag{16.33}$$

- The Landé g factors for the proton and deuteron are: $g_p = 5.586$ and $g_d = 0.857$.
- The magnetic moments of the proton, the deuteron, and triton are positive. Therefore, for a positive beam (such as on the high-energy side of a tandem Van-de-Graaff accelerator) the sense of spin rotation in these cases is the same as that of the magnetic deflection. For the polarized beams from negative-ion sources the opposite is true, which also has to be taken into account in Wien filters that have been used as spin-rotation devices without beam deflection.

16.8 Spin Rotation in Beamlines and Precession in a Wien Filter

- The spin Larmor precession and the deflection in magnetic fields, which is described by a *cyclotron motion* with $\omega_C = (q/m)B$, are proportional to each other and coupled together. The change of the polarization direction relative to the beam is the difference between the angles of precession and of deflection.
- The preparation and change of the polarization from the source to the target is preferably described in a beam-fixed coordinate system with a y axis vertical in space, a z axis attached to the beam direction (which may change in beam-deflection devices) and the x axis forming a right-handed system with both. An azimuthal angle ϕ of the polarization vector is counted starting from the x axis.
- In a deflection magnet (with B field in y direction) the precession occurs in the x–z plane and the change of the spin polar angle $\Delta\beta$ (measured from the z axis) with fixed azimuthal angle ϕ is (for positive beams):

$$\Delta\beta = \Delta\theta_L - \Delta\theta_C = \left(g\frac{m}{2m_p} - 1\right)\Delta\theta_C, \tag{16.34}$$

i.e. for protons

$$\Delta\beta = 1.793 \cdot \Delta\theta_C, \tag{16.35}$$

for deuterons

$$\Delta\beta = -0.143 \cdot \Delta\theta_C, \tag{16.36}$$

and for negative beams

$$\Delta\beta = -\Delta\theta_L - \Delta\theta_C = \left(-g\frac{m}{2m_p} - 1\right)\Delta\theta_C. \tag{16.37}$$

16.8.2 Spin Rotation in a Wien Filter

For the setting of the polarization at the source a Wien (velocity) filter that is rotatable around the beam axis on the source is especially suited. With the ion beam of velocity $\vec{v}$ in z direction, an electric field $\vec{E}$ in x, and a magnetic field $\vec{B}$ in y direction the filter transmits ions fulfilling the condition

$$v = \frac{E}{B} \tag{16.38}$$

for an ideal reference beam, i.e. one on the central z axis. For an extended beam with finite emittance (i.e. with particles having transverse momentum components) the above condition cannot be fulfilled for all particles simultaneously. This results in some (small) spreading of final spin directions in the beam, i.e. depolarization. This effect can be reduced by having a beam cross-over in the center of the device.

The changes of the spin orientation by deflecting fields and precession in a Wien filter are generally described by three Euler angles. However, the direction of a spin vector is completely determined by two parameters, i.e. in polar coordinates a polar angle β and an azimuthal angle ϕ, see also Sect. 5.4.3. Thus, one Euler angle is redundant (i.e. can only enter as a phase factor) and the other two are uniquely connected with β and ϕ. The description of the quantum-mechanical rotation of spinors via action of rotation matrices on spin operators (such as represented by Pauli matrices) is entirely equivalent to a classical 3×3 rotation matrix acting on angular momentum (spin) vectors.

If we describe the orientation of a spin unit vector in the beam-fixed coordinate system x, y, and z defined above, by polar coordinates

$$\hat{S} = \begin{pmatrix} \sin\beta\cos\phi \\ \sin\beta\sin\phi \\ \cos\beta \end{pmatrix} \quad (16.39)$$

then after a general rotation by polar and azimuthal angles α and ψ we have

$$\begin{pmatrix} \hat{S}_{x'} \\ \hat{S}_{y'} \\ \hat{S}_{z'} \end{pmatrix} = \begin{pmatrix} \sin\beta'\cos\phi' \\ \sin\beta'\sin\phi' \\ \cos\beta' \end{pmatrix}$$

$$= \begin{pmatrix} \cos\alpha\cos\psi & \cos\alpha\sin\psi & \sin\alpha \\ -\sin\psi & \cos\psi & 0 \\ \sin\alpha\cos\psi & \sin\alpha\sin\psi & \cos\alpha \end{pmatrix} \cdot \begin{pmatrix} \sin\beta\cos\phi \\ \sin\beta\sin\phi \\ \cos\beta \end{pmatrix}. \quad (16.40)$$

For an accelerator system where all rotations occur around the y axis, i.e. where $\psi = 0$, the rotation matrix simplifies to

$$\begin{pmatrix} \cos\alpha & 0 & \sin\alpha \\ 0 & 1 & 0 \\ -\sin\alpha & 0 & \cos\alpha \end{pmatrix}. \quad (16.41)$$

By applying the inverse of this matrix

$$\begin{pmatrix} \cos\alpha & 0 & -\sin\alpha \\ 0 & 1 & 0 \\ \sin\alpha & 0 & \cos\alpha \end{pmatrix} \quad (16.42)$$

to a desired arbitrary spin orientation at the target the necessary setting of the Wien filter magnetic field and azimuthal orientation can be calculated. Under the action of a vertical field B of length L along z the spin (polarization) vector of the particles precesses in the x–z plane by a polar angle proportional to $|B|$ and inversely proportional to v

$$\beta_{\text{prec}} = \frac{g_I \mu_N}{\hbar} \frac{(B \cdot L)_{\text{eff}}}{v}. \quad (16.43)$$

In the general case the azimuthal orientation of the field $\vec{B}$ determines the azimuthal angle of the polarization around the z axis. The Wien filter $\vec{E}$ field strength

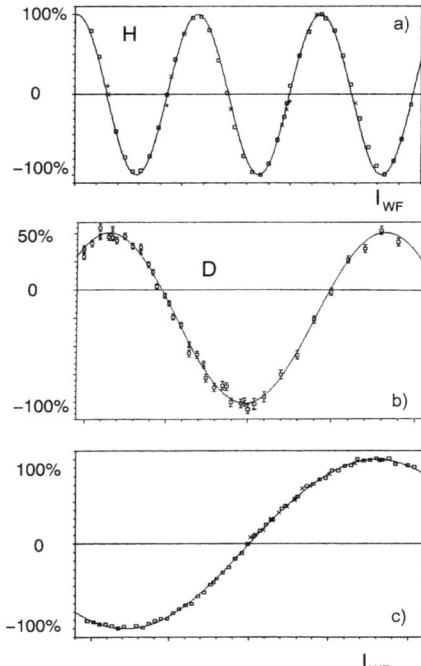

Fig. 16.45 Precession curves of the polarization (values were normalized to 100 %) as functions of the current of the Wien-filter magnetic field (the corresponding electric field is then fixed for a given particle velocity and is found by maximum beam transmission). (**a**) Vector polarization of protons, (**b**) Tensor polarization of deuterons, (**c**) Vector polarization of deuterons. Measurements were performed with the Cologne LSS at the FN tandem VdG accelerator

is adjusted for maximum transmission of the polarized beam. At the same time the device acts as a velocity filter (and mass filter for ions of the same energy). Any desired spin orientation at the target position may be achieved.

Figure 16.45 shows precession curves of the vector polarization of protons and deuterons as well as of the tensor polarization of the deuterons. The polarizations were measured with polarimeters using the reactions $^3\text{He}(d, p)^4\text{He}$, $^4\text{He}(p, p)^4\text{He}$, and $^4\text{He}(d, d)^4\text{He}$ after the acceleration by a tandem Van-de-Graaff accelerator and display the variation of the polar angle β of the polarization.

16.9 Exercises

16.1. In an old-fashioned tandem Van-de-Graaff accelerator the terminal electrode is a cylinder with a diameter of 0.8 m that is charged by the electric charges transported to its inner surface by a rubber-like belt. The accelerating structure is concentrically surrounded by a cylindrical pressure vessel of 3.5 m diameter filled with an insulating gas such as SF_6, CO_2, N_2 or a mixture of them, at several bar pressure. The breakthrough electric field strength is $E_{\max} = 1.7 \cdot 10^7$ V/m.

(a) How high is the breakthrough terminal voltage?

(b) What is the current that the belt (width: 60 cm) can transport if it is running with a velocity of 55 km/h and has the same surface breakthrough field strength of $E_{\max} = 1.7 \cdot 10^7$ V/m?

16.2. Find out about the average neutron separation energy for medium-heavy nuclei. Above which terminal voltage of a tandem VdG accelerator radiation shielding against neutrons must be in effect? If you had to provide shielding around an experimental setup, how would you proceed? How about γ radiation? Why are heavy-ion reactions less dangerous than proton reactions, and why are deuteron-induced reactions especially dangerous?

16.3. Which energy is required to pass the Coulomb threshold of the scattering of the identical particles ^{12}C on ^{12}C?

16.4. Which terminal voltage of a tandem Van-de-Graaff accelerator is necessary if the (in view of high beam intensity) favorable 3$^+$ charge state is to be used?

16.5. Accelerators must be calibrated to deliver beams with precisely known energies. Normally secondary standards are used which rely on some high-precision measurements. An example is using sharp resonances such as the isospin-forbidden state in ^{13}N, see Sect. 3.3.2. In order to check the linearity of the calibration and the "differential hysteresis" (calibration differences between energy changes up or down) other calibration points have to be established.

(a) For a VdG accelerator, threshold reactions may be used (requiring neutron detection and count-rate extrapolation to the thresholds) such as the (p, n) reactions on ^{13}C, ^{19}F, ^{27}Al, and ^{65}Cu. Using mass tables, find the proton lab. threshold energies.

(b) The energy resolution of a beam from a tandem VdG accelerator can be determined by measuring the broadening of a narrow resonance with known width. An example is discussed in Sect. 3.3.2 and Fig. 3.4. Justify the form of the relation between the experimental resonance width $\Gamma_{\exp}$, the accelerator's energy resolution Γ_{Beam}, the possible energy straggling in the target Γ_{Target}, and the true width of the resonance Γ_{Res}

$$\Gamma^2_{\exp} = \Gamma^2_{\text{Beam}} + \Gamma^2_{\text{Target}} + \Gamma^2_{\text{Res}}. \quad (16.44)$$

Calculate the beam resolution by assuming that it is entirely determined by the energy-defining slits behind the 90° analyzing magnet. Assume e.g. a proton beam energy of 14 MeV, a distance from magnet to slit of 3 m, and a slit width of ±0.5 mm.

16.6. A classical cyclotron is to accelerate protons to 45 MeV. How large is the minimum diameter of the homogeneous magnetic field if the maximum field strength is $B = 1$ T?

16.7. A tandem VdG accelerator reaches a maximum terminal voltage of 11 MeV.

(a) What energy (in MeV) can be obtained for a beam of ^{32}S in the 4$^+$ charge state?

(b) In order to deflect such a beam in a 90° analyzing magnet (which is used to determine and stabilize the beam energy via a pair of analyzer slits connected to a differential amplifier circuit) a homogeneous magnetic field B acting over a path length L is required. How high is B if $L = 1.8$ m?

(c) A Wien filter (also: velocity filter) consists of an electrical field E_x crossing a magnetic field B_y under 90°, both perpendicular to the direction z of a passing charged-particle beam. What would be the value of B for the above sulfur beam, if E has the (realistic) value of 2 kV/m?

(d) Can the Wien filter be used to separate masses?

16.8. Calculate the deflection of a ^{1}H atom of thermal (room temperature) velocity in a magnetic dipole field with the constant gradient of $dB/dz = 1$ T/cm over a length of 20 cm at a distance from the magnet exit of $d = 50$ cm

(a) due to the electron's magnetic moment?
(b) due to the nuclear magnetic moment?
(c) For both cases: If we assume a Maxwell-Boltzmann velocity distribution in the beam (this is not quite realistic!), how do the separations achieved relate to the beam-spot widths (FWHM) in view of creating highly polarized partial beams? What, if the beam "temperature" were 4 K?

References

[ABR58]   A. Abragam, J. Winter, Phys. Rev. Lett. **1**, 375 (1958)
[ALV46]   L.W. Alvarez, Phys. Rev. **70**, 799 (1946)
[BRE40]   G. Breit, E. Teller, Astrophys. J. **91**, 215 (1940)
[BRU70]   H. Brückmann, D. Finken, L. Friedrich, Nucl. Instrum. Methods **87**, 155 (1970) and in [MAD71], p. 823 (1971)
[CHA13]   A.W. Chao, K.H. Mess, M. Tigner, F. Zimmermann (eds.), *Handbook of Accelerator Physics and Engineering* (World Scientific, Singapore, 2013)
[CLA56]   G. Clausnitzer, R. Fleischmann, H. Schopper, Z. Phys. **144**, 336 (1956)
[CLA59]   G. Clausnitzer, Z. Phys. **153**, 600 (1959)
[CHR50]   N.C. Christofilos, U.S. Patent No. 2,736,766 (1950)
[COC32]   J.D. Cockroft, E.T.S. Walton, Proc. R. Soc. Lond. A **136**, 619 (1932)
[COW32]   J.D. Cockroft, E.T.S. Walton, Proc. R. Soc. Lond. A **137**, 229 (1932)
[COU52]   E. Courant, M.S. Livingston, H. Snyder, Phys. Rev. **88**, 1190 (1952)
[DON65]   B.L. Donnally, W. Sawyer, Phys. Rev. Lett. **15**, 439 (1965)
[FIT58]   W.L. Fite, R.T. Brackmann, Phys. Rev. **112**, 1141 (1958)
[GER22]   W. Gerlach, O. Stern, Z. Phys. **9**, 353 (1922)
[GIE86]   Ann. Rep., Strahlenzentrum Univ. Giessen, p. 51 (1981)
[GLA68]   H. Glavish, Nucl. Instrum. Methods **65**, 1 (1968)
[HAE68]   W. Haeberli, Nucl. Instrum. Methods **62**, 355 (1968)
[HAE82]   W. Haeberli, M.D. Barker, C.A. Gossett, D.G. Mavis, P.A. Quin, J. Sowinski, T. Wise, Nucl. Instrum. Methods **196**, 319 (1982)
[HAL80]   K. Halbach, Nucl. Instrum. Methods **169**, 1 (1980)
[HGS12]   H. Paetz gen. Schieck, *Nuclear Physics with Polarized Particles*. Lecture Notes in Physics, vol. 842 (Springer, Heidelberg, 2012)
[HIN97]   F. Hinterberger, *Physik der Teilchenbeschleuniger und Ionenoptik* (Springer, Heidelberg, 1997)

[ISI24] G. Ising, Ark. Mat. Fys. **18**(30), 1–3 (1924)
[KIB68] J.L. McKibben, G.P. Lawrence, G.G. Ohlsen, Phys. Rev. Lett. **20**, 1180 (1968)
[KIE66] L.J. Kieffer, G.H. Dunn, Rev. Mod. Phys. **38**, 1 (1966)
[KNU70] L.D. Knutson, Phys. Rev. A **2**, 1878 (1970)
[LAM50] W.E. Lamb Jr., R.C. Retherford, Phys. Rev. **79**, 549 (1950)
[LAT12] M. Lattuada, Nucl. Phys. News **22**, 5 (2012)
[LAW31] E.O. Lawrence, D. Sloan, Phys. Rev. **38**, 2022 (1931)
[LAW32] E.O. Lawrence, M.S. Livingston, Phys. Rev. **40**, 19 (1932)
[LEE11] S.Y. Lee, *Accelerator Physics*, 3rd edn. (World Scientific, Singapore, 2011)
[MAD71] H.H. Barschall, W. Haeberli (eds.), *Proc. 3rd Int. Symp. on Polarization Phenomena in Nucl. Reactions, Madison, 1970* (University of Wisconsin Press, Madison, 1971)
[MIK13] A. Mikirtychiants et al., Nucl. Instrum. Methods Phys. Res. A **721**, 83 (2013)
[OHL67] G.G. Ohlsen, J.L. McKibben, Los Alamos Scientific Lab. Rep. LA-3725 (1967)
[PRA74] P. Pradel, F. Roussel, A.S. Schlachter, G. Spiess, A. Valance, Phys. Rev. A **10**, 797 (1974)
[ROU77] F. Roussel, Phys. Rev. A **16**, 1854 (1977)
[RUD60] H. Rudin, H.R. Striebel, E. Baumgartner, L. Brown, P. Huber, Helv. Phys. Acta **34**, 58 (1961)
[SON67] P.G. Sona, Energ. Nucl. **14**, 295 (1967)
[VAS00] A. Vasiliev, H. Paetz gen. Schieck et al., Rev. Sci. Instrum. **71**, 3331 (2000)
[VDG31] R.J. Van de Graaff, Phys. Rev. **38**, 1919A (1931)
[WID28] R. Wideröe, Arch. Elektrotech. **21**, 387 (1928)
[WIE93] H. Wiedemann, *Particle Accelerator Physics* (Springer, Heidelberg, 1993)
[WIL01] E. Wilson, *An Introd. to Particle Accelerators* (Oxford University Press, New York, 2001)
[WIS06] T. Wise et al., Nucl. Instrum. Methods A **556**, 1 (2006)
[ZEL05] A. Zelenski et al., Nucl. Instrum. Methods A **538**, 248 (2005)

Chapter 17
Detectors, Spectrometers, and Electronics

Nuclear radiation cannot be felt nor seen. Already W.C. Röntgen detected X-rays by noticing effects on photographic material wrapped densely in black paper. Rutherford relied on seeing tiny flashes of light when α's impinged on a screen of ZnS, which was the beginning of scintillation detectors using phototubes with secondary-electron multipliers. Hess in 1912 detected cosmic rays by measuring the dependence of ionization in air as function of the height above ground in an electrometer. The nature of cosmic rays was then revealed by tracks in ionization chambers and stacks of photoplates. Finally Geiger and Müller developed the ionization chamber into the *proportional counter* and the *Geiger-Müller counter*. From these "ionization" detectors a number of modern high-energy detectors such as TPC's (*time projection chambers*) emerged. "Visual" detectors such as photoplates or Wilson's *cloud chamber*, and later the *spark chamber* and the famous *bubble chamber* not only led to important discoveries such as the positron or the particles with strangeness, but helped to convince the public of the "reality" of radiation. An enormous push was provided by developments in solid-state physics leading to compact Si detectors mainly for charged particles and efficient large Ge detectors mainly for the detection of γ rays. Due to their nature neutrons require more complicated detection schemes.

These detector developments cannot be discussed here in detail, and therefore only the principles of modern detector types widely used in nuclear physics will be explained here. For more details see the widely accepted books by Siegbahn [SIE68], Knoll [KNO10], and Leo [LEO94].

17.1 Ionization Chambers

Already the decay of charge in an electrometer indicated ionizing radiation. First quantitative measurements were made possible by measuring the weak electric pulse induced in the circuit connected to the electrodes containing the ionizing region. Using a thin wire as one electrode, where the high electric field leads to gas multiplication of charge carriers, thus forming an avalanche, resulted in much increased

Fig. 17.1 Typical scintillation detector setup. Voltages up to a few kV are applied to the cathode (sometimes to the anode) and output pulses in the volt range may be registered

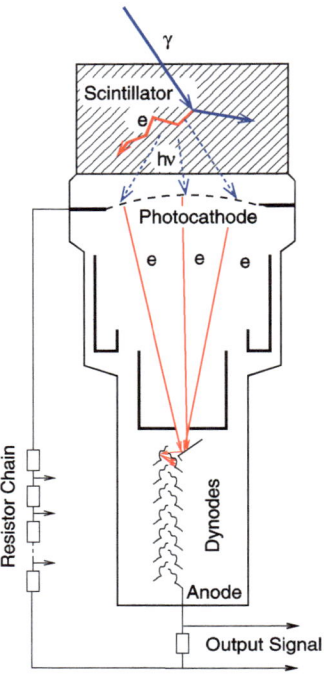

sensitivity of the counting tube. This device can be used basically in two modes depending on the applied voltage: At lower voltages the magnitude of the electric pulse is proportional to the energy deposited by the ionizing particle (*Proportional counter*). At higher voltages a discharge develops that has to be stopped either electronically or by adding a *quenching gas* such as CO_2. This mode (*Geiger-Müller counter*) registers "events" with high sensitivity and is used in radiation monitors. These detection principles have been developed into a number of modern detectors, especially in medium and high-energy applications, and where position sensitivity over wide areas is required (e.g. the multi-wire proportional chamber *MWPC*). A typical application is that as a detector in the focal plane of a magnetic spectrograph in low- and medium-energy nuclear physics, see Sect. 17.3.3.

17.2 Scintillation Detectors

Figure 17.1 depicts a typical scintillation detector setup consisting of a scintillator (inorganic or organic solid, liquid, or gas) suitable for different applications, a photocathode producing electrons from scintillator light, a photomultiplier tube consisting of a chain of electrodes, on which secondary-electron multiplication occurs, and a network to apply a high voltage to the cathode and at the same time to collect the output electron pulse for each event.

17.3 Solid-State Detectors

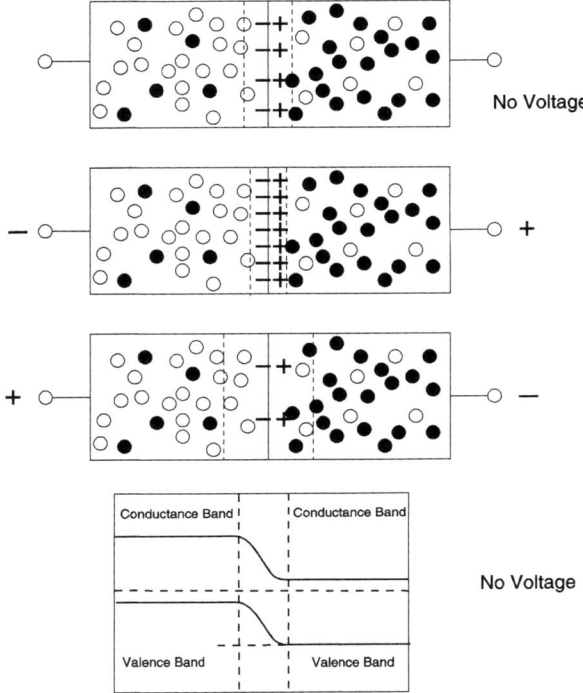

Fig. 17.2 Basic principle of sold-state detectors. By joining a p-type and an n-type semiconductor material a depletion layer is formed by charge-carrier diffusion (*upper frame*). By applying a voltage across the junction this layer becomes smaller or wider depending on polarity (*second* and *third frames*). In one case a conduction current flows, in the other there is only a very small reverse current but basically an insulating depletion layer is formed, in which charges from an ionizing nuclear particle are collected by the electric field existing across the layer. The *lowest frame* shows the situation without external voltage in the band-structure model of the two semiconductors. Applied voltages would either move the two bands (the conduction and the valence band) closer together or away from each other. Some doping reduces the gap that has to be "crossed" by the energy of the ionizing particles

17.3 Solid-State Detectors

For charged as well as for γ particles the development of solid-state technologies has brought enormous progress. It is mainly the compact size, but also the excellent energy resolution of these detectors, which led to their universal use in low-energy nuclear physics. Figure 17.2 illustrates the basic principle of such devices, i.e. that of a reversed-bias p-n junction, in which the electric field across a depletion layer with its very low electric conductivity allows the quantitative collection of electron-hole pairs created by the ionizing particles. The ionization energy, i.e. the energy to lift an electron from the valence into the conduction band creating an electron-hole pair is 3.66 eV for Si and 2.5 eV for Ge. Assuming Poisson counting statistics this explains the high resolution possible with semiconductors as compared to

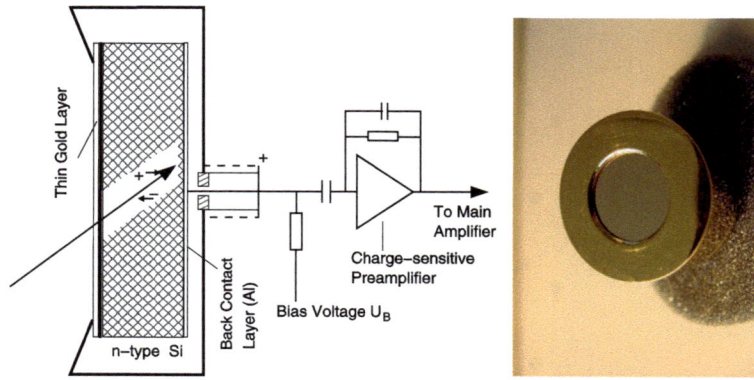

Fig. 17.3 Scheme and view of a typical surface-barrier detector with the front gold layer electrically connected with the housing as positive electrode

ionization-chamber type gas detectors where ΔE is $\approx$30 eV. The very different physical processes, by which γ radiation and charged particles interact with matter has led to very different designs of semiconductor detectors for both cases. It has, however, become customary to use Si detectors mainly for charged-particle spectroscopy (sometimes also for high-resolution X-ray detection) and Ge detectors for γ spectroscopy. Si technology in intermediate and high-energy applications is now often used for start detectors/triggers in large detector arrays with track reconstruction etc.

17.3.1 Si Detectors

The most common detector type for charged particles is the surface-barrier Si detector. The doped Si wafer is covered with a thin metal (gold) layer and a very thin oxide layer in between, thus forming a *Schottky p–n* diode. The depletion layer is therefore very asymmetric such that over the entire area of the wafer (a few cm^2) the particles can enter the depletion zone. If the particles are stopped in this depletion zone the total charge collected is proportional to the energy deposited, and the detectors are 100 % efficient: The number of output pulses is equal to the number of incident particles, which is important for absolute cross section measurements. Therefore for an energy measurement the incident particles have to be stopped within the depletion zone. The depth of the depletion zone goes as $\sqrt{U_B}$ with U_B the bias voltage. Typical maximum depths are $\approx$2 mm, in which protons of 20 MeV will be stopped. Figure 17.3 shows a typical surface-barrier detector of *pn*-type Si and a thin gold layer forming the asymmetric depletion layer. The theoretical resolution of Si detectors e.g. for protons is a few keV. In practical situations such as in an accelerator experiment in a scattering chamber with modest cooling of the detectors, with the usual amplifier chain and with thermal and other noise factors etc., 15 to 20 keV

17.3 Solid-State Detectors

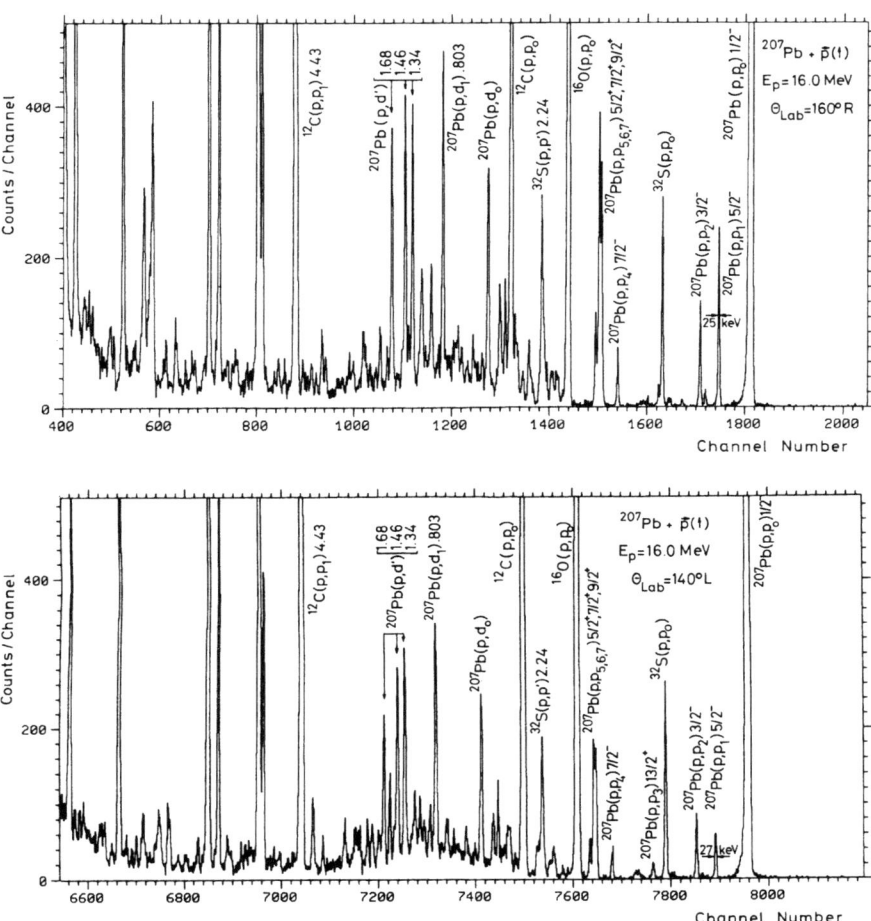

Fig. 17.4 Spectra from scattering (polarized) protons of 16 MeV at 160° and 140° from a thin ^{207}Pb target obtained with two standard surface-barrier detectors. A resolution of about 25 keV was obtained. The very low background allowed the measurement of the weak transitions to the first and second excited states of ^{207}Pb corresponding to shell-model hole states of ^{207}Pb [LAT79]

can be obtained. An example is shown in Fig. 17.4. For higher energies stacks of such detectors may be used or *Li drifted Si detectors*. A combination of a thin first detector, in which a small energy amount ΔE is deposited and a second stopping detector forming a *detector telescope* is used for particle identification according to the Bethe-Bloch formula for the differential energy loss of charged ions of charge z in matter of charge number Z and their velocity v

$$-\frac{dE}{dx} = e^4 \frac{z^2 Z N_A \rho}{m_e v^2 A u} \left[\ln \frac{2 m_e v^2}{I} - \ln(1 - \beta^2) - \beta^2 \right], \qquad (17.1)$$

where $\beta = v/c$, $I \approx (10 \text{ eV}) \cdot Z$ the average ionization potential of the stopping medium, for which extensive tabulations exist, N_A = Avogadro's number, ρ = the mass density of the material.

17.3.2 Particle Identification

For unambiguous identification of particles, especially in complex reactions with many exit channels, as is often the case for heavy-ion reactions, it is necessary (and sufficient) to know the mass and the charge number Z. Thus, the product of ΔE and v^2 is $\propto M_p z^2$, the particle's mass and squared charge, which can be used to identify it in complicated spectra and to improve the separation of particle species in high-sensitivity experiments such as AMS spectroscopy, see Sect. 20.1. Modern methods—in low-, medium-, and especially in high-energy nuclear or particle physics make use of event-by-event analysis of events. At least two different types of signals are required: an energy signal E, a ΔE signal, in addition a third information such as a time-of-flight signal help disentangle complex output. Magnetic deflection can be an additional piece of information on the momentum as well as the charge sign of the particles (e.g. when antiparticles are involved). Already a three-particle breakup reaction (see Sect. 9.2.6) in low-energy nuclear physics will profit from such information. Figure 17.5 gives an account of how for a complicated breakup reaction event-by-event processing, setting windows and cuts in energy and time-of-flight spectra can single out one reaction channel and remove background. In the relatively new field of exploring "exotic" nuclei and measuring their properties mainly by γ spectroscopy after fusion-evaporation reactions the high number of exit channels requires efficient means of identifying the final product nuclei by mass, momentum, and charge. Besides ΔE and energy/momentum the time-of-flight can be measured, e.g. by using pulsed incident beams and timing-transmission detectors to trigger and cut event by event in appropriate spectra. An impressive example is shown in the following Fig. 17.6. In intermediate- and high-energy physics an almost universal ΔE detector is the *Time Projection Chamber, TPC*, allowing not only particle identification via dE/dx, but also complete 3D tracking of many output events (multiplicities can be many thousands at the LHC). Si detectors have new and attractive applications in large high-energy detector systems such as the LHC detectors as start detectors with very good spatial and time resolution.

17.3.3 Magnetic Spectrographs

Another way of obtaining high-resolution charged-particle spectra is using a magnetic spectrograph in combination with some position-sensitive detection system. Precise ion-optical calculations have led to designs that fulfill a number of conditions: besides good energy resolution double focusing (i.e. focusing of particles

17.3 Solid-State Detectors

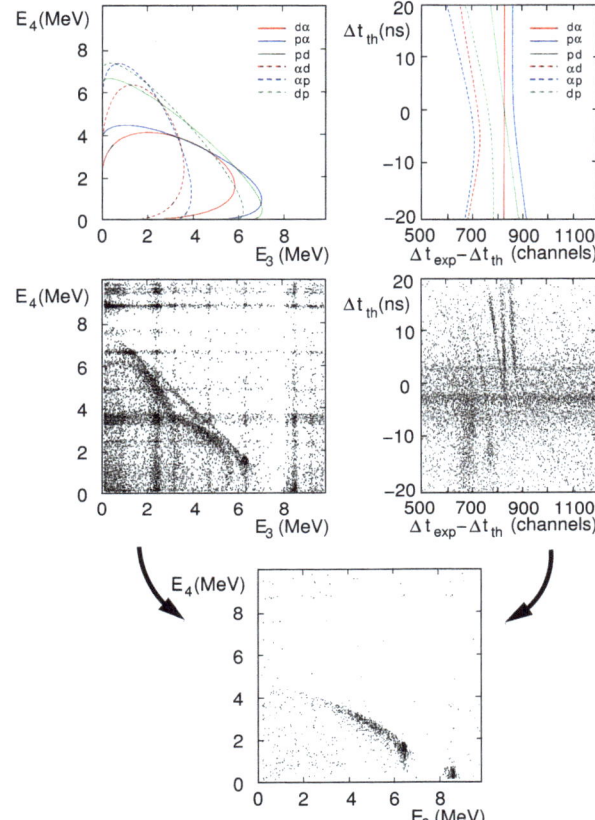

Fig. 17.5 *Left*: Raw E3/E4 coincidence spectrum of several reaction channels of $^6\text{Li} + p \rightarrow d\alpha, p\alpha, pd, \alpha d, \alpha p$, and dp coincidences. The six possible, partly overlapping, kinematical curves (theoretical curves *on top*) are almost invisible. Successive background reduction steps are: a cut on the time peak in the timing output signal of the electronics removing accidental coincidences, a cut around the kinematical curve of interest in the energy spectrum, and a cut on the time-of-flight difference (theoretical *on top*) intensity distribution of interest, leading to the pure $p\alpha$ intensity distribution (*bottom*) [NIE86]

emitted under different angles) to ensure constant efficiency across the entire spectrum necessary for absolute cross section measurements. The most-used design has been the *Q3D* design consisting of one quadrupole and three dipole magnets. For the MeV energies to be measured especially the magnets are heavy machinery, and in order to measure angular distributions of reaction products the entire system has to be easily rotatable around the target, which can be achieved e.g. with air cushions or rail systems. Energy resolutions and absolute energy calibration values in the low-keV range have been obtained. Figure 17.7 shows the Q3D spectrograph at the LMU tandem laboratory in Munich.

17.3.4 Ge Detectors

The interactions of γ rays with (detector) matter are governed by "all-or-nothing" processes, i.e. the photon is either absorbed or scattered (out of a good collimated-geometry setup) or unaffected, for details see Sect. 17.7. Thus, whereas with charged

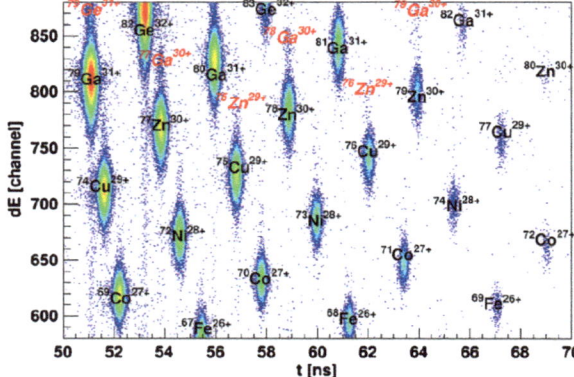

Fig. 17.6 ΔE vs. time-of-flight spectrum of final neutron-rich nuclei. They were produced by a 100 MeV/A ^{86}Kr beam from the NSCL K500/K1200 cyclotrons, impinging on a Be target and producing many different radioactive nuclei. These were selected and identified through a system of a fragment separator, a time-of-flight mass measurement beamline of 58.7 m length, and a magnetic spectrograph [MAT12]

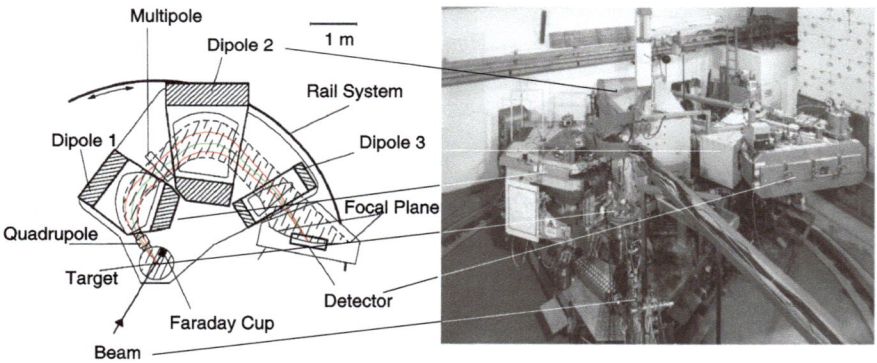

Fig. 17.7 Scheme with magnets and ion trajectories of the LMU Munich Q3D spectrograph (Courtesy R. Hertenberger, LMU Munich)

particles there are successive small energy losses down to zero energy and subsequently a finite range in a detector, for γ's there is an exponential decrease of *intensity* and no finite range but a mean absorption length. The energy is deposited in secondary particles (electrons and positrons), and in order to collect as much energy as possible from the photons the detectors have to be as large as possible (this is true also for γ scintillation detectors).

Historically Ge detectors have been fabricated much in the same way as Si ones: an asymmetric p–n junction with a large-volume doped region and a thin metal front contact. With much improved purities of large-volume Ge material (hyper-pure) Ge (HPGe) crystals that are intrinsically "depleted" without doping or Li drifting

Table 17.1 Neutron energy classification

Neutron Class	Subgroup	Energy Range	Source
Slow	Ultracold (UC)	<0.2 μeV	
	Very Cold	0.2 to 50 μeV	
	Cold	0.05 to 25 meV	Reactor or Spallation Source + (Ultra-)Cold Moderator
	Thermal	≈0.025 eV	Reactor + Moderator
Epithermal		25 meV to 500 keV	
Resonance		1 eV to 100 eV	
Fast		>500 keV	Nuclear Reactions
Very Fast		>5 · 10^2 MeV	Spallation Sources

have become customary. The latest development are Ge detectors that are position-sensitive (*tracking detectors*), providing improved angular resolution. Due to the nature of γ interactions the detectors are never 100 % efficient and they must be calibrated using also sophisticated multiple-scattering Monte-Carlo codes (such as GEANT4).

17.4 Neutrons

Table 17.1 shows the traditional classification of neutron according to their energies, partly following Ref. [NIC05]. The methods of production and detection of neutrons at the different energy regions are quite different and cannot be discussed here in detail. Only a short survey will be given below.

17.4.1 Production of Neutrons

Neutrons (better: neutron beams) can be produced either in nuclear reactors or in nuclear reactions initiated by charged particles, especially also in dedicated accelerator-based spallation neutron sources (SNS). In the first instance there exist special research reactors with high accessibility to the high neutron flux inside or near the core. Examples for such facilities are the Laue-Langevin Institute at Grenoble or the reactor FRM-II at TUM (Munich) with neutron fluxes of the order of 10^{14} s^{-1} cm^{-2}. Neutron energies reach from almost 0 (*ultracold or UC*) to a few MeV. Special methods have to be applied to obtain neutrons with well-defined energies or well-collimated beams of neutrons, or polarized neutrons, in order to perform reactions that are equivalent to mirror reactions with charged particles. Such "parallelism" is necessary to evaluate the effects of the Coulomb force and additional charge-symmetry breaking effects.

Typical neutron-producing reactions are (α, n), (γ, n), (p, n), and (d, n). The (α, n) reaction has been widely used as neutron sources by combining an α emitter such as ^{226}Ra, ^{239}Pu, or ^{226}Po with e.g. ^{9}Be, and initiating the ^{9}Be$(\alpha, n)^{12}$C ($Q = 5.7$ MeV) reaction. By surrounding the source with hydrogeneous material (paraffin, plastic or water) neutrons thermalized to $E_n = 0.025$ eV are obtained. Similarly, with a strong and energetic γ source such as ^{24}Na and ^{9}Be or ^{2}H as targets, near-monochromatic neutrons may be produced.

Accelerator-produced neutrons in the MeV energy range can be obtained from a host of different reactions. Most prominent are the ^{2}H$(d, n)^3$He and the ^{3}H$(d, n)^4$He reactions. Their high cross sections at low energies have made them useful tools in many low-energy installations. The choice of energies by varying the accelerator energies and emission angles together with proper collimation schemes provides (near-)monoenergetic neutron beams with good geometry for nuclear reactions.

Spallation neutron sources based proton beams from 1 GeV LINACS produce neutrons with energies in the hundreds of MeV range that can be used for fundamental research as well as for many industrial and scientific applications e.g. in solid-state physics. The SNS produce very high neutron fluxes. A planned European facility, the ESS, will start in 2019 with 1.5 MW and aims at 6 MW by 2025. The SNS developments are in line with efforts to develop nuclear-waste disposal facilities by combining a spallation neutron source with a subcritical reactor (*ADS transmutation*), see Fig. 19.6.

Detailed accounts of all the methods specific to production and handling of neutron beams are beyond the scope of this text, and the special literature should be consulted. For older monographies see e.g. [BAR58, MAR60, MAR63, BEC64]. Neutrons interact with all four fundamental forces and these interactions are accessible to experimental investigations. The many and increasing uses of neutrons (especially slow neutrons) as research objects complement in many ways the low- and high-energy research into fundamental questions of the standard model and beyond (e.g. details of the weak interaction, of symmetry breakings (parity, time-reversal, etc.)) and of Big Bang Nucleosynthesis. A more recent survey on this subject is given in Ref. [NIC05]. A good source of "neutron methods" in compact form is Ref. [MAR70].

Methods of detection of neutrons are briefly summarized below.

17.5 Neutron Detectors

The detection of neutrons has to rely on indirect methods. In the context of nuclear reactions neutrons play special roles.

17.5 Neutron Detectors

17.5.1 Neutron-Induced Reactions

- Neutrons are not subject to the Coulomb force in the entrance channel. Therefore, their cross sections do not become very small at very low energies.
- Neutron-induced reactions can be considered as the isospin-symmetric counterparts of proton reactions. By studying compound resonances for the two channels states with similar hadronic structures can be observed, their Coulomb-energy differences can be measured and possible effects of isospin breaking can be investigated. Actually the discovery of the neutron in 1932 by Chadwick [CHA32a, CHA32b] immediately cleared the composition of nuclei and the similarity between neutrons and protons caused Heisenberg [HEI32a, HEI32b, HEI32c] to propose the concept of isospin. The neutrons may be "injected" into a target nucleus in two ways: directly using neutrons from a nuclear reaction or a reactor, but also by one-neutron transfer reactions such as (d, p) stripping, see Chap. 10. The excitation energies in the compound or final-nucleus systems are quite different due to the negative neutron-separation energy.
- Neutrons must be measured in all energy regions in order to determine the incident flux in a reaction, necessary for absolute cross sections. Therefore, an entire arsenal of different methods depending on neutron energies is available here.

17.5.2 Neutrons as Reaction Products

Outgoing neutrons from nuclear reactions normally have energies in the keV/MeV range, for which the standard methods consist in using some recoil material combined with scintillators and phototubes. Neutron spectroscopy relies on quite different methods depending on energy.

17.5.3 Different Neutron Detection Methods Depending on Neutron Energies

Table 17.2 lists in compact form the different methods of neutron detection as function of the neutron energies. Like photons neutrons are not "stopped" in material (as charged particles with a definite *range*) but lose their energy by several processes ending up in charged particles that in turn are registered in the usual way. Such processes may be collisions with atoms resulting in recoiling charged nuclei (e.g. protons) that can be detected in a scintillator, or nuclear reactions with charged ejectiles. In any case, the absolute determination of neutron fluxes and thus detector efficiencies, and of neutron energies is difficult and requires careful calibrations taking into account multiple-scattering events by suitable correction methods and codes.

Outgoing neutrons from nuclear reactions normally have energies in the keV/MeV range, for which the standard methods consist in using some recoil material combined with scintillators and phototubes. Historically fast neutrons were registered

Table 17.2 Neutron detection methods

Method	Reaction	Energy Region (eV)	Features
Boron-filled Counters	$^{10}B(n,\alpha)^7Li + 2.78$ MeV		BF_3 gas-filled
Recoil Proton Scintillation	np Scattering Recoils	Fast $<2\cdot 10^{-3}$	Plastic or Liquid
n Activation	Thermal	≈0.025	Reactor + Moderator
Fission Chamber	Epithermal	>0.5	
Semiconductor	np Recoils	Fast	Hydrogeneous Converter

in proportional counters filled with a gas such as $^{10}BF_3$, in which the reaction $^{10}B(n,\alpha)^7Li$ produces charged particles that—after electron multiplication—result in an electric pulse. The very limited sensitive volume and other properties of these detectors made them obsolete as soon as suitable scintillators had been developed that—in conjunction with photomultipliers—provided large sensitive volumes, often as solids and liquids high densities and the possibility to shape the detectors according to experimental requirements (a more recent example are scintillating fibres that combine the functions of scintillator and light guide with position sensitivity). Here only a few prominent types of modern neutron detectors will be discussed.

Plastic Scintillators Plastic scintillators consist of organic material, into which a scintillating material and a wavelength shifter are mixed. The latter is necessary because the maximum quantum efficiency of photomultiplier cathodes is in the visible region of the spectrum. In addition it should have the following properties:

- Optical transparency;
- Machinability and polishable surfaces;
- High quantum efficiency;
- High density;
- Discrimination between neutrons and γ's;
- Fast response for high count-rate capability.

The typical output spectrum of recoil protons in a hydrogeneous material such as polyethylene is a flat continuum resulting from the (almost) isotropic np cross section and scattering under all possible angles. The maximum energy is that of the incident neutrons.

Liquid Scintillators Scintillators filled with organic liquids are very much like solid organic scintillators in an appropriate solvent, often with a wavelength shifter to adapt the light spectrum to the sensitivity function of the multiplier. They can be made quite large and are therefore often used for cost reasons. Most liquid scintillators, sensitive to neutrons, are also sensitive to γ''s, and some kind of *pulse-shape discrimination* has to be applied to separate the two (the rise-time characteristic is different for the two).

Gaseous Scintillators Among the possible gaseous scintillators ^{3}He is most prominent. Its use is based on the reaction ^{3}He$(n, p)^3$H.

Owing to the continuous spectra in most neutron detectors the precise energy determination requires additional provisions. One good method is the measurement of the time-of-flight of the neutrons, e.g. with a start trigger signal from the pulsed accelerator beam and a stop signal from the neutron detector or triggering by an associated γ event. With fast scintillators ns time resolution can be obtained. For slow neutrons mechanical velocity selectors ("Fizeau"-type) have been used.

17.6 Polarized Neutrons

Slow neutrons can be polarized by the interaction of their nuclear spin with the aligned spins of magnetized iron (*Bloch Effect*.) High polarization values have been obtained after transmission of neutrons through a slab of magnetically saturated iron of sufficient thickness. If the cross sections for the interaction of neutron spins with the spins of the magnetized material are different for spin up $\sigma\uparrow$ and spin down $\sigma\downarrow$ the transmissions $T^\uparrow$ and $T^\downarrow$ can be described by

$$T^{\uparrow/\downarrow} = \exp(-\rho\sigma^{\uparrow/\downarrow}L), \tag{17.2}$$

where ρ is the spin density of the material and L is the interaction target thickness. The polarization is

$$P_n = \frac{\exp(-\rho\sigma^\uparrow L) - \exp(-\rho\sigma^\downarrow L)}{\exp(-\rho\sigma^\uparrow L) + \exp(-\rho\sigma^\downarrow L)}. \tag{17.3}$$

A large amount of neutron research is done with slow (from thermal to ultracold) neutrons. For these polarization methods, different from those for protons, have to be used. The small magnetic moment excludes Stern-Gerlach methods of mechanical spin-state separations. But the strong spin dependence of low-energy scattering and absorption cross sections can be used for efficient polarizers.

A very modern method for very-low energy neutrons is to use reflection from magnetized mirrors (*multi-layer super mirrors*), which have different critical angles for total reflection of the neutrons depending on the spin direction. Neutron polarizations >90 % have been achieved at flux densities of $\approx 2 \cdot 10^8$ cm^{-2} s^{-1} mA^{-1} at the "FUNSPIN" Polarized Cold Neutron Facility at PSI, Villigen, Switzerland.

17.6.1 Magnetized Materials

In fully magnetized iron the spins of almost all conduction electrons are aligned along the magnetic field. The two spin states of the neutron interact differently with these spins. Thus the transmission cross sections are very different

$$\sigma\left(n^\uparrow + \text{Fe}\right) \gg \sigma\left(n^\downarrow + \text{Fe}\right). \tag{17.4}$$

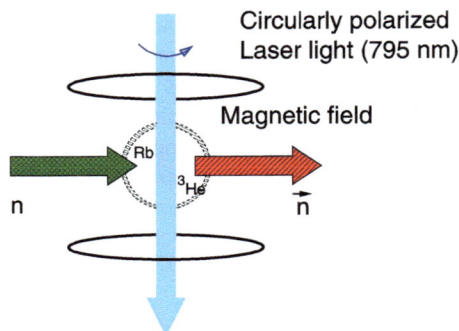

Fig. 17.8 SEOP neutron spin-filter scheme with optical-pumping device as target for unpolarized neutrons resulting in highly polarized outgoing neutrons. The polarization of Laser-pumped Rb atoms is transferred to ^{3}He by gas collisions. The target density may be several bar

The transmitted neutrons are therefore highly polarized, and a similar transmission device can serve as a polarization analyzer.

17.6.2 Polarized ^{3}He as Spin Filter

The neutron absorption by ^{3}He is highly spin-dependent. A beam of unpolarized neutrons (i.e. from a reactor) passing through polarized ^{3}He gas acquires spin polarization, i.e. the reaction acts as a spin filter. The transmission cross section for that half of the neutron beam with spin up (relative to a quantization axis defined by a magnetic field) is almost zero, that for the spin-down neutrons is very high, resulting in very high polarization (see Chap. 5 and [HGS12]). At resonance in ^{4}He near 25 meV the cross section $\sigma^\downarrow$ is many thousand times the cross section $\sigma^\uparrow$. If, in Eq. (17.3), we insert the spin-state cross sections and replace the exponential expression by the tanh function we obtain for the polarization of the outgoing neutrons as function of the ^{3}He polarization

$$p_n = -\tanh(nL\sigma p_{3\text{He}}), \qquad (17.5)$$

where n is the density of ^{3}He atoms and L the gas-cell diameter, typically 5–10 cm.

The methods (MEOP: Optical pumping with metastability exchange; or SEOP: Optical pumping with spin exchange with laser-polarized Rb atoms) of producing highly polarized ^{3}He with acceptable densities have been perfected in recent years. Therefore a glass cell containing optically pumped polarized ^{3}He gas, if used as a neutron target, is a good neutron polarizer. Figure 17.8 shows the setup schematically. Polarization values of $p_{3\text{He}} \approx 0.65$ have been reached. Fast polarized neutrons (at MeV energies) may be produced in nuclear (d,n) reactions such as the fusion reaction ^{2}H(^{2}H,n)^{3}He, provided there is a spin-dependent interaction. Especially efficient for producing polarized neutrons is polarization transfer, i.e. inducing the reaction by beams with high polarization from a polarized-ion source at the required energy and under 0° (see Chap. 5 and [HGS12]).

Special Dual Role of Neutrons The neutron is a useful probe for different interactions (in nuclear reactions as well as in the interaction with solid and liquid

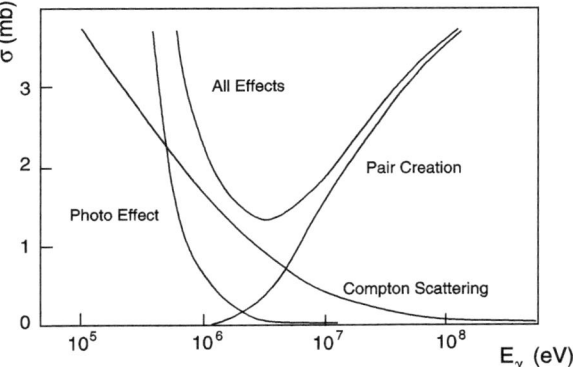

Fig. 17.9 Energy dependence of the cross sections of the three main interactions of γ's with matter

samples), but its structure and decay are interesting objects of fundamental investigations. A few examples:

- The β decay of the neutron is a key observable for the weak interaction. Its decay spectrum is investigated to get a hold on the electron-neutrino mass.
- The search for its electric dipole moment is a key to a possible time-reversal violation.
- The wave nature of the neutron enables the study of interference and diffraction of neutrons.
- Neutron radiography and activation analysis are important methods in art history, archaeology, and national security matters.

17.7 γ Spectroscopy

The three main processes, by which γ's are interacting with matter are

- Photo effect: the photon is absorbed and loses its energy to an atomic electron according to Einstein's equation

$$h\nu = T_{\text{electron}} + W \tag{17.6}$$

where W is the work function of the detector material. The fully stopped electron produces a line in a γ spectrum (full-energy peak).
- Compton Scattering: The photon is scattered from an electron in a characteristic angular distribution described by the Klein-Nishina formula, creating a recoil electron, which produces light (in a scintillator) and/or secondary ion pairs in semiconductor. Like in classical mechanics the photon collides and is scattered into all possible angles, similarly for the recoil electron. Therefore, when weighted with the differential cross section of Compton scattering a continuous spectrum from 0 up to a maximum energy ending below the full-energy peak results. The maximum recoil-electron energy (the *Compton edge*) is given by

Fig. 17.10 Schematic spectrum from one γ of energy >1.022 MeV in a detector of finite size in a γ-reflecting environment (e.g. large Z shielding etc.). The different contributions are indicated. Their relative contributions depend on energy, detector material, detector size and shape, shielding etc.

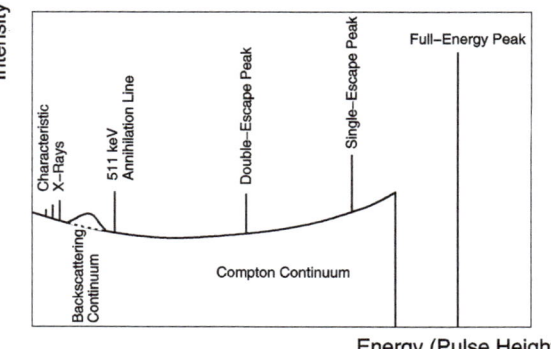

Fig. 17.11 View of MINIBALL in the setup phase as an illustration

$$h\nu_{\text{out}} = \frac{h\nu_{\text{in}}}{1 + \frac{h\nu_{\text{in}}}{m_0 c^2}(1 - \cos\theta)} \quad (17.7)$$

for $\theta = 180°$ where m_0 is the electron's rest mass.

- Pair Formation: Above a minimum energy of 1.22 MeV electron-positron pairs may be created, which again produce light and/or ion pairs. The positron quickly annihilates with an electron and creates a characteristic line in a spectrum at 511 keV. If the positron escapes the finite detector volume its energy is missing in the spectrum and a photopeak with energy E_0—0.511 keV (*escape peak*), when both (the e^+ and e^-) escape, one with E_0—1.22 MeV (*double-escape peak*) are created.

The three processes have very different energy dependences shown in Fig. 17.9. The many different effects following from the interaction of just one γ result in a

17.7 γ Spectroscopy

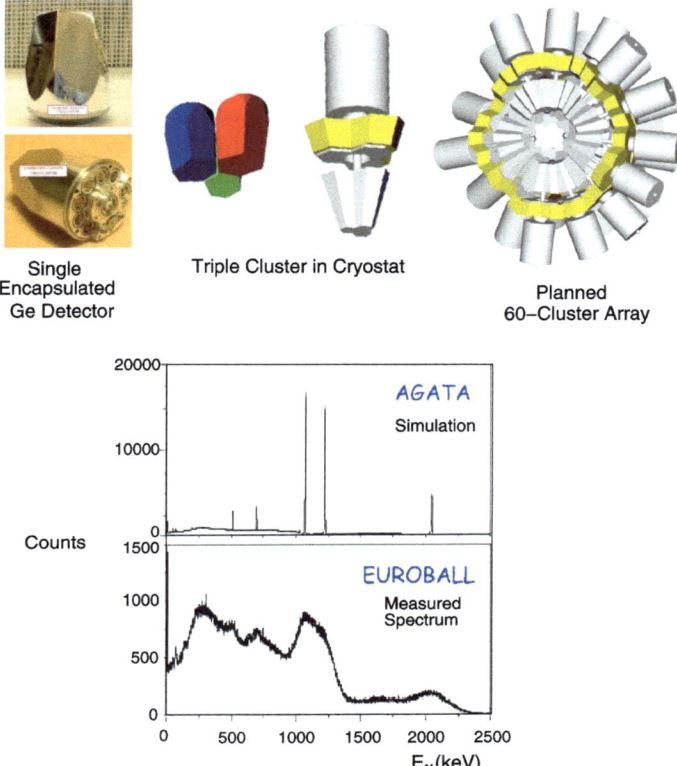

Fig. 17.12 View of different parts of the 180-detector AGATA setup and expected resolution around 1 keV in comparison with that of EUROBALL. The spatial resolution is from 3–7 mm and the efficiency from 20 to 40 %, depending on energy. From the homepage of the AGATA collaboration 2013

complicated spectrum, as shown schematically in Fig. 17.10. It is clear that already a spectrum from two γ's with different energies (e.g. the two energies of a ^{60}Co source at 1.33 MeV and 1.17 MeV) is complicated. Much more so if we have very many γ transitions as is typical for nuclei excited by Coulomb excitation or fusion-evaporation reactions (see Chap. 13). Therefore, efficient methods to suppress especially the Compton continuum have been developed. One method is to surround the Ge detector with another detector in coincidence and adding up all events generated in both systems, which ideally reduces the Compton continuum and enhances the full-energy peak.

The limited efficiency of γ detectors can be counteracted by registering events in the full solid angle of 4π, i.e. by constructing detector "balls" (such balls have names like EUROBALL, MINIBALL, GAMMASPHERE etc. and have been compton suppressed), see e.g. Refs. [EBE08, WAR13]. Figure 17.11 gives a view of MINIBALL, a smaller 4π gamma multi-detector setup during assembly with cluster detectors with γ-ray tracking capability. New techniques allow tracking of γ rays, i.e. a position-sensitive reconstruction of the path together with identification of the nature of each event. This allows the summation of all relevant energies and no special Compton suppression is necessary. Such 4π detectors with up to 180 (planned) detectors on a sphere use HPGe detectors only and will have high (relatively) efficiency together with unprecedented energy resolution on the order of 1 keV, depending on energy. A European collaborative project of this kind is AGATA (for an authoritative reference on the latest developments see [AKK12]) and a similar one in the US is GRETINA/GRETA. Figure 17.12 shows a feature survey of the AGATA project under construction. An AGATA demonstrator with a smaller number of detectors has recently been implemented. For more details, see also Ref. [GRE13].

References

[AKK12] S. Akkon et al. (several 100 authors). Nucl. Instrum. Methods Phys. Res. A **668**, 26 (2012)
[BAR58] H.H. Barschall, Detection of neutrons, in *Handbuch der Physik*, vol. 45, ed. by S. Flügge (Springer, Berlin, 1958), p. 437
[BEC64] K.H. Beckurts, K. Wirtz, *Neutron Physics* (Springer, Berlin, 1964)
[CHA32a] J. Chadwick, Nature **129**, 312 (1932)
[CHA32b] J. Chadwick, Proc. Roy. Soc. A **136**, 692 (1932)
[GRE13] F.C.L. Crespi et al., Nucl. Instrum. Methods Phys. Res. A **705**, 47 (2013)
[EBE08] J. Eberth, J. Simpson, Prog. Part. Nucl. Phys. **60**, 283 (2008)
[HEI32a] W. Heisenberg, Z. Phys. **77**, 1 (1932)
[HEI32b] W. Heisenberg, Z. Phys. **78**, 156 (1932)
[HEI32c] W. Heisenberg, Z. Phys. **80**, 587 (1932)
[HGS12] H. Paetz gen. Schieck, *Nuclear Physics with Polarized Particles*. Lecture Notes in Physics, vol. 842 (Springer, Heidelberg, 2012)
[LAT79] G. Latzel, H. Paetz gen. Schieck, Nucl. Phys. A **323**, 413 (1979)
[LEO94] W.R. Leo, *Techniques for Nuclear and Particle Physics Experiments* (Springer, Heidelberg, 1994)
[KNO10] F. Knoll, *Radiation Detection and Measurement*, 4th edn. (Wiley, New York, 2010)
[MAR60] J.B. Marion, J.J. Fowler (eds.), *Fast Neutron Physics, Part I: Techniques* (Wiley-Interscience, New York, 1960)
[MAR63] J.B. Marion, J.J. Fowler (eds.), *Fast Neutron Physics, Part II: Experiments and Theory* (Wiley-Interscience, New York, 1963)
[MAR70] P. Marmier, E. Sheldon, *Physics of Nuclei and Particles*, vol. II, 811 ff. (Academic Press, New York, 1970)
[MAT12] M. Matoš et al., Nucl. Instrum. Methods Phys. Res. A **696**, 171 (2012)
[NIC05] J.S. Nico, W.M. Snow, Fundamental neutron physics. Annu. Rev. Nucl. Part. Sci. **55**, 27 (2005)
[NIE86] P. Niessen, Diploma Thesis, University of Cologne (1985)
[SIE68] K. Siegbahn (ed.), *α-, β-, γ-Ray Spectroscopy* (North-Holland, Amsterdam, 1968)
[WAR13] N. Warr et al., Eur. Phys. J. A **49**, 40 (2013)

Part III
Applications of Nuclear Reactions and Special Accelerators

Chapter 18
Medical Applications

18.1 Particle (Hadron) Tumor Therapy

The leading principle of radiation therapy is to destroy tumor cells as efficiently as possible while sparing healthy surrounding tissue cells as much as possible. The classical methods of destroying tumor cells (more specifically: the DNA in the cells) besides chemotherapy have been the irradiation by γ rays from intense radioactive sources such as Cs or from bremsstrahlung produced by electron beams typically from betatrons. The interaction with tissue is characterized by the exponential decrease of the intensity with depth due to absorption by the "all-or-nothing" character of the photo, Compton and pair-creation events. Thus, the damage to tissue of the human body is maximal at the skin. Rotation of the patient (or the radiation source) around the tumor improves the ratio of tumor damage to that of surrounding tissue. Nevertheless severe side-effects are common in these "classical" treatments.

Particle tumor therapy is based on interactions of charged particles with the electron shell of the atoms or molecules inside human cells (more specifically with the DNA) by ionization and excitation. Due to the large mass ratio between the ions and the electrons these interactions consist in many small energy losses in the collisions with very small angular deflections that add up stochastically to three effects:

- Finite *average* range of the particles in matter depending on energy.
- For a beam of many particles: A scatter of ranges about this average, i.e. *energy and range straggling* with gaussian behavior.
- Gaussian widening of an initially collimated beam by *angle straggling*.

However, these straggling effects are small as compared to the well defined energy, range, or beam shape. In addition, the energy loss (the "differential ionization") increases with decreasing energy, i.e. during the stopping process and has a maximum near the end of the path (Bragg curve [BRA05]—this behavior is already well described by the semi-classical theory of Bethe and Bloch [BET30, BLO33, FAN63], see many textbooks on nuclear physics). Modern computing codes such as GEANT4 [AGO03, GEA], SRIM [ZIE] and FLUKA [FLU05] take all details of the slowing-down in matter into account and are widely applied. Figure 18.1 shows a typical

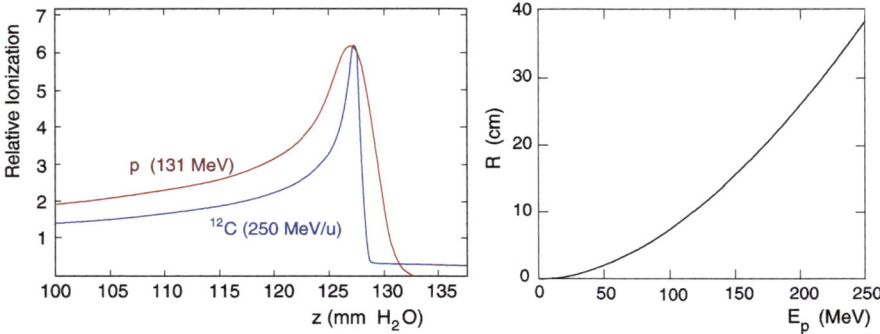

Fig. 18.1 Typical Bragg curve of the medically applied relative dose, in a water phantom, for protons of 131 MeV with a range of about 12.7 cm, compared to the ionization by a beam of ^{12}C ions of 250 MeV/u (*left*). The energies have been adjusted to show the Bragg peaks at the same depth. The depth distribution of the proton peak, which is caused by range straggling $\propto M^{-1/2}$ is wider than that of the ^{12}C peak, which, in turn, shows some remaining ionization behind the Bragg peak. On the *right* the relation of the range of protons in water to the initial proton energy in MeV. The range is often defined as the depth of penetration where the dose behind the Bragg peak decreases to 80 %

Bragg curve for protons and ^{12}C ions and the range-energy relation of protons in water. Water phantoms are generally used to model the human tissue. This makes charged-particle beams of sufficient energy appear ideally suited to treat circumscribed tumors in the depth of the human body where they deposit the maximum of their energy while the surrounding tissue gets much less ionization. The methods of delivering the charged-particle energy have been developed largely in analogy to nuclear physics methods. The beams can be shaped accurately in three dimensions, laterally by ion-optical elements such as magnetic or electrostatic lenses and diaphragms, longitudinally by choosing the appropriate energies. The necessary ranges require medium-energies that can be produced by LINACS, synchrocyclotrons, or synchrotrons. Mainly protons and ^{12}C ions have been used so far successfully. The shapes of their ionization vs. distance (Bragg) curves differ slightly. Proton ranges of tens of cm are obtained with energies of several hundred MeV, similarly for heavy ions, when measured in MeV/nucleon. One problem associated with the stopping of charged particles in tissue may be secondary neutral particles such as neutrons or γ's, for which the advantages of charged particles such as controlled spatial confinement are not equally valid.

So far more than 90,000 patients worldwide have been treated by proton therapy methods, more than 6,500 by ^{12}C ion therapy. The recent developments include methods to spread the beam Bragg region evenly over a given tumor volume that often has an irregular shape. This can be approached by straggling the beam on a scatterer, but a better method is guiding the focused beam with optical elements over the volume with proper adjustment of the energies (raster scan). Gantry constructions, which allow irradiation from all directions without loss of beam quality and with synchronous adaptation of the beam spot to the treatment volume have become

18.2 Isotope Production

Table 18.1 Most common general radionuclides

Radionuclide	Half-life (h)	γ energy (keV)	Imaged organ
^{201}Tl	73.0	80	Heart
^{111}In	67.2	240	Infection
^{67}Ga	78.3	100–300	Abdomen
^{123}I	13.2	160	Thyroid

operational (e.g. at the HIT (Heidelberg Ion Therapy) center), with ^{12}C as well as proton beams, [HEI12], or CNAO, Pavia [ROS11, ROS13]. Rotation of the beam around the tumor together with raster scanning over the 3D volume of the tumor allows maximal efficiency together with minimal damage to surrounding tissue.

18.2 Isotope Production

A large number of different (radioactive) isotopes are used in medical diagnostics and therapies. Examples are shortlived nuclides such as β^+ emitters like ^{11}C, ^{13}N, ^{15}N, and ^{17}F used in PET scanners (PET = positron emission tomography). The increased metabolism of tumors leads to enrichments of molecules containing the isotopes in primary tumors and metastases, which can be spotted inside the body with high accuracy and sensitivity not obtainable by other methods. A number of radio-pharmaceuticals are used for diagnosis and treatment of many different diseases such as heart conditions etc.

- Scintigraphic methods use predominantly ^{99m}Tc ($T_{1/2}$ = 6 h) as a tracer isotope. Table 18.1 lists a few common radio-pharmaceuticals (out of more than 30 medically useful isotopes). For the production of the radionuclides mostly cyclotrons with proton beams of up to 40 MeV and beam currents of up to several hundred µA are in use.
- Positron emission tomography (PET) is the most detailed method to especially spot metastases in the human body that may not be seen by other methods such as nuclear magnetic resonance tomography (MRT) or computer tomography (CT). It is based on the enrichment of certain elements in tumor tissue, emission of two γ's head-on after emission of a positron. The tracer nuclide of suitable half-life has to be inserted into a suitable biomedical compound. More than 500 different PET compounds have been identified, but only a small number is used in practice. Very common is ^{18}F deoxycluose (FDG) which is used in oncology, neurology, and cardiology. The nuclides are mostly produced with (compact) cyclotrons (<20 MeV protons or <10 MeV deuterons with beam currents of ≈100 µA). Typical tracers are listed in Table 18.2.

Additional useful references are [GAN88, TRS08]. It should be mentioned that many dedicated facilities are operating worldwide but that research efforts towards

Table 18.2 Most common PET radionuclides

Radionuclide	Production reaction	Half-life	Method
^{11}C	^{14}N$(p,\alpha)^{11}$C	20.4 m	Cyclotron
^{13}N	^{16}O$(p,\alpha)^{13}$N	10 m	Cyclotron
^{15}O	^{14}N$(d,n)^{15}$O	2 m	Cyclotron
^{18}F	^{18}O$(p,n)^{18}$F	110 m	Cyclotron
^{82}Rb		1.2 m	"^{82}Rb Generator"
	EC decay of ^{82}Sr	25.35 d	Spallation $E_p = 40\text{--}90$ MeV

smaller ("table-top") and automated accelerators for nuclide production, using superconducting magnets or even laser-beam acceleration techniques, are underway in order to provide e.g. more wide-spread PET services or other technical applications.

18.3 Exercises

18.1 Considering the Bethe-Bloch equation (17.1) describing the process of *energy loss* of charged particles in matter, and the *attenuation* of uncharged particles (γ's and neutrons) by "all-or-nothing" interactions in matter: How do the dose-depth distributions differ in the two extreme cases? What are the medical consequences, i.e. to which kinds of cancers are both methods best suited?

18.2 How would you proceed to produce suitable beams of γ's, neutrons, protons, or heavy ions? How can the geometries (in three dimensions) of the different kinds of beams be shaped to optimize the pertinent treatment?

18.3 Beams of π^- have been used for tumor therapies. How can they be produced (types of primary particles, their energies, intensities, types of accelerators)? What happens in the Bragg-peak region, i.e. near the end of their path through tissue (think of their negative charge and of muonic atoms for comparison)? What is the final fate of the pions as hadronic antiparticles and what are the additional medical effects, as compared to protons or heavy ions?

18.4 For tumor diagnostics by positron-emission tomography (PET) β^+ emitters such as ^{18}F (half-life $T_{1/2} = \ln 2/\lambda = 110$ m) are used. It can be produced by a compact cyclotron with a proton beam of typically 10 MeV and currents of up to 100 µA in the ^{18}O$(p,n)^{18}$F reaction (Q-value $= -2.437$ MeV). For one PET diagnosis a quantity of 20 mg is required.

 (a) Confirm the Q-value from the masses involved (use nuclear-mass tables or compilations such as the Nuclear Wallet Card, for references see Sect. 22.1).
 (b) What is the lab. threshold energy of this reaction (with the target at rest)? Discuss the shape of the integrated cross-section excitation function of this reaction as shown in Fig. 18.2 (cf. Sect. 4.1).

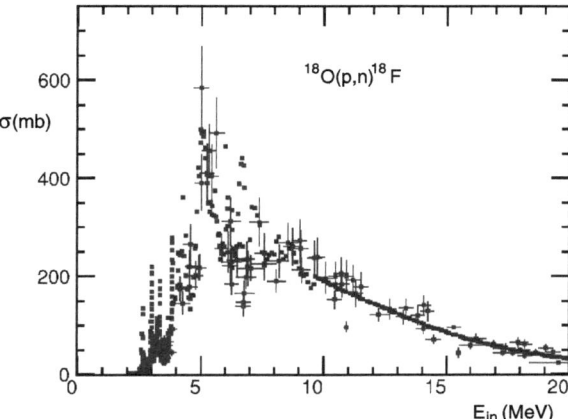

Fig. 18.2 Excitation function of the integrated cross section of the reaction $^{18}O(p,n)^{18}F$ producing the PET tracer isotope ^{18}F. From [NNDC]

(c) The production rate R (in s^{-1}) of the final nuclide, when using a thick target, is described by the formula

$$R \equiv \frac{dn}{dt} = n \cdot \Phi \left(1 - e^{-\lambda t}\right) \int_{E_{\text{in}}}^{E_{\text{fin}}} \frac{\sigma(E)}{dE/dx} dE, \qquad (18.1)$$

with n the number of target nuclei per cm^{-2}, Φ the flux of incident particles per s, $\sigma(E)$ the reaction cross section as function of energy in mb, and dE/dx the stopping power (the differential energy loss) at the energy E in the target material. Discuss the structure of this formula that describes the process of saturation of the production rate. What is the final saturation rate R_0? What are reasonable irradiation times in terms of the half-life?

(d) Try to evaluate R approximately for a proton beam of 10 MeV, a current of $I = 10$ µA, and a thick liquid target of 100 % enriched $H_2^{18}O$ at normal density. Use all information on energy loss and range from tables or polynomial fits to data, e.g. [ZIE85, ZIE] and approximate the cross section suitably. Programs such as GEANT4, FLUKA, and SRIM can perform the task more accurately.

(e) What is the irradiation time for $R = R_o/2$? What is the activity of the entire sample after that time and how many treatments are possible?

18.5 Suppose a superconducting mini-cyclotron with a magnetic field of 5 T could be built. What would be its diameter for 15 MeV protons (deuterons) (see Sect. 16.2.2)?

References

[AGO03] S. Agostinella et al., Nucl. Instrum. Methods Phys. Res. A **506**, 250 (2003)
[BET30] H. Bethe, Ann. Phys. **397**, 325 (1930)

[BLO33] F. Bloch, Z. Phys. A **81**, 363 (1933), Ann. Phys. **408**, 285 (1933)
[BRA05] W. Bragg, Philos. Mag. **10**, 318 (1905)
[FAN63] U. Fano, Annu. Rev. Nucl. Sci. **13**, 1 (1963)
[FLU05] FLUKA, A. Ferrari et al., CERN-2005-10, INFN/TC_05/11, SLAC-R-773 (2005)
[GAN88] D. Gandarias-Cruz, K. Okamoto, Status on the Compilation of Nuclear Data for Medical Radioisotopes Produced by Accelerators, Rpt. INDC (NDS)-209/GZ, IAEA, Vienna (1988)
[GEA] http://geant4.web.cern.ch/geant4/UserDocumentation/UsersGuides/PhysicsReferenceManual/fo/PhysicsReferenceManual.pdf
[HEI12] HIT Center Heidelberg (2012)
[NNDC] Natl. Nucl. Data Center, EANDC, Nucl. Reactions, http://www.nndc.gov
[ROS11] S. Rossi, Eur. Phys. J. Plus **126**, 78 (2011)
[ROS13] S. Rossi, Nucl. Phys. News **23**, 27 (2013)
[TRS08] Cyclotron Produced Radionuclides: Principles and Practice, IAEA, Technical Rep. Series 465, Vienna (2008)
[ZIE] J.F. Ziegler, SRIM, http://www.srim.org/
[ZIE85] J.F. Ziegler, J.P. Biersack, U. Littmark, *The Stopping and Ranges of Ions in Matter*, vol. 1 (Pergamon, New York, 1985)

Chapter 19
Nuclear-Energy Applications

19.1 Fusion-Energy Research

19.1.1 Fusion Basics

Fusion-energy research concerns mainly nuclear reactions in the low-energy regime of up to a few hundred keV and mainly the determination of cross sections. The energy range of interest overlaps largely with that of nuclear astrophysics, see the preceding Chap. 14 whereas the relevant nuclear reactions are limited to few-nucleon fusion reactions i.e. those of the four- and five-nucleon systems. In cases where the goal is to determine the exact reaction mechanism also measurements using polarized particles are important. Very specific questions deal with possibilities using polarized particles to enhance the reaction rates, to suppress certain reaction channels (e.g. outgoing neutrons), and to give ejectiles preferential emission directions, see e.g. [HGS10]. Non-linearities between reaction rates and ignition criteria could make the use of polarized fusion fuels even more attractive.

Besides the cross sections and reaction parameters (see Figs. 14.1 and 14.6) the power densities produced by different reactions in typical reactor situations are important. Figure 19.1 illustrates this and shows the prominent role of the ^{3}H$(d,n)^4$He reaction.

Here a few quantities relevant for fusion (as well as for nuclear astrophysics) at low energies will be recapitulated. These are the basic cross sections, the reaction parameter, which determines the reaction rates in a thermonuclear reactor, and the power density, which depends also on the reaction-particle densities.

19.1.2 Nuclear Cross Sections

The basic cross sections of the relevant nuclear low-energy reactions have been shown in Fig. 14.1. The resonant behavior of the five-nucleon reactions is clearly

Fig. 19.1 Relative power density of relevant fusion reactions as functions of energy

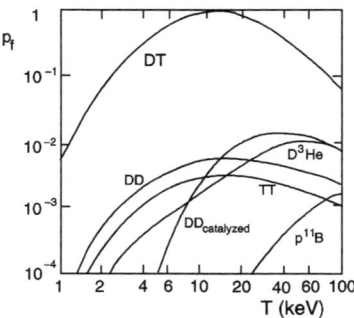

visible whereas other reactions such as the $D + D$ reactions appear non-resonant. It is also evident how the cross section towards lower energies is entirely dominated by the Coulomb penetrability.

In order to separate the influence of the Coulomb penetrability from the nuclear-reaction part it is customary to introduce the astrophysical S-factor $S(E)$, which is defined, using the Sommerfeld parameter with Z_1, Z_2 the charge numbers and μ the reduced mass of the entrance-channel particles as

$$\eta_S = \frac{Z_1 Z_2 e^2}{\hbar v} = Z_1 Z_2 \left(\frac{e^2}{\hbar c}\right)\frac{c}{v} = Z_1 Z_2 \frac{\alpha}{\beta} = \sqrt{\frac{\mu}{2E_{\text{c.m.}}}} \frac{Z_1 Z_2 e^2}{\hbar}. \quad (19.1)$$

For s-waves only and assuming a point-Coulomb interaction of the bare nuclei it is

$$S(E_{\text{c.m.}}) = \sigma_{\text{tot}}(E_{\text{c.m.}}) \cdot E_{\text{c.m.}} \cdot e^{2\pi \eta_S}. \quad (19.2)$$

For purely s-wave, non-resonant, reactions in a limited range of low energies such as the $D + D$ reactions the S-factor is smooth and practically energy-independent, higher partial waves cause an increase, and resonances show the typical excursions. The effects of electron screening at very low energies has been discussed in Sect. 14.2.

The quantities relevant for fusion-energy studies thus are the integrated (or total) cross section σ and, derived from this, the S-factor, the reaction coefficient (or reaction parameter) $\langle \sigma v \rangle$, and the (relative) power density P_f.

19.2 Five-Nucleon Fusion Reactions

The important reactions to be discussed here are:

- $d + {}^3\text{H} \to n + {}^4\text{He} + 17.58$ MeV
- $d + {}^3\text{He} \to p + {}^4\text{He} + 18.34$ MeV

The two mirror reactions have some very pronounced features: At the low energies discussed here both proceed via strong s-wave resonances (at deuteron lab. energies of 107 keV for ^{3}H$(d, n)^4$He, and 430 keV for ^{3}He$(d, p)^4$He, respectively). These resonant states are quite pure $J^\pi = 3/2^+$ states with possibly very little admixture of a $J^\pi = 1/2^+$ s-wave and/or higher-wave contributions.

19.2.1 "Polarized" Fusion

Increasing energy demand in view of limited supply, as well as environmental and nuclear-safety concerns leading to increased emphasis on renewable energy sources such as solar or wind energy are expected to focus public and scientific interest increasingly also on fusion energy. With the decision to build ITER (low-density magnetic confinement, MCF) and also continuing research on (high-density) inertial-confinement fusion (ICF, cf. the inauguration of the laser fusion facility at the Lawrence Livermore National Laboratory) prospects of fusion energy have probably entered a new era. The idea of "polarized fusion", i.e. using spin-polarized particles as nuclear fuel was developed long ago ([KUL82, KUL88], and for more recent developments see [POL99, HGS10, HGS12]). It offers a number of modifications as compared to conventional unpolarized fusion. The main features are:

- Neutron management: replacement or reduction of neutron-producing reactions in favor of charged-particle reactions.
- Handling of the emission direction of reaction products.
- Increase of the reaction rate.

Some of these improvements may lead to lower ignition thresholds and to more economical running conditions of a fusion reactor due to less radiation damage and activation to structures and especially the blanket, necessary to convert the neutron energy to heat, or may lead to concepts of a much simpler and longer-lasting blanket. At the same time its realization will meet additional difficulties, for which solutions have to be studied. Some of these are:

- Preparation of the polarized fuel, either in the form of intense beams of polarized ^{3}H, D, or ^{3}He atoms or as pellets filled with polarized liquid or solid.
- Injection of the polarized fuel.
- Depolarization during injection or during ignition.

As an example of a recent efforts to address some of these questions we cite Refs. [HON07, DID11, DID11a]. The energy range, in which the relevant fusion reactions will take place is <100 keV where the Coulomb barrier strongly suppresses charged-particle cross sections. This is the reason why necessary experimental polarization data with sufficiently high precision such as spin-correlated cross sections have not been measured. Existing reaction analyses and predictions for polarized fusion relied on existing world data sets of other (simpler) data. On the other hand, sufficiently microscopic and therefore realistic theoretical predictions (such as for the three-nucleon system) are just beginning to become available for the four-nucleon systems at the required low energies [DEL10]. An interesting question is whether

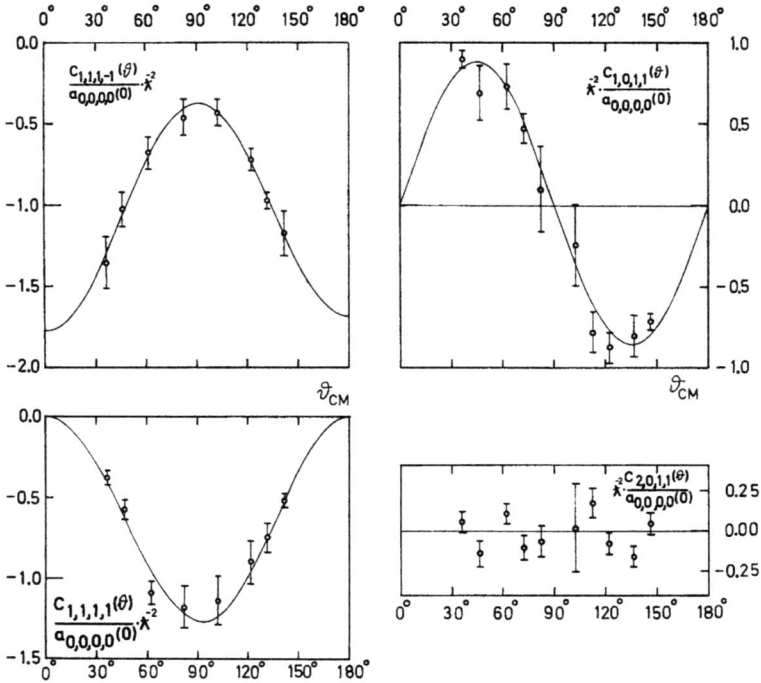

Fig. 19.2 Spin-correlation measurement of the $^3\text{He}(\vec{d}, p)^4\text{He}$ reaction at $E_d(lab) = 430$ keV. This energy corresponds to that of the s-wave $J^\pi = 3/2^+$ resonance. The *lines* are least-squares Legendre fits. By permission of Birkhäuser Verlag, Basel

the recently discussed electron-screening enhancement ([ROL95, HUK08] and references therein) of the very-low energy cross sections has any bearing on polarized fusion.

It should be mentioned here that in the past polarization observables played a decisive role in elucidating the reaction mechanisms of few-body reactions as well as the nuclear structure of few-body nuclei, especially in the two- to six-body systems. At present only four- or five-nucleon systems are considered for fusion energy.

This has been a long-time point of discussion, mainly because of the reactions being very good absolute tensor-polarization analyzers, provided they proceed only through the s-wave, $J^\pi = 3/2^+$ state. Experimental evidence shows that other contributions are small (of the order of a few %). An example of the $^3\text{He}(d, p)^4\text{He}$ reaction on resonance is an early spin-correlation measurement [LEE71a, LEE71b] supporting this assumption, see Fig. 19.2. For a recent discussion of this reaction at low energies see e.g. Refs. [GEI99, BRA04]. The results for the mirror reaction $^3\text{H}(d, n)^4\text{He}$ are similar.

The relatively good knowledge about these two reactions allows the conclusion that with polarized beams and targets an enhancement of the fusion yield close to a factor of 1.5 may be expected. A simple hand-waving statistical argument shows that

19.3 Four-Nucleon Reactions

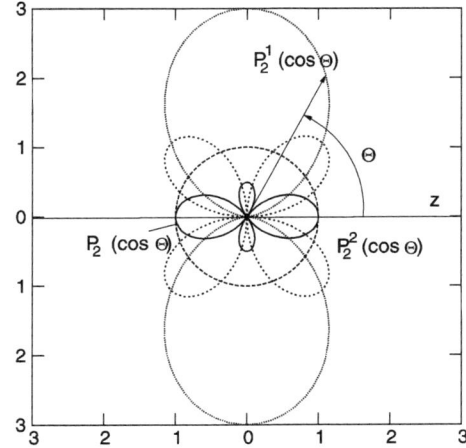

Fig. 19.3 Typical emission anisotropies of neutrons/protons from the s-wave analyzing power T_{20}, as well as the vector-vector and tensor-vector spin correlation cross sections from $C_{10,10}$, $C_{20,11}$, and $C_{11,11}$ along and perpendicular to the z axis of the reactions $^3\mathrm{H}(d,n)^4\mathrm{He}$ or $^3\mathrm{He}(d,p)^4\mathrm{He}$

the reactions, if they go entirely through the $3/2^+$ state and with the entrance channel prepared in a stretched configuration, as compared to the unpolarized entrance channel with a purely statistical spin configuration, yields just this enhancement.

Another interesting feature of polarizing the fuel for the two five-nucleon reactions is the possibility of controlling the emission directions of the neutrons (protons) and also the α particles. The angular distributions for unpolarized nuclei at the resonance energies are isotropic in the c.m. system (with pure s-wave assumed) resulting in a small anisotropy in the laboratory system. The angular distribution of the tensor analyzing power T_{20} follows $\frac{d\sigma}{d\Omega} \propto P_2(\cos\theta)$ and that of the spin-correlation cross section, if all nuclei were vector-polarized along the magnetic field (e.g. of a tokomak reactor) also $C_{10,10} \propto P_2(\cos\theta)$. Figure 19.3 shows the $P_2(\cos\theta)$ anisotropy. Other spin correlations have other anisotropies: $C_{11,11} \propto P_2^2(\cos\theta)$, and $C_{11,20} \propto P_2^1(\cosh\theta)$. Thus neutrons could be diverted away from the walls and alpha particles directed into the plasma enhancing the plasma temperature. It has also been argued [KNI86] that for a tokomak near ignition the ignition conditions change non-linearly with the reactivity thus leading to increases of gains by more than 50 % using polarized fuel.

19.3 Four-Nucleon Reactions

The most important four-nucleon fusion reactions are the $D + D$ reactions, which in a plasma also inevitably accompany the more important five-nucleon reactions discussed above:

- $d + d \to n + {}^3\mathrm{He} + 3.268$ MeV
- $d + d \to p + {}^3\mathrm{H} + 4.033$ MeV.

Whereas the situation of the five-nucleon systems is relatively clear-cut the four-nucleon systems and especially the two DD reactions have a number of problems in their description, especially in view of "polarized fusion". Different from the five-nucleon case the non-resonant reaction mechanism is very complicated (at least 16 complex matrix elements including S, P, and D waves have to be considered with spin-flip transitions from the entrance to the exit channel, which contribute even at low energies). One consequence of participating P waves is that they are the only reactions with appreciable vector- (besides tensor-)analyzing power even down to 20 keV lab. energy, which makes them very useful analyzer reactions at these energies (see also [HGS10, HGS12]). In a semi-classical picture this is made plausible with the large extension of the deuteron wave function and therefore large interaction distance of the two deuterons.

19.3.1 Suppression of Unwanted DD Neutrons

Aneutronic fusion may have a number of advantages (not the least unimportant *economic* ones) over the use of neutron-producing reactions. At an advanced stage the ^{3}H(d, n) reaction could be replaced by the ^{3}He(d, p) reaction. However, DD neutrons would remain. It has been suggested by theoretical approaches that DD neutrons could be reduced substantially by polarizing the deuterons, thus forming a quintet ($S = 2$) state. The main argument was that quintet states in the entrance channel would require spin-flip transitions, which are Pauli-forbidden in first order. However, this argument would be invalid if the reactions proceeded via the D state of the deuteron, and so far the (indirect) experimental evidence does not support this conjecture, see e.g. [HGS10]. For all of these reasons the prediction of suppression or enhancement of the $D + D$ reactions is not possible by considering spin coupling only but requires detailed theoretical and experimental studies. All more recent and more modern studies point to relatively small, if not zero suppression or even some enhancement below 100 keV. A direct spin-correlated cross section measurement is still lacking, but is highly desirable.

Evidence for Suppression? Lacking a direct spin-correlation experiment at very low energies, several indirect approaches have been taken, two of which are:

- Parametrization of world data by a multi-channel R-matrix analysis [HOF87].
- Köln parametrization of world data of the ^{2}H$(d, n)^3$He and ^{2}H$(d, p)^3$H reactions by direct T-matrix analysis below 1.5 MeV including S, P, and D waves (16 complex matrix elements) [LEM90, LEM93, GEI95].

Both approaches allow predictions of any observable of the DD reactions, also of the *quintet suppression factor QSF and similar suppression factors for other spin configurations*, as defined below. Since these analyses the $D + D$ data base has not experienced much improvement by new data. New unpolarized differential cross section data for both reactions [LEO06] and two polarization-transfer measurements for ^{2}H$(d, p)^3$H [KAT01, IMI06] in the energy range discussed here should be cited. However, it is not expected that these additional data would substantially change the predictions summarized in Fig. 19.4.

19.3 Four-Nucleon Reactions

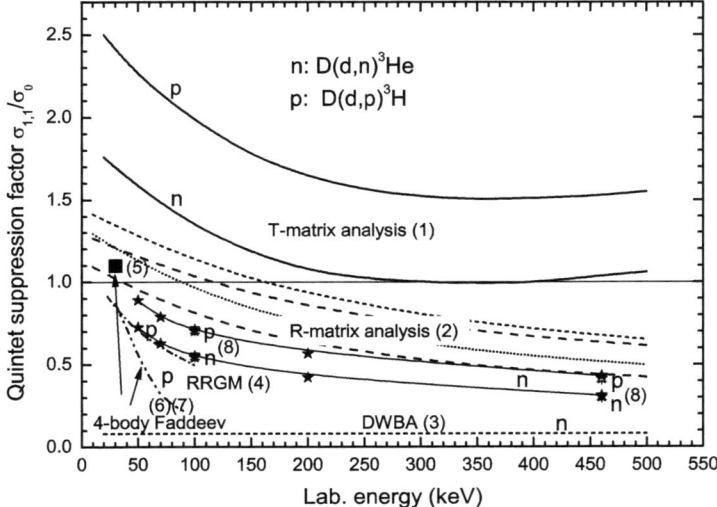

Fig. 19.4 Quintet suppression factor as predicted by various theoretical and from two experimental approaches using world data of DD reactions. The relevant references (numbers in parentheses in the figure) are: (1): [LEM93] (*thick solid lines*), (2): [HAL84, FLE94], (3): [ZHA85, ZHA86], (4): [FIC83, HOF84, HOF86b], (5): [UZU99], (6): [UZU02], (7): [ZHA99], and (8): [DEL07, DEL10] (*stars* and *thin solid lines*). The predictions of Refs. [UZU99, UZU02, DEL07, DEL10] are from microscopic Faddeev-Yakubovsky calculations

Definition of QSF In order to quantify the extent, to which DD neutrons may be suppressed by polarizing the fusion fuel nuclei the "Quintet Suppression Factor (QSF)" is defined as

$$\text{QSF} = \frac{\sigma_{1,1}}{\sigma_0}, \tag{19.3}$$

where

$$\sigma_0 = \frac{1}{9}(\underbrace{2\sigma_{1,1}}_{\text{Quintet}} + \underbrace{4\sigma_{1,0}}_{\text{Triplet}} + \underbrace{\sigma_{0,0} + 2\sigma_{1,-1}}_{\text{Singlet}}) \tag{19.4}$$

is the total (integrated) cross section, to which the four independent channel-spin cross sections $\sigma_{1,1}$ (spin-quintet configuration), $\sigma_{1,0}$ (spin triplet), $\sigma_{0,0}$, and $\sigma_{1,-1}$ (two spin-singlet terms) contribute with their statistical weights.

In Fig. 19.4 all results for the QSF from different theoretical predictions as well as from the two data parametrizations for both DD reactions are shown. The theoretical approaches reach from DWBA calculations to very recent microscopic calculations including the Coulomb force [DEL07, DEL10] and vary widely. However, these latest and most advanced calculations lend confidence to the idea that substantial suppression occurs only in the higher energy range, i.e. above the region of the Gamow peak, at which fusion-energy production will preferably take place.

19.3.2 Possible Reaction-Rate Enhancement for the DD Reactions by Polarization?

Whereas the appreciable enhancement of the reaction rate for the two principal five-nucleon reactions can be considered certain this effect has to be investigated for the two $D + D$ fusion reactions. Although these reactions will not be first choice for a fusion reactor, needing higher temperatures, they should be considered for more developed concepts. They would not need either ^{3}H or ^{3}He, both of which would have to be "bred" artificially in contrast to ^{2}H, which may be extracted from seawater in sufficient quantities. The lower energies of the ejectiles of these reactions (neutrons and protons) may also have advantages or disadvantages compared to those of the ^{3}H and ^{3}He reactions.

Because the $D + D$ reactions inevitably accompany the main fusion reactions a possible rate enhancement (or attenuation) has to be weighed against a possible rate suppression (or enhancement) by polarizing the fuel.

19.4 Other Fusion Reactions

One other interesting option of a neutron-lean reaction should be mentioned, the reaction

$$^{11}B + p \to 3\alpha. \tag{19.5}$$

This reaction is probably not a first choice because it has its maximum fusion cross section near 600 keV.

19.5 Present Status of "Polarized" Fusion

In view of the wide range of theoretical predictions and the lack of direct experimental evidence e.g. for the QSF it seems mandatory to perform a direct $D + D$ spin-correlation experiment in the energy range from 10 to 100 keV. The number of correlation coefficients, however, is quite formidable. The general cross section for the reaction of a spin-1 polarized beam with a polarized spin-1 target contains—besides the unpolarized cross section—analyzing powers of beam and of target in addition to the 32 spin-correlation terms. Parity conservation has been taken into account. This number is reduced to 20 independent correlation terms due to the identity of the incident projectiles, see [HGS10, HGS12]. The simplest correlation experiment is that with both deuterons polarized in the z (the beam) direction

$$[\sigma(\theta,\phi)]_{\Phi=0} = \sigma_0(\theta)\big[1 + C_{z,z}(\theta)p_z q_z + C_{zz,zz}(\theta)p_{zz}q_{zz}\big]. \tag{19.6}$$

Simplifications are achieved by selection of polarization components along single coordinates and the choice of pure vector or tensor polarizations at the source (terms such as $C_{z,zz}$ or $C_{zz,z}$ as well as the analyzing powers $A_z(b)$, $A_x(b)$, $A_z(t)$, and $A_x(t)$ are forbidden under parity conservation).

The main difficulties with spin-correlation measurements at these low energies are:

- The low cross sections;
- The use of solid polarized targets can be excluded because it appears impossible to make them sufficiently thin. Therefore only two interacting polarized beams may be employed resulting in low target densities and small yields;
- The use of (compressed) polarized gas at these low energies meets the difficulties of the need for a container including very thin and, at the same time, strong windows of polarization-conserving materials.

Thus, the only sensible experimental arrangement for measuring spin correlations for the $D + D$ reactions is using an intense atomic beam of polarized deuterons as target that is crossed by another atomic or ion beam of polarized deuterons. Alternatively, one could think of building a low-energy storage-ring device in analogy to COSY-Jülich where multiple target passes would compensate for the low cross sections. However, the technical and financial requirements on such a device seem prohibitive. The high forward multiple-scattering cross section e.g. requires extremely good vacuum. With existing (decommissioned) polarized-ion sources an experiment can be set up with relatively modest efforts such that acceptable count rates result. Figure 19.5 sketches such an experimental setup. Besides and after clearing the nuclear-physics questions concerning the low-energy DD reactions many other problems such as preparation of polarized fusion fuel, its injection into magnetic fields and the conservation of polarization have to be investigated. Similar to beam-beam collisions in high-energy colliders the reaction rate into 4π solid angle is

$$\dot{n}_{\text{out}} = \dot{n}_{\text{in}} \cdot \sigma_{\text{int}} \cdot t = L \cdot \sigma_{\text{int}}, \qquad (19.7)$$

where the luminosity is defined as $L = \dot{n}_{\text{in}} \cdot t$ ($\dot{n}_{\text{in}}$ is the flux of polarized ions from a polarized-ion source, incident on the target region, and t the current of "target" polarized-beam particles from an ABS). It is left to the reader to evaluate the parameters of this experiment in the following exercises. It will be obvious that the experiment is difficult and should be run automatically. A project along these ideas is underway [GAT12].

19.6 New Calculations for Few-Body Systems

A phenomenon studied only rather recently is the enhancement of cross sections of few-nucleon reactions at the very-low-energy range. Although for fusion-energy production this may be a favorable feature, for nuclear astrophysics of the Big-Bang scenario this constitutes a problem because the extrapolation of the S-factor to even lower energy ranges than measured is uncertain as long as there is no reliable theoretical guideline helping to extract the bare nuclear cross sections needed for

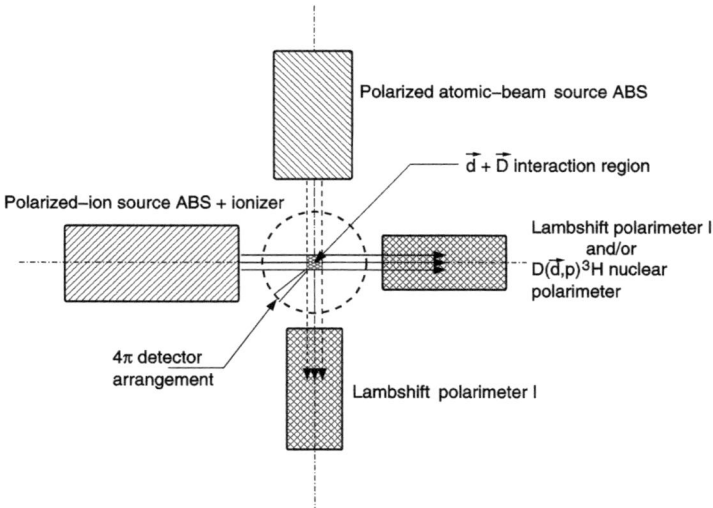

Fig. 19.5 Scheme of a possible spin-correlation experiment with an atomic $\vec{D}$ beam crossing a $\vec{d}$ ion beam of 10 to 100 keV. A granular 4π system of detectors surrounds the interaction region

astrophysics. Therefore, exact predictions of the pure nuclear cross sections from microscopic theories are urgently needed.

19.6.1 Theoretical Approaches

So far—following the progress in theoretical descriptions of the three-nucleon systems—predictions for the four-nucleon systems have been based on resonating-group and Faddeev-Yakubovsky calculations. The high-accuracy nucleon-nucleon interaction data have been used as input. Recent improvements have been achieved by including the Coulomb interaction in a satisfactory way [DEL07, DEL10] and by using the effective-field theoretical approach. One quantity of interest for polarized fusion, the predicted quintet-suppression factor, is depicted in Fig. 19.4 and shows the trend with energy of all modern investigations and only weak quintet-state suppression of the DD neutrons at the relevant low energies. Other predictions or data parametrizations show also weak or no suppression except for the higher lab. energy range above ≈ 100 keV.

The five-nucleon systems, which have been thoroughly investigated experimentally have so far not been treated theoretically by truly microscopic methods. The calculations were either in the framework of resonating-group methods (RGM) or in R-matrix parametrization of experimental data at higher energies, i.e. above the astrophysically interesting energies but encompassing the $J = 3/2^+$ resonance region. Recently *ab initio* many-body calculations using again the nucleon-nucleon interaction but also three-body forces as input have had stunning successes in describing low-lying states of a number of light nuclei [PIE01, EPE09], e.g. for ^{10,11}B and ^{12,13}C.

Two successful methods are the *Green's Function Monte Carlo (GFMC)* and the *No-core shell model (NCSM)* methods. Recently, these have been applied to nuclear (astrophysically relevant) reactions such as ^{3}H$(\alpha, \gamma)^7$Li, ^{3}He$(\alpha, \gamma)^7$Be, and ^{7}Be$(p, \gamma)^8$Be [NA09]. Now a very recent letter [NAV11a] presented the first *ab initio* many-body calculation of the ^{3}H$(d, n)^4$He and ^{3}He$(d, p)^4$He fusion reactions in the framework of the *ab initio* NCSM/RGM approach, see e.g. [NAV11b] and references therein, in an effort to unify the description of the (bound-state) nuclei involved and the (scattering-state) reaction mechanism, starting from the chiral N^{3}LO NN interaction. S-factors in the energy range from almost zero to 2 MeV across the resonance region were calculated and—in view of the approximations used, e.g. the NNLO force—show quite satisfactory agreement with the data, especially in the resonance regions, but for the ^{3}He$(d, p)^4$He reaction the data in the resonance region have larger discrepancies than for the ^{3}H$(d, n)^4$He reaction. Since this is only a first shot much better agreement can be expected in the future. For details including the wealth of all available data and their comparison with the calculations the reader is referred to the original article.

19.7 New Aspects of Polarized Fusion

19.7.1 Effect of Electron Screening

In Fig(s). 14.4 the effect of electronic screening is clearly visible for the reaction ^{3}He$(d, p)^4$He, but not so evident for ^{3}H$(d, n)^4$He. Of course, purely nuclear calculations do not describe this part of the S-factor, and a consistent theory of the screening is still missing. Precise extrapolations of the nuclear S-factor together with precise measurements of the cross sections, which are difficult due to the low energies could help to pin down the screening details, maybe also including possible polarization effects on the screening. The cross section enhancement could be potentially useful for fusion-energy applications although the energy range of the enhancement is below that where fusion reaction yields have their maxima.

19.7.2 Rate Enhancement and Electron Screening

From relatively simple considerations the conclusion was drawn that for the five-nucleon particle reactions ^{3}H$(d, n)^4$He and ^{3}He$(d, p)^4$He an enhancement of the fusion cross sections and reaction rates of up to a factor $f = 1.5$ (for the case of a pure transition through the $J = 3/2^+$ resonance state) can be expected when both incident reaction partners are fully spin-polarized. Below the resonance energy the amount of enhancement is not so clear because other partial-wave states may have a stronger influence on the cross sections, relative to the dominant $3/2^+$ state. In addition, electron screening modifies the cross section, and the effect of the polarization on this modification/enhancement is unclear and should be measured. Figures 14.4 show the experimental low-energy behavior of the two five-nucleon reactions together with recent "ab initio" calculations, see Sect. 14.2. For the $D + D$ reactions strong screening has also been measured, see [RAI02].

19.7.3 Pellet Implosion Dynamics

In the case of inertial fusion using the compression of a fuel pellet by laser or ion beams seems to follow a dynamics, which can additionally increase the gain by polarizing the fuel by an additional factor. According to recent references [TEM12] numerical simulations show that, depending on the value of f and on a number of special conditions of pellet design etc., an additional enhancement factor seems to arise, which—in the case discussed—changed the situation from "no ignition" to "ignition". Conversely, it is shown that the ignition conditions (temperature and density) can be somewhat relaxed when using polarized fuel. Further studies of this effect are important and necessary.

19.7.4 Technical Questions

The provision of either solid or liquid highly polarized pellets for inertial fusion may be easier to achieve than very intense polarized beams to be injected in magnetic-fusion devices. The hopes expressed in Ref. [PHT82] of having "amperes of polarized nuclei at acceptable power cost" have not been borne out. Perhaps polarized molecules could be produced in large quantities. These ideas, however, need thorough and expensive investigations in the future. A recent proposal by [NIK11] is to separate brute-force polarized hydrogen molecules in the ortho spin-state with total spin $I = 1$ by a strong superconducting Stern-Gerlach magnet with the expectation to obtain beam intensities (or densities) higher by an order of magnitude as compared to the atomic-beam intensities from ABS that seem to have reached some saturation. At the proposed temperature of the molecules of $T = 20$ K about 99.8 % of the molecules are in the para state (the para state has $I = 0$) and the ortho molecules have to be "filtered" out from this background. Because the magnetic interaction is with the nuclear magnetic moment only, the deflection is very small, see [FRI33]. The polarization will depend on the degree of selectivity of one of the nine hyperfine states against different background. For the measurement of the polarization of the molecules the Lambshift polarimeter developed recently is an ideal instrument, see [ENG03].

19.7.5 Preservation of Polarization on Injection

The preservation of the fuel polarization either during the transport of the polarized particles to the reactor, during injection and in the plasma collisions has been discussed already quite early. Kulsrud [KUL82] concluded from quite general considerations that depolarization would not be a problem in magnetic-confinement fusion, and More [MOR83] came to similar conclusions for the case of inertial confinement. The authors of Refs. [DID11, DID11a] recently proposed an experiment to investigate this question in a polarized HD molecular target where the idea of observing the γ's from the reaction $^2\text{H} + {}^1\text{H} \rightarrow {}^3\text{He} + \gamma$ was discarded in favor of the $D + D$ reaction neutrons because of their much higher cross section.

19.8 Future of Polarized Fusion

Due to the slow, but visible progress in the physics and technologies of fusion energy, it is appropriate to focus again on the old ideas of using polarized fuel in MCF as well as ICF devices and the advantages this could offer. For the five-nucleon reactions we have quantitative results at hand, but for the $D + D$ reactions the complicated reaction mechanism requires renewed and increased theoretical and experimental efforts to decide on reaction-rate suppression or enhancement, or effects on emission directions. The questions of production of high-density highly polarized beams and targets, of polarization preservation on injection and ignition, and others are largely unanswered and open interesting fields of research in the future.

19.9 Transmutation of Nuclear Waste

Producing energy with nuclear reactors based on fission inevitably produces nuclear waste in the form of medium-mass fission products and trans-uranium nuclei. Many are highly radioactive and long-lived, many also toxic, and must be isolated from the environment for very long times. Even if all nuclear power plants could be shut down immediately the threat will remain with us for a long time. The problem of where and how to store the waste safely has nowhere been solved satisfactorily.

The proposition to "transmute" the nuclear waste into more short-lived and less toxic species using methods developed for nuclear reactions appears attractive. It was mainly promoted by Nobel-prize winner Carlo Rubbia at CERN, cf. [RUB01] who already in 1993 proposed the scheme of a subcritical reactor as energy amplifier. The necessary components for ADS (*Accelerator Driven Systems*) are a medium-energy ($\approx$1 GeV high-current (typically 30 mA) accelerator (LINAC) whose intense proton beam could *spallate* transuranium nuclei into smaller debris, transmute existing fission fragments into shorter-lived species, and at the same time the remaining "spallation neutrons" (a total of about 30 neutrons per proton) could

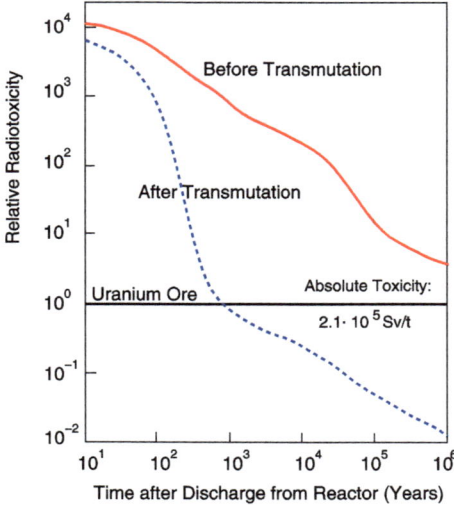

Fig. 19.6 Radiotoxicity of transuranium elements as functions of time with and without transmutation using fast neutrons from a spallation facility combined with a subcritical reactor (ADS system). Also shown is the level of natural uranium ores that is taken as measure for comparison, see e.g. [RUB01]

drive a subcritical breeder reactor. The reactor appears necessary, also for economical reasons, because the spallation neutrons alone would not suffice, but the subcritical ($k = 0.98$) reactor would act as an *anergy amplifier* that could produce energy for the LINAC etc. by fission, entirely controlled by the LINAC beam. Up to 99 % of transuranium isotopes would be removed by fission and up to 95 % of long-lived fission products such as ^{129}I, ^{99}Tc etc. could be "transmuted" by resonant capture of fast neutrons (*TARC process*) into shorter-lived species. The reactor would be of an "inherently safe" design with cooling by convection only. Figure 19.6 shows the predicted effect of transmutation on the radiotoxicity over time of spent nuclear fuel with and without transmutation.

A comprehensive text on ADS is a DOE "White Paper" [ABD10]. Two European projects have been started to build a demonstrator ADS, GUINEVERE [GUI12] and MYRRHA [MYR12] including a fast-neutron reactor at Mol, Belgium. At the same time a high-intensity LINAC is being developed. Even after countries such as Germany decided to phase out nuclear energy production, the enormous reduction of the mass and activities of the remaining highly active waste by transmutation may alleviate the answers to the unsolved questions of final storage of radioactive waste.

19.10 Exercises

19.1 Design a spin-correlation experiment of the aneutronic fusion reaction $^3\vec{\text{He}}(\vec{d}, p)^4\text{He}$ with a strong resonant cross section of $\sigma_{\text{tot}} = 800$ mb at $E_{\text{lab}}(d) = 430$ keV. Assume that by optical pumping a target density corresponding to 1 bar at room temperature of 50 % polarized ^{3}He gas has been achieved and a polarized deuteron beam of 50 µA from an ABS source can be focused into the gas cell.

Table 19.1 Cross sections of the spin-correlation cross section experiment of the $^2\text{H}(\vec{d},n)^3\text{He}$ reaction at three incident energies

E_{in} (keV)	σ_{int} (b)
10	$8.4 \cdot 10^{-6}$
50	$4.4 \cdot 10^{-3}$
100	$1.5 \cdot 10^{-2}$

(a) What is the total proton rate?

(b) What is the rate into a solid-state detector of 1 cm² area at 20 cm distance from the target center (disregard any ϕ dependence of σ)? Is it necessary to take into account an anisotropy of the proton emission?

(c) How long is the measurement time for an absolute precision of 0.05 of $A_y(\theta)(d)$ (target unpolarized), $A_y(^3\text{He})$ (beam unpolarized), and of the spin-correlation coefficient $C_{y,y}(\theta)$ (both polarized)?

19.2 Evaluate the parameters of the $D+d$ spin-correlation experiment of Sect. 19.5. For simplicity assume that the overlap target region is a cube with 1 cm side lengths, see also Fig. 19.5.

(a) Assume a deuteron current from the polarized-ion beam source of 20 µA. What is the equivalent $\dot{n}_{in}$?

(b) The intensity of the D atomic beam is $5 \cdot 10^{16}$ s^{-1} at a temperature of $T = 100$ K. What is the equivalent average beam velocity $\langle v \rangle$?

(c) Calculate L and t.

(d) With the cross-section data of Table 19.1 calculate the reaction rates for three energies of the d beam.

(e) To measure angular distributions of the polarization observables with e.g. 100 detectors each covering $\Delta\Omega = 4\pi/100$ the time $T_{5\%}$ is necessary to reach a statistical uncertainty of N, the number of registered counts, $\Delta N/N$ of 5 %. For polarization observables to reach sufficient accuracy with different polarization setups an additional factor of four is expected. These times are pure experiment running times. Calculate the 5 % error time $T_{5\%}$ in s for one detector and the realistic measurement time $T_{5\%} \cdot 4$ in d.

References

[ABD10] H.A. Abderrahim et al. Rep. FERMILAB-FN-0907-DI, http://lss.fnal.gov/archive/test-fn/0000/fermilab-fn-0907-di.pdf (2010)

[BRA04] D. Braizinha, C.R. Brune, A.M. Eiró, B.M. Fisher, H.J. Karwowski, D.S. Leonard, E.D. Ludwig, F.D. Santos, I.J. Thompson, Phys. Rev. C **69**, 0246008 (2004)

[DEL07] A. Deltuva, A. Fonseca, Phys. Rev. C **76**, 021001(R) (2007)

[DEL10] A. Deltuva, A. Fonseca, Phys. Rev. C **81**, 054002 (2010)

[DID11] J.-P. Didelez, C. Deutsch, J. Phys. Conf. Ser. **295**, 012169 (2011)

[DID11a] J.-P. Didelez, C. Deutsch, Laser Part. Beams **29**, 169 (2011)

[ENG03]   R. Engels, R. Emmerich, J. Ley, G. Tenckhoff, H. Paetz gen. Schieck, Rev. Sci. Instrum. **74**, 345 (2003)
[EPE09]   E. Epelbaum, H.-W. Hammer, U.-G. Meißner, Rev. Mod. Phys. **81**, 1773 (2009)
[FIC83]   D. Fick, H.M. Hofmann, Phys. Rev. Lett. **55**, 1650 (1983)
[FLE94]   K.A. Fletcher, Z. Ayer, T.C. Black, R.K. Das, H.J. Karwowski, E.J. Ludwig, G.M. Hale, Phys. Rev. C **49**, 2305 (1994)
[FRI33]   R. Frisch, O. Stern, Z. Phys. **85**, 4 (1933)
[GAT12]   PolFusion collaboration, Gatchina (2012)
[GEI95]   O. Geiger, S. Lemaître, H. Paetz gen. Schieck, Nucl. Phys. A **586**, 140 (1995)
[GEI99]   W.H. Geist, C.R. Brune, H.J. Karwowski, E.J. Ludwig, K.D. Veal, G.M. Hale, Phys. Rev. C **60**, 054003 (1999)
[GUI12]   GUINEVERE, http://hal.in2p3.fr/in2p3-00722493
[HAL84]   G. Hale, G. Doolen, LA-9971-MS, Los Alamos (1984)
[HGS10]   H. Paetz gen. Schieck, Eur. Phys. J. A **44**, 321 (2010)
[HGS12]   H. Paetz gen. Schieck, *Nucl. Phys. with Polarized Particles*. Lecture Notes in Physics, vol. 842 (Springer, Heidelberg, 2012)
[HOF84]   H.M. Hofmann, D. Fick, Phys. Rev. Lett. **52**, 2038 (1984)
[HOF86b] H.M. Hofmann, D. Fick, Phys. Rev. Lett. **57**, 1410 (1986)
[HOF87]   H.M. Hofmann, in *Proc. of Models and Methods in Few-Body Physics*, ed. by L.S. Ferreira, A.C. Fonseca, L. Streit. Lecture Notes in Physics, vol. 273, Lisboa, Portugal, 1986 (Springer, Berlin, 1987), p. 243
[HON07]   A. Honig, A. Sandorfi, in *Proc. 17th Int. Spin Phys. Symp. (SPIN2006)*, ed. by K. Imai et al.. AIP Conf. Proc., vol. 915, Kyoto, 2006 (AIP, New York, 2007), p. 1010
[HUK08]   A. Huke, K. Czerski, P. Heide, G. Ruprecht, N. Targosz, W. Żebrowski, Phys. Rev. C **78**, 015803 (2008)
[IMI06]   A. Imig, C. Düweke, R. Emmerich, J. Ley, K.O. Zell, H. Paetz gen. Schieck, Phys. Rev. C **73**, 024001 (2006)
[KAT01]   T. Katabuchi, K. Kudo, K. Masuno, T. Iizuka, Y. Aoki, Y. Tagishi, Phys. Rev. C **64**, 047601 (2001)
[KNI86]   R.J. Knize, in *Proc. 6th Int. Symp. on Polarization Phenomena in Nucl. Phys.*, ed. by M. Kondo et al., Osaka, 1985. Suppl. J. Phys. Soc. Jpn., vol. 55 (1986)
[KUL82]   R.M. Kulsrud, H.P. Furth, E.J. Valeo, M. Goldhaber, Phys. Rev. Lett. **49**, 1248 (1982)
[KUL88]   R.M. Kulsrud, Nucl. Instrum. Methods A **271**, 4 (1988)
[LEE71a] Ch. Leemann, H. Bürgisser, P. Huber, U. Rohrer, H. Paetz gen. Schieck, F. Seiler, Helv. Phys. Acta **44**, 141 (1971)
[LEE71b] Ch. Leemann, H. Bürgisser, P. Huber, U. Rohrer, H. Paetz gen. Schieck, F. Seiler, Ann. Phys. (N.Y.) **66**, 810 (1971)
[LEM90]   S. Lemaître, H. Paetz gen. Schieck, Few-Body Syst. **9**, 155 (1990)
[LEM93]   S. Lemaître, H. Paetz gen. Schieck, Ann. Phys. (Leipz.) **2**, 503 (1993)
[LEO06]   D.S. Leonard, H.J. Karwowski, C.R. Brune, B. Fisher, E.J. Ludwig, Phys. Rev. C **73**, 045801 (2006)
[MOR83]   R.M. More, Phys. Rev. Lett. **51**, 396 (1983)
[MYR12]   MYRRHA, http://myrrha.sckcen.be/en
[NA09]   P. Navrátil, S. Quaglioni, I. Stetcu, B.R. Barrett, J. Phys. G **36**, 083101 (2009)
[NAV11a] P. Navrátil, S. Quaglioni, arXiv:1110.0460v1 [nucl-th] (2011)
[NAV11b] P. Navrátil, S. Quaglioni, Phys. Rev. C **83**, 044609 (2011)
[NIK11]   D.M. Nikolenko, I.A. Rachek, Yu.V. Shestakov, D.K. Toporkov, *Proc. XIVth Int. Workshop on Polarized Sources, Targets, and Polarimeters*, St. Petersburg (2011)
[PHT82]   Physics Today, AIP New York, 8/(1982)
[PIE01]   C. Pieper, V.R. Pandharipande, R.B. Wiringa, J. Carlson, Phys. Rev. C **64**, 014001 (2001)
[POL99]   M. Tanaka (ed.), *Proc. RCNP Workshop on Spin Polarized Nucl. Fusions (POLUSION99)* (RCNP, Osaka, 1999)
[RAI02]   F. Raiola et al., Eur. Phys. J. A **13**, 377 (2002)

[ROL95]	C. Rolfs, E. Somorjai, Nucl. Instrum. Methods Phys. Res. B **99**, 297 (1995)
[RUB01]	C. Rubbia et al., The European Roadmap for Developing Accelerator Driven Systems (ADS) for Nuclear Waste Incineration, ENEA (2001), and http://www.oecd-nea.org/pt/docs
[TEM12]	M. Temporal, V. Brandon, B. Canaud, J.P. Didelez, R. Fedosejevs, R. Ramis, Nucl. Fusion **52**, 103011 (2012)
[UZU99]	E. Uzu, S. Oryu, M. Tanifuji, in Ref. [POL99], p. 30
[UZU02]	E. Uzu, nucl-th/0210026 (2002)
[ZHA85]	J.S. Zhang, K.F. Liu, G.W. Shuy, Phys. Rev. Lett. **55**, 1649 (1985)
[ZHA86]	J.S. Zhang, K.F. Liu, G.W. Shuy, Phys. Rev. Lett. **57**, 1410 (1986)
[ZHA99]	J.S. Zhang, K.F. Liu, G.W. Shuy, Phys. Rev. C **60**, 054614 (1999)

Chapter 20
Other Important Applications of Nuclear-Reaction Techniques

20.1 Archaeology, Geology, and Art

The determination of the age of archaeological and geological samples as well as of works of art has developed substantially in recent years. Older methods using e.g. mass spectrometry to determine $^{14}C/^{12}C$ ratios have been superseded by the use of accelerators, partly in dedicated facilities. Tandem Van-de-Graaff machines are especially suited because of their extremely high sensitivity and low background contamination allowing the age measurements on carbon samples back to <100000 years corresponding to such ratios on the order of 1 in 10^{15}. One decisive advantage is the very small amount of material ($\approx$mg quantities) required, which keeps the destruction of the samples to a minimum.

20.1.1 Archaeology and Age Determination

The required installations are similar to nuclear tandem VdG labs, except that medium energies normally suffice, but high beam currents are desirable. Especially very good mass separation and quantitative detection of the carbon ions are necessary. Other mass-14 specimens are ^{14}N, which, however does not form a negative ion, and CH_2, which is dissociated in the stripping process. High mass and energy resolution are obtained already after the sputter ion source before injection into the tandem as well as after acceleration by electrostatic and magnetic fields. Detection systems use ion-mass and energy analysis by time-of-flight and ΔE–E telescopes. It is necessary to know the transmission through the entire system for all relevant nuclides.

A problem is that the natural ^{14}C of the atmosphere has never been constant, especially during the period of nuclear-bomb testing (the reference year BP = 0 is defined as the year 1950 AD). Therefore, the relation between the $^{14}C/^{12}C$ ratio and the time when the carbon intake stopped has to be calibrated by other means. These are e.g. tree-ring dating (e.g. the *German oak and pine chronologies*) [SPU98], back

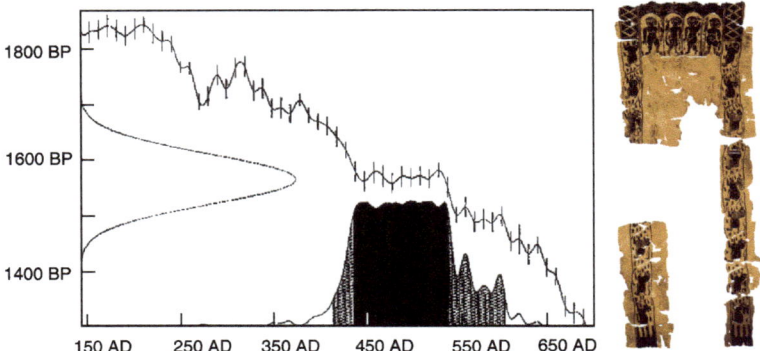

Fig. 20.1 ^{14}C age determination of a piece of coptic fabric. The measured carbon age is 1568 ± 45 years. After calibration with the calibration function shown time intervals of from 434 to 535 AD with 1σ C.L. and from 411 to 599 AD with 2σ C.L. for the manufacturing of the fabric result. Courtesy Deutsches Textilmuseum (DTM) Krefeld, inv. no. 12479. Photograph ©Dieter Gasse (DTM). The dating was done at the Erlangen ^{14}C AMS laboratory [APS13]. See also Ref. [APS07]

to ≈ 12 ky BP), U-Th dating of corals (back to ≈ 24 ky BP), and marine varves (back to ≈ 52.8 ky BP), see Refs. [INT98, INT04, INT09, RAM12] and references therein. An example is shown in Fig. 20.1. Depending on the shape of the calibration function unique age assignments cannot always be made but e.g. two probable assignments with their probabilities instead. The very high prices of older and younger pieces of art have incited large-scale forgeries especially of paintings. The truly high skills of art forgers make the certification of art works based only on stylistic criteria quite uncertain and error-prone (and increasingly so with increasing age of the objects). Thus, objective methods are often necessary such as the determination of the age of the canvas or wooden panels or of the composition of the dyes and paints used. In a number of cases the chemical composition can be determined by physical methods such as PIXE (= proton-induced x-ray emission), in others Rutherford backscattering (RBS) or neutron-activation analysis are required.

20.2 Materials Analysis and Modification

A number of methods using either protons or neutrons have been developed for the use of elemental analysis of samples in technical fields as well as archaeology and history of art. Among the few most important methods are

- RBS: Rutherford backscattering,
- PIXE: proton induced X ray analysis,
- NAA: neutron activation analysis,
- Neutron radiography, and
- PGAA: prompt γ activation analysis.

20.2 Materials Analysis and Modification

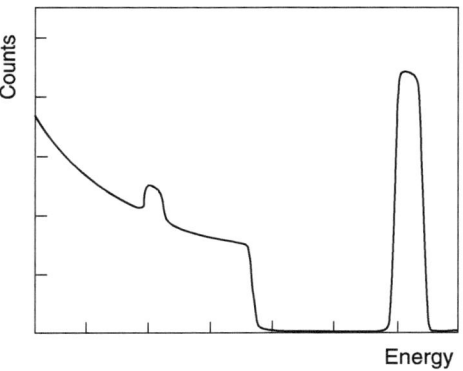

Fig. 20.2 Schematic RBS spectrum of α's on a bulk material with intermediate mass number (continuum) and a thin surface layer of a metal oxide (i.e. with a smaller and a larger mass number resulting in two peaks)

Whereas the first two scan the surface of objects because of the small ranges in matter the others can be used to get information on the bulk of the samples. The five methods will be addressed briefly here.

20.2.1 RBS

The method of surface analysis is based on Rutherford scattering e.g. with α's, see Sect. 2.2 but normally on a thick substrate with some surface structure or layer such that only backscattering can occur. The substrate, due to continuous energy loss produces a continuum up to the maximum lab energy of backscattered projectiles and, in addition one or more peaks from the thin surface layer. From kinematics the Z of the element, the layer thickness and some surface-structure information such as roughness can be deduced. Typical energies are several hundred keV and varying backscattering angles are used. Figure 20.2 shows schematically the shape of an RBS spectrum of α's backscattered at a typical angle from a bulk material (continuum) and a thin surface layer of a metal oxide (two peaks).

20.2.2 PIXE

Protons impinging on material surfaces undergo not only nuclear and Rutherford scattering but can excite inner atomic shells causing the emission of characteristic X rays. These are specific for each element and can therefore be used to identify the elemental composition of samples. These could be e.g. a small sample of pigment from a painting identifying a forgery by a pigment that was not used before a certain period. Since antiquity the white pigment white lead ($2PbCO_3 \cdot Pb(OH)_2 = C_2H_2O_8Pb_3$) was used by artists, whereas zinc-white (ZnO) was only introduced around 1840, and titanium-white (TiO_2) around 1918. Another example would be dust from a filter, through which air has been blown and deposited material from

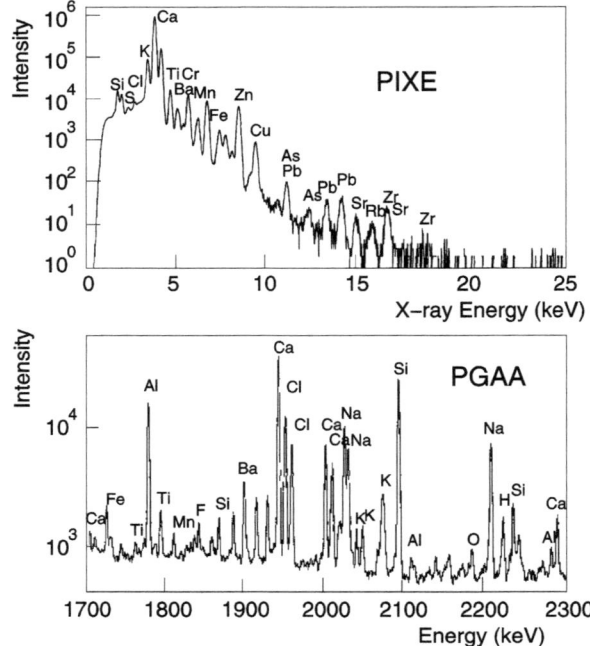

Fig. 20.3 A PIXE and a PGAA spectrum of the same piece of glass. The background in the PIXE spectrum are bremsstrahlung X rays. Courtesy of P. Kudějová [KUD05]

the environment, e.g. along highways to check on the vehicles' exhaust (Pb, soot etc.). Figure 20.3 shows a PIXE and a PGAA spectrum of the same piece of green glass [KUD05] and illustrates the different and partly complementary information provided by both methods.

20.2.3 Neutron Radiography

Cold to thermal neutron beams from nuclear reactors are used for radiography, already on an industrial scale. The scattering and absorption properties of the neutrons are different and often complementary to X-rays, which interact most strongly with high-Z material. Neutrons interact most strongly with hydrogen, i.e. also with compounds containing hydrogen (organic liquids in metal containers, biological material, explosives etc.). Analog as well as digital methods of image formation are being used. Figure 20.4 shows—in contrast to X-ray absorption and in part also complementary for different elements—the "un-systematic" response of different materials (elements, isotopes) to thermal neutron irradiation and a biological example (it could have been a technical device where cracks in the material would show up, or a hidden explosive device).

20.3 Exercises

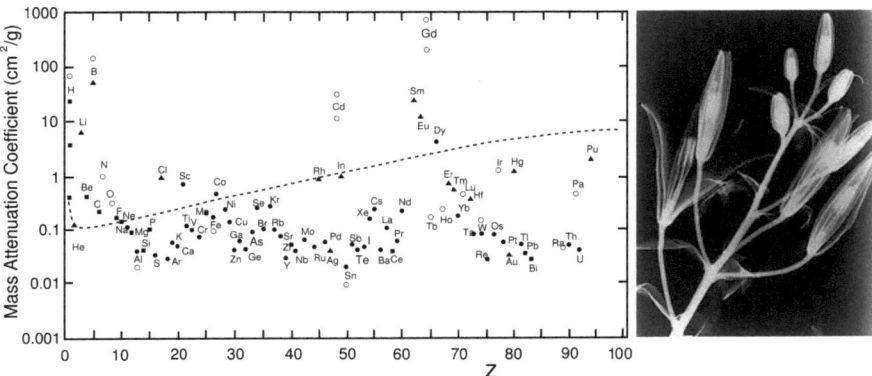

Fig. 20.4 The response to thermal neutrons of elements (isotopes) along the periodic table, expressed as mass attenuation coefficient (*left*, and a biological example of a radiography (*right*). From [CEA13]. The symbols are: Scattering + absorption (*full circles*), mainly scattering (*squares*), mainly absorption (*triangles*), absorption only (*circles*). The *dashed line* is the mass attenuation coefficient of 125 keV X-rays. After [PNR92]

20.2.4 Materials Modifications

Only a few short remarks can be devoted to these applications of accelerators and nuclear methods despite partly widespread use.

- Ion implantation is used to dope pure Si wafers with foreign atoms like boron, phosphorus, and arsenic with concentrations from 10^{10} to 10^{15} atoms/cm^2 to produce n- or p-type semiconductor materials. The energy of the beams from high-current accelerators determines the depth of implantation. Applications are, among others, semiconductor detectors as well as complex integrated circuits.
- Surface treatments of medical implants, ball bearings and machining tools by implanting nitrogen, boron, carbon titanium etc. ions with high-current, low-voltage accelerators in order to improve the mechanical properties and reduce wear.
- In view of terrorism and for contraband and drug detection accelerator-based pulsed-neutron radiographic systems (PFNA) and methods are being developed that could allow to scan entire trucks by (n, γ) for certain materials (e.g. is a high carbon to oxygen ratio characteristic for drugs, high nitrogen content for explosives etc.) The fusion neutrons are produced by a deuteron beam from a low-energy accelerator and the γ's are detected by large arrays of suitable detectors, thus creating large-area or even tomographic images.

20.3 Exercises

20.1. Remnants of a wooden boat were found in the mud of the Baltic Sea. A small piece of a plank was prepared and inserted into the target pill of the sputter source of an AMS accelerator. A ratio of $N(^{14}C)/N(^{12}C)$ of $1.221 \cdot 10^{-12}$ was measured. When was the tree cut down, which served to make the plank? Assume an original ratio at that time—as confirmed by tree-ring calibration—of $1.5 \cdot 10^{-12}$ and $T_{1/2}(^{14}C) = 5730 \pm 40$ a, as well as the simple radioactive-decay law. Which age error results from the error of the halflife of ^{14}C alone?

20.2. Older methods used mass spectrometers to determine mass ratios of nuclei decaying into daughter nuclei. In a rock sample of 10 g with a uranium content of 4 ppm a ^{4}He quantity of $5 \cdot 10^{-3}$ cm^3 was measured. How long ago did the rock solidify?

20.3. Another method for radioactive materials is measuring the activity per mass unit ("specific activity") of an old sample vs. that of a recent one. As an example take a wooden beam from an old settlement. You measure an activity of ^{14}C of 1.7 decays/s whereas the activity in a recent sample is 2.5/s. How old is the old wooden beam (assuming it was cut directly from a freshly felled tree)?

20.4. With your knowledge of Rutherford scattering (see Sect. 2.2), construct qualitatively (and possibly quantitatively) an RBS spectrum of 5.5 MeV α's (such as from an ^{241}Am source), backscattered under $\theta = 170°$ from a thin Al foil ("thin" is to mean that the energy loss in the foil is small relative to the incident energy). Assume, in addition, that the Al has a thin oxide layer on its surface and, also, some carbon buildup from cracking of the residual oil vapor by the beam which continues to grow with time (see e.g. the spectra of Fig. 17.4). How does the spectrum change with time? With increasing levels of sophistication, use kinematics and energy-loss calculations (Bethe-Bloch) and/or suitable tabulations, graphs, internet etc. Because there is no perfect vacuum (evaluate the approximate number density of residual-gas molecules at a "vacuum" pressure of $1 \cdot 10^{-6}$ mbar): which measures could you try to reduce carbon buildup?

References

[APS07] A. Paetz gen. Schieck, in *Proc. Conf. on Meth. of Dating Ancient Textiles of the 1st Millenium AD from Egypt and Neighbouring Countries*, ed. by A. De Moor, C. Fluck. Antwerp 2095 (Lannoo Publishers, Tielt, 2007), p. 167
[APS13] A. Paetz gen. Schieck, private communication
[CEA13] CEA Saclay
[INT98] M. Stuiver, P.J. Reimer, E. Bard, J.W. Beck, G.S. Burr, K.A. Hughen, B. Kromer, G. McCormac, J. Van der Plicht, M. Spurk, Radiocarbon **40**, 1041 (1998)
[INT04] P.J. Reimer et al., Radiocarbon **46**, 1029 (2004)
[INT09] P.J. Reimer, Radiocarbon **51**, 1111 (2009)
[KUD05] P. Kudějová, Ph.D. dissertation, University of Cologne (2005)
[PNR92] J.C. Domanus (ed.), *Practical Neutron Radiography*, Int. Advances in NDT, vol. 16 (Kluwer Academic, Dordrecht, 1992)
[RAM12] C.B. Ramsey et al., Science **338**, 370 (2012)
[SPU98] M. Spurk, M. Friedrich, J. Hofmann, S. Remmele, B. Frenzel, H.H. Leuschner, B. Kromer, Radiocarbon **40**, 1107 (1998)

Chapter 21
Trends and Future Developments of Nuclear Reactions

At least two dominant trends can be observed in nuclear-reaction physics:

- Studies of "exotic nuclei"
- Studies of low-cross section reactions.

Technological advances can be expected in the field of accelerators, but not very quickly: One is the *wakefield acceleration* of ions and energetic beams directly from the interaction of high-power lasers with the atoms in target foils, which might also produce polarized ions ("tilted-foil" method).

Theoretical work on nuclear structure and nuclear reactions that increasingly relies on the use of very large computing resources has made enormous progress in recent years and will continue to do so.

21.1 Exotic Nuclei

All nuclei far from the line of stability could be named "exotic". These encompass nuclei at the lifetime limits along the proton and neutron driplines and halo nuclei as well as superheavies on the way to a new island of stability beyond element 118. The short-lived β^+ and β^- unstable nuclei which can be classified by their varying isospin quantum number $T = -1/2(N - Z)$ are of interest for many reasons:

- It is interesting to follow the development of the shell-model energy gaps with varying neutron or proton numbers.
- The structure of these nuclei and their function in nuclear reactions may serve as tests of advanced theoretical approaches such as no-core shell model and similar *ab initio* calculations.
- The contributions of these nuclei as intermediate steps in the different chains of nucleosynthesis must be known.

The challenge is of course to devise accelerators for beams, with which these nuclei may be produced either as compound or as final nuclei, which also means accelerators for radioactive beams. A number of such facilities is already working or in the planning stage, see Chap. 15.

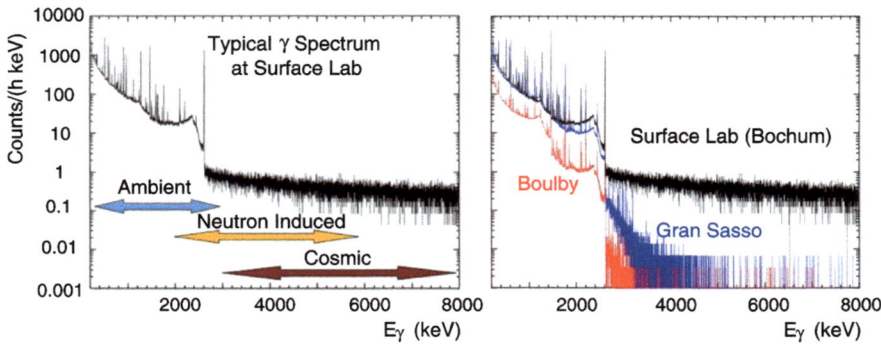

Fig. 21.1 Characteristic background γ spectra taken at a surface laboratory, with the main contributions from different sources, and at two different underground laboratories, using the same detector, the Boulby mine (UK) and LUNA/LNGS (Gran Sasso, Italy) showing background suppression by up to three orders of magnitude, see [ALI12]

21.2 Low-Cross Section Reactions

In order to study nuclear reactions (e.g. relevant to nucleosynthesis), which, due to the Coulomb barrier may have very small cross sections, two developments are essential: Good statistics in reasonable measurement times require beam intensities as high as possible, for which suitable accelerators and ion sources have to be provided.

With the expected small count rates in the presence of background from terrestrial and cosmic-ray sources (consisting mainly of muons) radiation shielding is necessary. This is best achieved by going underground. The number of already existing underground facilities complete with accelerators etc. such as LUNA at the Gran Sasso, Italy, will certainly increase [BRO10]. Figure 21.1 shows the dramatic effect of shielding on the background in environmental γ spectra. Similar effects are expected for charged-particle spectra from solid-state detectors. An example showing surface-laboratory background compared to an anti-coincidence spectrum that simulates the LUNA underground situation is given by Fig. 21.2. The facility for the PANDAX experiment in China has the so far lowest muon background and will be used for the search of *Weakly Interacting Massive Particles* (WIMPs), the particles connected to the *Cold Dark Matter* constituting $\approx$22 % of the mass of the universe, using nuclear reaction methods—elastic scattering recoils imparted to nuclei such as Xenon.

21.3 Accelerator Developments

The sheer size of highest-energy accelerators such as the LHC at Geneva, SLAC at Stanford, or DESY at Hamburg indicates a limit for future accelerators with even higher energies. But the possible reduction of the size of smaller accelerators could have an enormous impact on accelerator applications, e.g. for particle-therapy instal-

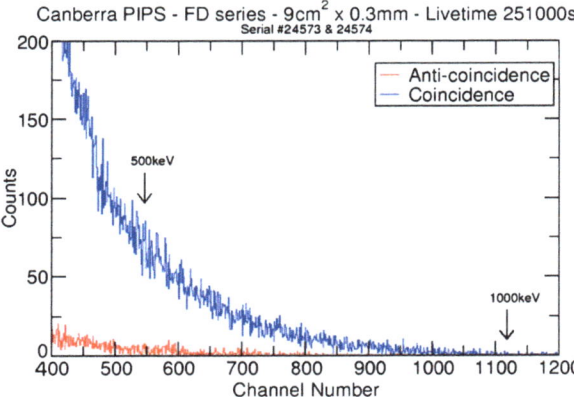

Fig. 21.2 Comparison of background in the upper Si solid-state detector of a stack of two 0.3 mm detectors taken at a surface laboratory in coincidence, with background from a simulation of the LUNA deep-underground situation with the two detectors in anti-coincidence [ALI11]

lations in hospitals. The limitation is mainly connected with the strength of electric accelerating fields that can be maintained across a gap of given length. Thus, strong research efforts are seeking alternative ways of achieving higher acceleration values. The idea of plasma acceleration was first published in 1979 ([TAJ79]) and first acceleration was observed in 1988 ([ROS88]). One concept is that of *wakefield* acceleration by a plasma "wave" produced with laser or charged-particle beams. Femtosecond TW titanium-sapphire lasers have been applied, as well as high-energy high-current beams at facilities such as SLAC (Stanford) and BNL (Brookhaven). Acceleration has been achieved with field gradients several orders of magnitude higher than in existing accelerators. For a recent survey see Ref. [JOS12] and references therein.

21.4 Theoretical Progress

Recently large collaborations of theorists have been formed in order to tackle a number of unsolved problems using large computers. Three regions of the nuclear chart can be defined, in which different goals and methods have been proposed. In the region of light nuclei no-core shell model calculations using the NN interaction or EFT as input have been quite successful. A brief survey on these projects was recently published in Nucl. Phys. News [FUR11].

References

[ALI12] M. Aliotta, Nucl. Phys. News **22**, 13 (2012)
[ALI11] M. Aliotta, Private communication (2012)
[BRO10] C. Broggini, D. Bemmerer, A. Guglielmetto, R. Menegazzo, LUNA: astrophysics deep underground. Annu. Rev. Nucl. Part. Sci. **60**, 53 (2010)
[FUR11] R. Furnstahl, Nucl. Phys. News **21**, 18 (2011)

[JOS12] C. Joshi, V. Malka, New J. Phys. **12**, 045003 (2012)
[ROS88] J.B. Rosenzweig, D.B. Cline, B. Cole, H. Figueroa, W. Gai, R. Konecny, J. Norem, P. Schoessow, J. Simpson, Phys. Rev. Lett. **61**, 98 (1988)
[TAJ79] T. Tajima, J.M. Dawson, Phys. Rev. Lett. **43**, 267 (1979)

Chapter 22
Appendices

22.1 Appendix A—Tables of Useful Numbers and Relations

All numerical values are taken from the CODATA(NIST) compilation as published in Nuclear Wallet Card [NWC11] and [RMP12]. All masses are rest masses.

Table 22.1 Fundamental constants

Vacuum velocity of light	c	$2.99792458 \cdot 10^8$	ms^{-1}
Elementary charge	e	$1.602176565 \cdot 10^{-19}$	C
Planck constant	h	$6.62606957 \cdot 10^{-34}$	J s
Reduced Planck constant	$\hbar = h/2\pi$	$1.054571726 \cdot 10^{-34}$	J s
Fine structure constant	$\alpha = \frac{e^2}{\hbar c}$	$7.2973525698 \cdot 10^{-3}$	
Avogadro constant	N_A	$6.02214129 \cdot 10^{23}$	mol^{-1}
Bohr magneton	$\mu_B = \frac{e\hbar}{2m_e}$	$9.27400968 \cdot 10^{-24}$	J T^{-1}
Electron magnetic moment	μ_e	-1.00115965218076	μ_B
Atomic mass unit	u	$1.660538921 \cdot 10^{-27}$	kg
	uc^2	931.494013	MeV
Mass of H atom	m_H	1.00782503207	u
Proton mass	m_p	1.007276466812	u
Neutron mass	m_n	1.00866491600	u
Deuteron mass	m_d	2.013553212	u
Electron mass	m_e	$5.4857990946 \cdot 10^{-4}$	u
Bohr radius	$a_0 = \hbar^2/m_e e^2$	$5.2917721092 \cdot 10^{-11}$	m
Nuclear magneton	μ_N	$5.05078353 \cdot 10^{-27}$	J T^{-1}
Proton gyromagnetic ratio	$\gamma_I = \frac{\mu_I}{\hbar I}$	$2.675222005 \cdot 10^8$	s^{-1} T^{-1}
g-factor neutron		-3.82608545	
g-factor proton		5.585694713	

Table 22.1 (Continued)

g-factor deuteron		0.8574382308	
g-factor electron		−2.00231930436153	
Boltzmann constant	k	$1.3806488 \cdot 10^{-23}$	J K^{-1}
Farady constant	$F = N_A e$	$9.64853365 \cdot 10^4$	C mol^{-1}

Table 22.2 Useful relations and ratios

Energy/Frequency	$E = h\nu$	1 MeV $\hat{=}$ $2.418 \cdot 10^{20}$ Hz
Energy/Wavelength	$E\lambda = hc$	$1.2398 \cdot 10^{-10}$ MeV cm
	$E\lambda = \hbar c$	197.326990 MeV fm
Energy/Velocity		
non-relativistic	$v = 1.3889\sqrt{E(\text{MeV})/m(u)} \cdot 10^7$	m/s
Energy/Momentum	$p = \frac{1}{c}\sqrt{2m_0c^2 E + E^2}$	
non-relativistic $E \ll m_0c^2$	$\approx \sqrt{2m_0 E}$	
	$= 43.162\sqrt{m_0(u)E(\text{MeV})}$	MeV/c
highly relativistic $E \gg m_0c^2$	$\approx E/c$	
De Broglie Wavelength	$\lambda = \hbar/p = \frac{\hbar}{\sqrt{2m_0 E + E^2/c^2}}$	
non-relativistic	$\approx \frac{\hbar}{\sqrt{2m_0 E}} = \frac{4.572}{\sqrt{m_0 E}}$	fm
relativistic	$\approx \frac{\hbar c}{E} = \frac{197.327}{E}$	fm
Sommerfeld Parameter	$\eta_S = Z_1 Z_2 e^2 / \hbar v$	
	$= 0.1575 Z_1 Z_2 [\mu(u)/E_{\text{c.m.}}(\text{MeV})]^{1/2}$	
Coulomb Barrier	$E_C = Z_1 Z_2 e^2 / R_C$	
Coulomb Radius [BAS80]	$R_C \approx [1.12(A_1^{1/3} + A_2^{1/3})$ $-0.94(A_1^{-1/3} + A_2^{-1/3}) + 3]$	fm

22.2 Appendix B—Practical Units

In practical work in nuclear and atomic physics a set of units is costumarily used that is adapted to the size and energies of atoms and nuclei. The following table lists a few of them.

Table 22.3 List of a few "practical" units of nuclear physics

Energy	1 eV	$1.602176565 \cdot 10^{-19}$ J
Energy/Temperature	1 eV	$1.1604519 \cdot 10^4$ K
Length	1 fm	$1 \cdot 10^{-15}$ m
	1 Å	$1 \cdot 10^{-10}$ m
Cross Section	1 b	$1 \cdot 10^{-28}$ m^2

22.3 Appendix C—Angular-Momentum Recoupling Coefficients

The coefficients used in the text are briefly defined here. The tedious summation over magnetic quantum numbers has been enormously alleviated by introducing these coefficients, which have a number of symmetry properties and rules and exist as tables or can simply be calculated in appropriate subroutines. Especially the algebra introduced by Racah (*Racah algebra* [RAC42]) has proved very useful. Further details can be found in the literature [BRI71, EDM60, SAT90, BLA52] and references therein. Caveat: Many different definitions, notations, and symbols have been published in the literature; we follow Ref. [BRI71].

22.3.1 Coupling of Two Angular Momenta

THe coupling of two angular momenta $a \equiv j_1, \alpha \equiv m_1$ and $b \equiv j_2, \beta \equiv m_2$ to $c \equiv J, \gamma \equiv M$ is described by *Clebsch-Gordan or Wigner coefficients or 3-j symbols*. They are defined by the linear transformation from the uncoupled to the coupled system

$$|JM\rangle = \sum_{m_1 m_2} |j_1 j_2 m_1 m_2\rangle \langle j_1 j_2 m_1 m_2 | JM\rangle \tag{22.1}$$

or in abbreviated form

$$|abc\gamma\rangle = \sum_{\alpha\beta} |ab\alpha\beta\rangle \langle ab\alpha\beta | c\gamma\rangle \tag{22.2}$$

Wigner's 3j symbol is related by

$$\langle ab\alpha\beta | c - \gamma\rangle = (-)^{a-b-\gamma}(2c+1)^{1/2} \begin{pmatrix} a & b & c \\ \alpha & \beta & \gamma \end{pmatrix}. \tag{22.3}$$

22.3.2 Coupling of Three Angular Momenta

The different ways of coupling three angular momenta $a, b,$ and c to $d, e,$ and f with respective magnetic quantum numbers $\alpha, \beta, \gamma, \delta, \epsilon,$ and ϕ are connected by a linear transformation mediated by *Racah coefficients W or 6-j symbols*

$$\langle ab\alpha\beta | c\gamma\rangle \langle ed\alpha + \beta, \gamma - \alpha - \beta | c\gamma\rangle$$
$$= \sum_f \langle bd\beta, \gamma - \alpha - \beta | f\gamma - \alpha\rangle \langle af\alpha, \gamma - \alpha | c\gamma\rangle$$
$$\times \left[(2e+1)(2f+1)\right]^{1/2} W(abcd; ef). \tag{22.4}$$

W is related to the 6-j symbol

$$\begin{Bmatrix} a & b & e \\ d & c & f \end{Bmatrix} = (-)^{a+b+c+d} W(abcd; ef). \tag{22.5}$$

Both coefficients can be expressed in terms of CG coefficients or 3-j symbols.

The coefficient Z (used e.g. in the Hauser-Feshbach cross section) is related to $W(abcd; ef)$ by [BLA52]

$$Z(abcd; ef) = i^{f-a+c} \big[(2a+1)(2b+1)(2c+1)(2d+1)\big]^{1/2}$$
$$\times \langle ac00|f0\rangle W(abcd; ef). \tag{22.6}$$

22.3.3 Coupling of Four Angular Momenta

A case, which occurs in nuclear reactions is that of (re-)coupling of four angular momenta including all summations over magnetic quantum numbers and important especially for the case of polarization observables. The coefficients, which are defined in analogy to the 6-j case are the *X coefficients (Fano)* or *9-j symbols*

$$X(abc, def, ghi) \equiv \begin{Bmatrix} a & b & c \\ d & e & f \\ g & h & i \end{Bmatrix}. \tag{22.7}$$

They can be contracted out of six 3-j symbols after summing over all (minus one) magnetic quantum numbers. Welton's general formula (7.1) contains four 9-j symbols.

22.3.4 Spherical Harmonics

The spherical harmonics are the angular-momentum eigenfunctions and solutions of the angular part of the Schrödinger equation. They are defined as

$$Y_{\ell m} = \left(\frac{2\ell+1}{4\pi}\right)^{1/2} \tag{22.8}$$

$$\times (-)^m \left[\frac{(\ell-m)!}{(\ell+m)!}\right]^{1/2} P_\ell^m(\theta) e^{im\phi}. \tag{22.9}$$

The P_ℓ^m are the *Associated Legendre Functions*, for $m = 0$ the *Legendre Polynomials* $P_\ell(\cos\theta)$.

22.4 Appendix D—General Resources

In this appendix a selection of references with resource material is collected. They concern:

- **Recommended General Texts on Nuclear Reactions** Refs. [BAS80, MAR70, SAT90],
- **Angular Momentum** Refs. [BLA52, BRI71, EDM60, RAC42], and
- **Data Resources** Refs. [ELL, GWU11, KNC, LED, NWC11, NCR, NDT, NDS, NNO13, NUD, RMP12, RPP08, AMS08, BER12, SAID13].

References

[AMS08] C. Amsler et al., Phys. Lett. B **667**, 1 (2008)
[BAS80] R. Bass, *Nucl. Reactions with Heavy Ions*. Texts and Monographs in Physics (Springer, Berlin, 1980)
[BER12] J. Beringer et al. (Particle Data Group), Phys. Rev. D **86**, 010001 (2012)
[BLA52] J.M. Blatt, L.C. Biedenharn, Rev. Mod. Phys. **24**, 258 (1952)
[BRI71] D.M. Brink, G.R. Satchler, *Angular Momentum* (Clarendon Press, Oxford, 1971)
[EDM60] A.R. Edmonds, *Angular Momentum in Quantum Mechanics* (Princeton University Press, Princeton, 1960)
[ELL] Energy Levels of Light Nuclei ($A = 3$ to 20), a series of compilations in Nuclear Physics A
[GWU11] http://gwdac.phys.gwu.edu
[KNC] Karlsruhe Nuclide Chart (2012)
[LED] C.M. Lederer, V.S. Shirley, *Table of Isotopes*, 8th edn. (Wiley, New York, 1999). Newer: R.B. Firestone, V.S. Shirley, also on CD-ROM
[MAR70] P. Marmier, E. Sheldon, *Physics of Nuclei and Particles, Vols. I + II* (Academic Press, New York, 1970)
[NWC11] J.K. Tuli, *Nucl. Wallet Card*, 8th edn. (Brookhaven Natl. Lab., Upton, 2011)
[NCR] Neutron Cross-Sections Reference Books, Formerly BNL-325, NNDC, BNL
[NDS] Nuclear Data Sheets, Academic Press
[NDT] Nuclear Data Tables (Q-Values, Reaction Lists etc.), Academic Press
[NNO13] Theor. High-Energy Phys. Group, Radboud University, Nijmegen, The Netherlands, http://nn-online.org
[NUD] NuDat Chart of Nuclides
[RAC42] G. Racah, Phys. Rev. **62**, 438 (1942)
[RMP12] P.J. Mohr, B.N. Taylor, D.B. Newell, Rev. Mod. Phys. **84** (2012)
[RPP08] Particle Data Group, Review of particle properties. Rev. Mod. Phys. **80**, 633 (2008)
[SAID13] SAID Partial-Wave Analysis Facility, Physics Dept., The George Washington University, CNS DAC Services, Center for Nuclear Studies, http://gwdac.phys.gwu.edu
[SAT90] G.R. Satchler, *Introd. Nucl. Reactions*, 2nd edn. (McMillan, London, 1990)

Index

A
Absorption, 133
Accelerators, 251
    Electrostatic
        Cockroft-Walton, 251
        Van de Graaff, 252
    Future Developments, 352
    History, ix
    RF, 255
        Betatron, Synchrotron, 258
        Cyclotron, 255
        LINAC, 255
    Underground Facilities, 352
Alternating-Gradient Focusing, 266
Analyzing Powers, 111, 113, 116
    Generalized, 116, 126
    Longitudinal, 118
Appendices
    Angular-Momentum Algebra, 357
    Fundamental constants, 355
    Practical Units, 356
    Resources, 359
    Useful Numbers and Relations, 355
Applications, 319
    Archaeology and Art, 345
        AMS dating, 345
    Materials Analysis and Modification, 349
        Neutron Radiography, 346
    Medical, 321
    Nuclear Energy, 327
    Transmutation, 339

B
Born Approximations, 175
    DWBA, 176
Bosons, 54

Breit-Rabi
    Diagram, 286, 289
    Formula, 270
    Zeeman States, 270

C
Charge, Current, or Matter Distributions
    Charge, 21
    Current, 33
    Matter, 33
    Sampling, 25
Charge Exchange
    Alkali, 280
    $H(2S) + Ar$, 291
    $H(2S) + I_2$, 291
    Resonant, 281
Chiral-Perturbation Theories, 145, 153
Conservation
    Angular Momentum, 43, 115
    Energy, 42
    Isospin, 48, 50
    Laws and Symmetries, 39
    Momentum, 42
    Parity, 44, 45, 85, 115
        Three-Particle Reactions, 117
Coordinate Systems, 118
    Analyzing Powers, 110
    Helicity, 116
    Notation, 110
    Polarization Transfer, 111
    Rotation, 97, 101
    Space-Fixed, 112
    Spin Correlations, 112
    Tensor Moments, 110
Correlations
    Angular, 195
        Spin Channels, 195

Channel-channel, 194
Coulomb
    Effects, 48
    Energy, 48
    potential, 17
Coulomb Barrier, 6, 187, 222, 226, 231, 239, 329, 356
Coulomb Force, 162
Coulomb Potential, 15
Coulomb Scattering
    Identical Particles, 54
Cross Section
    Electron Loss in I_2, 291
    Electron-Impact Ionization, 279
    General, Polarized, 125
    General Appearance, 61
    Hauser-Feshbach, 199
    Metastable Production in Cs, 287
    Neutral Cs on H or D, 281
    Unpolarized, 125, 131

D
Deflection Function, 13, 223
Density Distributions
    Charge, 31
    Current, 33
    Matter, 33
Density Matrix, 83, 85
    Entrance Channel, 105, 106
    Exit Channel, 106
    Expansion, 85, 95
        Cartesian or Spherical Tensors, 107
    Incident, 85
    Rotation, 84, 97, 99
    $S = 1$, 95
    $S = 1/2$, 94
Density Operator, 78
Detectors, 301
    Detector Telescope, 305
    History, xiii
    Neutron Detectors, 310
    Scintillation Detectors, 302
    Solid-State Detectors, 303
        Ge, 307
        Si, 304
Direct Reactions
    Spectroscopic Factor, 173, 178
Doorways
    Fission Isomers, 213
    HI Molecular Resonances, 207, 225
    IAS, IAR, 213
    Transfer Reactions, 226
Driplines, 241

E
Effective-Field Theories EFT, 8, 35, 52, 145, 153, 161, 353
Elastic Scattering, ix
Electron Scattering, 29
Ensemble, 77, 88
    Average, 79
    Unpolarized, 88
Ericson Fluctuations, 192
Escape Width, 207, 215
Exotic Nuclei, 241, 351

F
Fermions, 54
Fermi's Golden Rule, 21, 57, 61, 106, 176
Few-Body Theories
    EFT or Chiral-Perturbation Calculations, 221
    Faddeev Calculations, 161, 221
    New Calculations, 335
        Faddeev-Yakubovsky, 162, 336
        Green's Function Monte Carlo,
        No-Core Shell Model, 337
Few-Nucleon Reactions
    Five-Nucleon Reactions, 328
    Four-Nucleon Reactions, 331
Few-Nucleon Systems, 152, 162
    Kinematics, 154
Fission, 187
    Fission Barrier, 187
Form Factor, 26
Fusion
    Cross Sections, 327
    Fusion Energy, 327
    "Polarized", 329
Fusion Reactions
    Rate Enhancement by Polarization, 334

G
Gamma
    Spectroscopy, 315
Giant Resonances, 210, 242

H
Halo Nuclei, 34
    Density Distribution, 35
Heavy Ions
    Elastic Scattering, 224
    Molecular Resonances, 207, 225
Heavy-Ion Reactions, 221

I
Integral Equations, 174
    Lippmann-Schwinger Equation, 175

Index 363

Interactions
  Nucleon-Nucleon, 145
Intermediate Structures, 207
  Fission Isomers, 213
  Giant Resonances, 210
  Isobaric Analog Resonances IAR, 213
  Molecular Resonances, 207
  n Scattering, 209
Invariances, 40
  CP, 55
  Gauge, 42
  Isospin, 48
  Parity, Space Reflection, 45
  Rotation, 116
  Time Reversal, 54
    Reciprocity, 56
Ion Sources, 266
  Polarized Beams, 269
  Unpolarized Beams, 266
Ionization
  Ionization Yield, 273
Ionizers, 279
Isobaric-Analog Resonances IAR, 51, 213
Isobaric-Analog States IAS, 213, 226
Isospin, 48, 311
  Barshay-Temmer Theorem, 50
Isospin Breaking, 51
Isotope Production, 323

L
Legendre Polynomials, 71, 128
Lenses, 260
  Electrostatic, 262
  Magnetic, 263
Lepton
  Elastic Scattering
    Rosenbluth Formula, 30
Level Densities, 185, 201, 216, 218, 227
Level Widths, 197
Liouville Theorem, 260
Low-Energy Fusion
  Electron Screening, 337

M
Magnet
  SC Stern-Gerlach, 338
Magnetic Moments
  Effective, 270
Magnets
  Stern-Gerlach
    Matching, 271
Materials Analysis
  Neutron Radiography, 348
  Proton-Induced X-Rays, 347
  Rutherford Backscattering, 347
Mixed State, 79
M or Transfer Matrix
  Spin $1/2 + 0$, 118
  Spin $1/2 + 1/2$ Elastic, 119

N
Neutrons, 309
  Detection, 310, 311
  Production, 309
  Spectroscopy, 311
NN force, 153
NN Scattering, 145
Nuclear Astrophysics, 231
Nuclear g Factors, 294
Nuclear Radii, 21
Nuclear Reactions
  Charged Particles, 128
  Compound-Nuclear, 181
  Direct, 167
    Rearrangement, 169
    Stripping, 171
  Models, 167
  Neutral Particles, 61, 125
  Optical Model, 29, 168, 210

O
Observables, 85
  Analyzing Power, 96, 97, 107, 108
  Cartesian, 117
  Definition, xiv
  Few-Nucleon, 160
  NN, 146
  One-Spin, 108
  Optical Model and DWBA, 176
  Polarization, 69, 97
  Polarization Transfer, 117
  PWBA, 177
  Spin Correlation, 109, 117
  Spin-Dependent, 151
  Three- and Four-Spin, 109
  Two-Spin, 109
  Unpolarized Differential
    Cross Section, 106
  Zero-Spin, 108
Occupation Numbers, 81
Operator
  Expectation Value, 77, 85
  Parity, 116
  Pauli, 107
  Rotation, 97, 98
  Spin, 81
  Time-Reversal, 115, 116
  Vector, Tensor, 91

Optical Pumping
  Alkali
    Rb, 283
Optical Theorem, 73

P
Parity
  Violation, 116
Partial Waves, 65, 128, 132, 146, 178, 203, 328
  Coulomb, 74, 125
  Expansion, 71, 125
  IdenticalParticles, 54
  Truncation, 20
Partial-Width Amplitude, 184
Particle Tumor Therapy, 321
Phase Shift, 70, 72, 148
Polarization Observables, 85
  General, 108
  Parity, 85
  Types, 108
Polarization Transfer
  Coefficients, 108, 109, 117
  Wolfenstein Parameters, 109, 118
Polarized-Ion Sources
  ABS
    ^{7}Li and ^{23}Na Beams, 283
    Beam Formation, 272
    Colliding-Beams Ionizers, CBS, 281
    Dissociator, 271
    ECR Ionizers, 280
    Electron-Bombardment Ionizers, 279
    Multipole Fields, 274
    Multipole Magnets of Halbach Design, 274
    Separation Magnets, 273
    Trajectories, 274
  Atomic Beam, ABS, 271
  Ground-State Atomic Beam, ABS, 271
  Lamb Shift, 284
    Ionization in Argon and Iodine, 291
    Metastables, 286
    Sona Scheme, 286, 289
    Spin Filter, 286, 290
  Optical Pumping, 283
  Spin Rotation
    Deflection Devices, 296
    Rotatable Wien Filter, 295
Positron Emission Tomography (PET), 323
Potential
  Coulomb, 15, 20, 168, 222, 225
    Extended Charge, 224
    Screened, 23, 233
  Hard-Sphere
    Neutrons, 139

Heavy-Ion, 222
Rectangular Well
  Neutrons, 140
Spin-Orbit, 168
Woods-Saxon, 31, 142, 168
Yukawa, 26, 52, 142
Pure State
  Superposition, 81

Q
QCD, 52, 153, 162

R
Radioactive Ion Beams, 35, 241
Reactions
  Fusion-Evaporation, 306
  Heavy-Ion, 221
Resonances
  Angular Distribution, 189
  Argand Plot, 189
  Breit-Wigner, 8, 51, 179, 181, 183, 186, 189, 192, 202, 215
  Delta, 190
  Overlapping, 192
  Scattering Phase, 189
  Single Isolated, 188
  Single-Particle Neutron, Shape, 209
RF Transitions, 275
  MFT, 278
  Quantum-Mechanical Treatment, 277
  Rotating Reference
    Quantum-Mechanical, 278
  SFT, 278
  Weak-, Medium-, and Strong-Field
    Transitions, 275
  WFT
    Quantum-Mechanical, 278
Rotation Function, 99
  Reduced, 98, 99
Rotations, 97
  Finite, 97
  Infinitesimal, 97
  Spherical Tensors, 86
Rutherford, 15
  Classical
    Cross Section, 15
    Scattering Distance, 16
    Trajectories, 17
  Cross Section
    Born Approximation and "Golden Rule", 21
    Quantum-Mechanically, 22
  Historic Significance, 19
  Scattering, 15

Index

S
Scattering
 Rutherford, 3
Scattering Amplitude, 68, 73, 74, 105, 137
Scattering Function, 73
Scattering Length, 137
Schrödinger Equation, 69, 135, 168, 174, 212, 358
S Matrix, 73
Spectrograph, 306
Spectroscopy
 Charged-Particles, 305
 Gamma, 315
 Neutron, 311
Spherical Harmonics, 100
Spherical Tensors, 94, 99
Spin Correlation, 107
 Coordinate System, 112
 Definition, 109
Spin Polarization, 77
Spin Precession, 97, 101, 294
 Beam Lines and Wien Filters, 293
 Curves, 297
 Larmor Frequency, 294
Spin Tensor Moments, 85
Spin Tensors
 Entrance and Exit Channels, 106
Spreading Width, 179, 207, 215, 216
State
 Zeeman, 114
Statistical Distributions
 χ^2, 194
 Porter-Thomas, 194
 Stochastic, 194
 Wigner, 194
Stern-Gerlach, 269, 273, 275, 283, 313
 Quadrupole, 271
 Sextupole, 271
Stopping Power and Range
 Bethe-Bloch Formula, 10, 306, 324
Stripping Reactions
 Analyzing Power, 172
 Cross-Section Maximum, 172
 Newns Absorption Model, 172, 179
Strong Focusing, 256, 258, 265
 Cyclotrons and Synchrotrons, 265
 Einzellens, 265
 Quadrupole Doublet, 265
Superheavies, 225, 244, 246, 351
Symmetries
 Charge
  Breaking, 52
 Conservation and Breaking, 39
 Exchange, 52, 146
 Isospin, 148
 Mirror, 115
 Parity, Time Reversal, Exchange, Isospin, 149
 Rotational, 115
Time-Reversal
 Time-Reversal Operator, 115
 Invariance, 117

T
Tensor Moments, 94
 Notation, 110, 126
 Nuclear Reactions, 101
 Parity Behavior, 116
 Rotation, 97, 99
 $S = 1$, 95
 $S = 1/2$, 94
Three-Body Forces
 Axilrod-Teller Force, 161
 Tucson-Melbourne Force, 162
T matrix, 174
Total Width, 207
 Coherence Width, 192
Transfer Reactions, 169
 Stripping Reactions, 226
Transformations
 Continuous, 41
 Discrete, 41
 Fourier, 26
 Parity, 116
 Properties, 91
 Rotation, 85
 Unitary, 277
Transmutation, 339
Transuranium, 188
Trends and Future Developments, 351

U
Uncertainty Relation, 13, 30, 149, 167

W
Wavelength
 De Broglie, 7, 14, 356
 Virtual Photons, 30
Wien Filter, 101, 294–296

Z
Zeeman
 Adiabatic Level Crossing, 286
 HFS Splitting, 285
 HFS States, 278
 Lifetime of $n = 2$ States, 286
 Occupation Numbers, 278
 Region of the HFS, 276
 Transitions Between States, 279

MIX
Papier aus verantwortungsvollen Quellen
Paper from responsible sources
FSC® C105338

If you have any concerns about our products,
you can contact us on
ProductSafety@springernature.com

In case Publisher is established outside the EU,
the EU authorized representative is:
**Springer Nature Customer Service Center GmbH
Europaplatz 3, 69115 Heidelberg, Germany**

Printed by Libri Plureos GmbH
in Hamburg, Germany